T0076266

Methods in Molecular Biology™

Series Editor
John M. Walker
School of Life Sciences
University of Hertfordshire
Hatfield, Hertfordshire, AL10 9AB, UK

For further volumes:
http://www.springer.com/series/7651

Vaccinia Virus and Poxvirology

Methods and Protocols

Second Edition

Edited by

Stuart N. Isaacs

University of Pennsylvania and the Philadelphia VA Medical Center,
Philadelphia, PA, USA

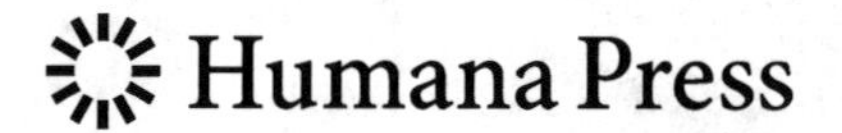

Editor
Stuart N. Isaacs
Department of Medicine
Division of Infectious Disease
University of Pennsylvania
and the Philadelphia VA Medical Center
502 Johnson Pavilion
Philadelphia, PA, USA

ISSN 1064-3745 ISSN 1940-6029 (electronic)
ISBN 978-1-61779-875-7 ISBN 978-1-61779-876-4 (eBook)
DOI 10.1007/978-1-61779-876-4
Springer New York Dordrecht Heidelberg London

Library of Congress Control Number: 2012937882

Printed on acid-free paper

Humana Press is a brand of Springer
Springer is part of Springer Science+Business Media (www.springer.com)

Dedication

To my Mom and Dad; and to my wife and children—Lauren and Jordan—who tolerate the crazy hours I work.

Preface

Nine years ago, the chapters for the first edition of Vaccinia Virus and Poxvirology were submitted for publication in the Methods in Molecular Biology Series. This second edition does not replace the first edition since essentially every chapter in this volume represents a protocol not covered in the first edition. To allow new readers to be aware of what topics were covered in the first edition, I include the following list of chapters from the first edition of Vaccinia Virus and Poxvirology, Methods and Protocols (Volume 269 in the Methods in Molecular Biology Series):

1. Working Safely with Vaccinia Virus: Laboratory Technique and the Role of Vaccinia Vaccination (Stuart N. Isaacs)
2. Construction and Isolation of Recombinant Vaccinia Virus Using Genetic Markers (María M. Lorenzo, Inmaculada Galindo, and Rafael Blasco)
3. Construction of Recombinant Vaccinia Virus: Cloning into the Thymidine Kinase Locus (Chelsea M. Byrd and Dennis E. Hruby)
4. Transient and Inducible Expression of Vaccinia/T7 Recombinant Viruses (Mohamed R. Mohamed and Edward G. Niles)
5. Construction of Recombinant Vaccinia Viruses Using Leporipoxvirus Catalyzed Recombination and Reactivation of Orthopoxvirus DNA (Xiao-Dan Yao and David H. Evans)
6. Construction of cDNA Libraries in Vaccinia Virus (Ernest S. Smith, Shuying Shi, and Maurice Zauderer)
7. Construction and Isolation of Recombinant MVA (Caroline Staib, Ingo Drexler, and Gerd Sutter)
8. Growing Poxviruses and Determining Virus Titer (Girish J. Kotwal and Melissa Abraham)
9. Rapid Preparation of Vaccinia Virus DNA Template for Analysis and Cloning by PCR (Rachel L. Roper)
10. Orthopoxvirus Diagnostics (Hermann Meyer, Inger K. Damon, and Joseph J. Esposito)
11. An In Vitro Transcription System for Studying Vaccinia Virus Early Genes (Steven S. Broyles and Marcia Kremer)
12. An In Vitro Transcription System for Studying Vaccinia Virus Late Genes (Cynthia F. Wright)
13. Studying Vaccinia Virus RNA Processing In Vitro (Paul D. Gershon)
14. Methods for Analysis of Poxvirus DNA Replication (Paula Traktman and Kathleen Boyle)
15. Studying the Binding and Entry of the Intracellular and Extracellular Enveloped Forms of Vaccinia Virus (Mansun Law and Geoffrey L. Smith)
16. Pox, Dyes, and Videotape; Making Movies of GFP Labeled Vaccinia Virus (Brian M. Ward)

17. Interaction Analysis of Viral Cytokine-Binding Proteins Using Surface Plasmon Resonance (Bruce T. Seet and Grant McFadden)
18. Monitoring of Human Immunological Responses to Vaccinia Virus (Richard Harrop, Matthew Ryan, Hana Golding, Irina Redchenko, and Miles W. Carroll)
19. Vaccinia Virus as a Tool for Immunologic Studies (Nia Tatsis, Gomathinayagam Sinnathamby, and Laurence C. Eisenlohr)
20. Mouse Models for Studying Orthopoxvirus Respiratory Infections (Jill Schriewer, R. Mark L. Buller, and Gelita Owens)
21. Viral Glycoprotein-Mediated Cell Fusion Assays Using Vaccinia Virus Vectors (Katharine N. Bossart and Christopher C. Broder)
22. Use of Dual Recombinant Vaccinia Virus Vectors to Assay Viral Glycoprotein-Mediated Fusion with Transfection-Resistant Primary Cell Targets (Yanjie Yi, Anjali Singh, Joanne Cutilli, and Ronald G. Collman)
23. Poxvirus Bioinformatics (Chris Upton)
24. Preparation and Use of Molluscum Contagiosum Virus (MCV) from Human Tissue Biopsy Specimens (Nadja V. Melquiot and Joachim J. Bugert)

In this second edition of poxvirus protocols, there are multiple new chapters covering various approaches for the construction of recombinant viruses. Other chapters focus on methods to isolate the various forms of infectious virus, methods to study the entry of poxviruses into cells, and various protocols covering in vivo models to study poxvirus pathogenesis. There are also chapters on studying cellular immune responses and generation of monoclonal antibodies to poxvirus proteins. This book also contains chapters to cover methods in poxvirus bioinformatics as well as various ways to study poxvirus immunomodulatory proteins. The protocols are designed to be easy to follow and the Note sections include both additional explanatory information and important insights into the protocols.

Since the last edition of this book, a number of important events related to poxvirology have occurred. Examples include the FDA approval of a culture-based live smallpox vaccine and the vaccination of large numbers of US military and relatively large numbers of US civilians. Novel anti-poxvirus therapeutics have been developed and have been used in emergency settings. I will not even attempt to summarize the scientific advances in poxvirology that have been made over this time period. Since the last edition of this book, there have been a number of retirements of prominent poxvirologists. So I would like to acknowledge the retirements of Joseph J. Esposito (Centers for Disease Control and Prevention, Atlanta), Richard (Dick) W. Moyer (University of Florida, Gainesville), and Edward G. Niles (SUNY School of Medicine, Buffalo). These fine scientists ran outstanding labs, made countless contributions to the poxvirus field, and throughout their careers helped create the community that those of us in poxvirology have enjoyed. While no longer lab-based, we are fortunate that they all remain active and continue to contribute to the scientific community. Since the last edition of this book, the poxvirus community sadly marked the deaths of colleagues. So I would like to acknowledge the passing of Riccardo (Rico) Wittek (April 26, 1944 to September 19, 2008) and Frank J. Fenner (December 21, 1914 to November 22, 2010). Their contributions to our field will not be forgotten.

Philadelphia, PA, USA *Stuart N. Isaacs*

Contents

Preface . . . . *vii*
Contributors . . . . *xi*

1 Working Safely with Vaccinia Virus: Laboratory Technique and Review of Published Cases of Accidental Laboratory Infections 1
Stuart N. Isaacs

2 In-Fusion® Cloning with Vaccinia Virus DNA Polymerase 23
Chad R. Irwin, Andrew Farmer, David O. Willer, and David H. Evans

3 Genetic Manipulation of Poxviruses Using Bacterial Artificial Chromosome Recombineering 37
Matthew G. Cottingham

4 Easy and Efficient Protocols for Working with Recombinant Vaccinia Virus MVA 59
Melanie Kremer, Asisa Volz, Joost H.C.M. Kreijtz, Robert Fux, Michael H. Lehmann, and Gerd Sutter

5 Isolation of Recombinant MVA Using F13L Selection 93
Juana M. Sánchez-Puig, María M. Lorenzo, and Rafael Blasco

6 Screening for Vaccinia Virus Egress Inhibitors: Separation of IMV, IEV, and EEV 113
Chelsea M. Byrd and Dennis E. Hruby

7 Imaging of Vaccinia Virus Entry into HeLa Cells 123
Cheng-Yen Huang and Wen Chang

8 New Method for the Assessment of Molluscum Contagiosum Virus Infectivity 135
Subuhi Sherwani, Niamh Blythe, Laura Farleigh, and Joachim J. Bugert

9 An Intradermal Model for Vaccinia Virus Pathogenesis in Mice 147
Leon C.W. Lin, Stewart A. Smith, and David C. Tscharke

10 Measurements of Vaccinia Virus Dissemination Using Whole Body Imaging: Approaches for Predicting of Lethality in Challenge Models and Testing of Vaccines and Antiviral Treatments 161
Marina Zaitseva, Senta Kapnick, and Hana Golding

11 Mousepox, A Small Animal Model of Smallpox 177
David Esteban, Scott Parker, Jill Schriewer, Hollyce Hartzler, and R. Mark Buller

12 Analyzing CD8 T Cells in Mouse Models of Poxvirus Infection 199
Inge E.A. Flesch, Yik Chun Wong, and David C. Tscharke

13 Generation and Characterization of Monoclonal Antibodies Specific for Vaccinia Virus 219
Xiangzhi Meng and Yan Xiang

14 Bioinformatics for Analysis of Poxvirus Genomes. 233
Melissa Da Silva and Chris Upton

15 Antigen Presentation Assays to Investigate Uncharacterized Immunoregulatory Genes 259
Rachel L. Roper

16 Characterization of Poxvirus-Encoded Proteins that Regulate Innate Immune Signaling Pathways 273
Florentina Rus, Kayla Morlock, Neal Silverman, Ngoc Pham, Girish J. Kotwal, and William L. Marshall

17 Application of Quartz Crystal Microbalance with Dissipation Monitoring Technology for Studying Interactions of Poxviral Proteins with Their Ligands. 289
Amod P. Kulkarni, Lauriston A. Kellaway, and Girish J. Kotwal

18 Central Nervous System Distribution of the Poxviral Proteins After Intranasal Administration of Proteins and Titering of Vaccinia Virus in the Brain After Intracranial Administration. 305
Amod P. Kulkarni, Dhirendra Govender, Lauriston A. Kellaway, and Girish J. Kotwal

Index *327*

Contributors

RAFAEL BLASCO • *Departamento de Biotecnología, Instituto Nacional de Investigación y Tecnología Agraria y Alimentaria (INIA), Madrid, Spain*
NIAMH BLYTHE • *Department of Microbiology and Infectious Diseases, Cardiff Institute of Infection and Immunity, Cardiff, UK*
JOACHIM J. BUGERT • *Department of Microbiology and Infectious Diseases, Cardiff Institute of Infection and Immunity, Cardiff, UK*
R. MARK BULLER • *Department of Molecular Microbiology and Immunology, St. Louis University Health Sciences Center, St. Louis, MO, USA*
CHELSEA M. BYRD • *SIGA Technologies, Inc., Corvallis, OR, USA*
WEN CHANG • *Academia Sinica, Institute of Molecular Biology, Taipei, Taiwan, ROC*
MATTHEW G. COTTINGHAM • *The Jenner Institute, University of Oxford, Oxford, UK*
MELISSA DA SILVA • *Biochemistry and Microbiology, University of Victoria, Victoria, BC, Canada*
DAVID ESTEBAN • *Biology Department, Vassar College, Poughkeepsie, NY, USA*
DAVID H. EVANS • *Department of Medical Microbiology and Immunology, Li Ka Shing Institute of Virology, University of Alberta, Edmonton, AB, Canada*
LAURA FARLEIGH • *Department of Microbiology and Infectious Diseases, Cardiff Institute of Infection and Immunity, Cardiff, UK*
ANDREW FARMER • *Clontech Laboratories, Inc., Mountain View, CA, USA*
INGE E.A. FLESCH • *Research School of Biology, The Australian National University, Canberra, ACT, Australia*
ROBERT FUX • *Institute for Infectious Diseases and Zoonoses, University of Munich LMU, Munich, Germany*
HANA GOLDING • *Division of Viral Products, Center for Biologics Evaluation and Research, Food and Drug Administration, Bethesda, MD, USA*
DHIRENDRA GOVENDER • *Faculty of Health Sciences, Division of Anatomical Pathology, Department of Clinical Laboratory Sciences, University of Cape Town, South Africa*
HOLLYCE HARTZLER • *Department of Molecular Microbiology and Immunology, St. Louis University Health Sciences Center, St. Louis, MO, USA*
DENNIS E. HRUBY • *SIGA Technologies, Inc., Corvallis, OR, USA*
CHENG-YEN HUANG • *Academia Sinica, Institute of Molecular Biology, Taipei, Taiwan, ROC*
CHAD R. IRWIN • *Department of Medical Microbiology and Immunology, Li Ka Shing Institute of Virology, University of Alberta, Edmonton, AB, Canada*
STUART N. ISAACS • *Division of Infectious Diseases, Department of Medicine, University of Pennsylvania and the Philadelphia VA Medical Center, Philadelphia, PA, USA*
SENTA KAPNICK • *Division of Viral Products, Center for Biologics Evaluation and Research, Food and Drug Administration, Bethesda, MD, USA*

LAURISTON A. KELLAWAY • *Division of Anatomical Pathology, Department of Clinical Laboratory sciences, Faculty of Health Sciences, University of Cape Town, South Africa*
GIRISH J. KOTWAL • *Kotwal Bioconsulting, LLC, Louisville, KY, USA; InFlaMed Inc, Louisville, KY, USA; Department of Microbiology and Biochemistry, University of Medicine and Health Sciences, Saint Kitts, West Indies*
JOOST H.C.M. KREIJTZ • *Institute for Infectious Diseases and Zoonoses, University of Munich LMU, Munich, Germany; Department of Virology, Erasmus MC, Rotterdam, The Netherlands*
MELANIE KREMER • *Institute for Infectious Diseases and Zoonoses, University of Munich LMU, Munich, Germany*
AMOD P. KULKARNI • *Division of Anatomical Pathology, Department of Clinical Laboratory sciences, Faculty of Health Sciences, University of Cape Town, South Africa*
MICHAEL H. LEHMANN • *Institute for Infectious Diseases and Zoonoses, University of Munich LMU, Munich, Germany*
LEON C.W. LIN • *Research School of Biology, The Australian National University, Canberra, ACT, Australia*
MARÍA M. LORENZO • *Departamento de Biotecnología, Instituto Nacional de Investigación y Tecnología Agraria y Alimentaria (INIA), Madrid, Spain*
WILLIAM L. MARSHALL • *Division of Infectious Disease, Department of Medicine, University of Massachusetts School of Medicine, Worcester, MA, USA*
XIANGZHI MENG • *Department of Microbiology and Immunology, University of Texas Health Science Center at San Antonio, San Antonio, TX, USA*
KAYLA MORLOCK • *Division of Infectious Disease, Department of Medicine, University of Massachusetts School of Medicine, Worcester, MA, USA*
SCOTT PARKER • *Department of Molecular Microbiology and Immunology, St. Louis University Health Sciences Center, St. Louis, MO, USA*
NGOC PHAM • *Division of Infectious Disease, Department of Medicine, University of Massachusetts School of Medicine, Worcester, MA, USA*
RACHEL L. ROPER • *Department of Microbiology and Immunology, Brody School of Medicine, East Carolina University, Greenville, NC, USA*
FLORENTINA RUS • *Division of Infectious Disease, Department of Medicine, University of Massachusetts School of Medicine, Worcester, MA, USA*
JUANA M. SÁNCHEZ-PUIG • *Departamento de Biotecnología, Instituto Nacional de Investigación y Tecnología Agraria y Alimentaria (INIA), Madrid, Spain*
JILL SCHRIEWER • *Department of Molecular Microbiology and Immunology, St. Louis University Health Sciences Center, St. Louis, MO, USA*
SUBUHI SHERWANI • *Department of Microbiology and Infectious Diseases, Cardiff Institute of Infection and Immunity, Cardiff, UK*
NEAL SILVERMAN • *Division of Infectious Disease, Department of Medicine, University of Massachusetts School of Medicine, Worcester, MA, USA*
STEWART A. SMITH • *Research School of Biology, The Australian National University, Canberra, ACT, Australia*
GERD SUTTER • *Institute for Infectious Diseases and Zoonoses, University of Munich LMU, Munich, Germany*
DAVID C. TSCHARKE • *Research School of Biology, The Australian National University, Canberra, ACT, Australia*

CHRIS UPTON • *Biochemistry and Microbiology, University of Victoria, Victoria, BC, Canada*
ASISA VOLZ • *Institute for Infectious Diseases and Zoonoses, University of Munich LMU, Munich, Germany*
DAVID O. WILLER • *Department of Microbiology, Mt. Sinai Hospital, Toronto, ON, Canada*
YIK CHUN WONG • *Research School of Biology, The Australian National University, Canberra, ACT, Australia*
YAN XIANG • *Department of Microbiology and Immunology, University of Texas Health Science Center at San Antonio, San Antonio, TX, USA*
MARINA ZAITSEVA • *Division of Viral Products, Center for Biologics Evaluation and Research, Food and Drug Administration, Bethesda, MD, USA*

Chapter 1

Working Safely with Vaccinia Virus: Laboratory Technique and Review of Published Cases of Accidental Laboratory Infections

Stuart N. Isaacs*

Abstract

Vaccinia virus (VACV), the prototype orthopoxvirus, is widely used in the laboratory as a model system to study various aspects of viral biology and virus–host interactions, as a protein expression system, as a vaccine vector, and as an oncolytic agent. The ubiquitous use of VACVs in the laboratory raises certain safety concerns because the virus can be a pathogen in individuals with immunological and dermatological abnormalities, and on occasion can cause serious problems in normal hosts. This chapter reviews standard operating procedures when working with VACV and reviews published cases on accidental laboratory infections.

Key words: Vaccinia virus, Biosafety Level 2, Class II Biological Safety Cabinet, Personal protective equipment, Smallpox vaccine, Complications from vaccination, Laboratory accidents

1. Introduction

Poxviruses are large DNA viruses with genomes of nearly 200 kb. Their unique site of DNA replication and transcription (1), the fascinating immune evasion strategies employed by the virus (2, 3), and the relative ease of generating recombinant viruses that express foreign proteins in eukaryotic cells (4, 5) have made poxviruses an exciting system to study and a common laboratory tool. Variola virus, the causative agent of smallpox, is the most

*The views expressed in this chapter are solely those of the author and do not necessarily reflect the position or policy of the Department of Veterans Affairs or the University of Pennsylvania.

Stuart N. Isaacs (ed.), *Vaccinia Virus and Poxvirology: Methods and Protocols*, Methods in Molecular Biology, vol. 890, DOI 10.1007/978-1-61779-876-4_1, © Springer Science+Business Media, LLC 2012

famous member of the poxvirus family. It was eradicated as a human disease by the late 1970s and now work with the virus is confined to only two World Health Organization-sanctioned sites under Biosafety Level 4 conditions. Thus, vaccinia virus (VACV) is more widely studied and has become the prototype member of the orthopoxvirus genus. VACV was used as the vaccine to confer immunity to variola virus and helped in the eradication of smallpox. In the USA, routine vaccination with the smallpox vaccine ended in the early 1970s. Since then, the Advisory Committee on Immunization Practices (ACIP) and the CDC have recommended that people working with poxviruses continue to get vaccinated (6–10). This recommendation for those working with VACV is based mainly on the potential problems that an unintentional infection due to a laboratory accident may cause. Rationale for this recommendation is furthered by the understanding that the strains of VACV used in the laboratory setting (e.g., Western Reserve (WR); see Note 1) are more virulent than the vaccine strain. Also, lab workers frequently handle virus at much higher titers than the dose given in the vaccine (see Note 2). There have been reports of laboratory accidents involving VACV (discussed later in this chapter), but a much greater number of such incidents likely go unreported. The total number of people working with VACV and the frequency with which they work with the virus is also unknown. Thus, for laboratory workers, both the full extent of the problem and potential benefit from the vaccine are not known. This chapter discusses laboratory procedures, personal safety equipment, and published laboratory accidents, all of which will serve as aides in preventing accidental laboratory infections, and highlights the need to work safely with the virus.

2. Materials and Equipment

1. Class II Biological Safety Cabinet (BSC).
2. Personal protective equipment.
3. Autoclave.
4. Disinfectants: 1% sodium hypochlorite, 2% glutaraldehyde, formaldehyde, 10% bleach, Spor-klenz, Expor, 70% alcohol.
5. Sharps container disposal unit.
6. Centrifuge bucket safety caps.
7. Occupational medicine access to the smallpox vaccine (see Note 3).

3. Methods

3.1. Laboratory and Personal Protective Equipment

The following section describes safety practices when working with fully replication-competent live VACV. Table 1 summarizes some published cases of laboratory accidents and will be used to highlight various aspects of working safely with the virus. In addition to fully replication-competent VACVs, there are very attenuated strains of VACV (e.g., MVA and NYVAC) that are unable to replicate and form infectious progeny virus in mammalian cells. These highly attenuated, non-replicating VACVs are considered BSL-1 agents (25). With that said, labs that work with both replication-competent and non-replicating viruses should be wary of potential contamination of stocks of avirulent virus with replication-competent poxviruses. This could result in an accidental laboratory infection as highlighted in Case 19 in Table 1. Since unintentional VACV infections most commonly occur through direct contact with the skin or eyes, the most important aspect of working safely with VACV is to use proper laboratory and personal protective equipment to help prevent accidental exposure to the virus. One of the first lines of defense against an accidental exposure is to always work with infectious virus in a BSC. A BSC is a requirement when working with VACV. The cabinet not only confines the virus to a work area that is easily defined and cleaned, but the glass shield on the front of the BSC also serves as an excellent barrier against splashes into the face. A BSC draws room air through the front grille, circulates HEPA-filtered air within the cabinet area, and HEPA filters the air that is exhausted. Thus, working in a BSC protects the worker and the room where VACV is being handled from the unlikely event of aerosolization of the virus (see Note 4).

An equally important line of defense against accidental exposure to the virus is wearing proper personal protective equipment. This includes gloves, lab coat, and eye protection. VACV does not enter intact skin, but gains access through breaks in the skin. Thus, gloves are critical (see Note 5). Accidental infections due to breaks in skin are highlighted by Cases 1, 5–7, 18, and 19 in Table 1 (see Figs. 1–3). Some of these accidents could have been prevented by use of personal protective equipment. While the front shield of the BSC serves as a first line of protection against splashes into the eye, it is also recommended that safety glasses with solid side shields be worn when working with VACV. Depending upon the work being done (e.g., handling high-titer purified stocks of VACV), one should consider additional eye protection like goggles or a full-face shield. This is important to consider because, as an immunologically privileged site (26), the eye can be susceptible to a serious infection even in those previously vaccinated (27). Finally, a lab coat or some other type of outer protective garment decreases the chance of contaminating clothing. If such a contamination occurs,

Table 1
Published cases of accidental laboratory infections with orthopoxviruses

Case no.	Journal, year (reference)	Age (years) or state (year) and underlying medical conditions	Exposure activity	Virus	Site and cause of infection	Prior vaccination status	Illness	Antibiotics/ surgery/antivirals	Resolution and follow-up	Figure
1	Nature, 1986 (11)	>31	Injecting mice	TK-minus WR strain (2×10^6 pfu/50 μl)	Cut on right ring finger	Vaccinated 30 years prior to exposure	4 days after exposure, finger was red and swollen and it progressed from base of finger nail to first joint; day 8, right axillary LN became swollen; no fever or malaise		10 days; worker developed antibodies to the recombinant VACV-expressed protein	
2	Lancet, 1991 (12)	London (1990)	Injecting mice	TK-minus WR strain	Needlesticks into the left thumb and left forefinger	Vaccinated 1 year prior to exposure	3 days after the needlestick, regions became itchy and by day 4 were red and papular. Days 5–6, the lesions were discharging serous fluid and reached a max diameter of 1 cm; kept in occlusive dressing and healed spontaneously		No antibody response to protein expressed by recombinant VACV, but potential evidence of T-cell response	

3	NEJM, 2001 (13)	28 (15 weeks pregnant w/h/o epidermolytic hyperkeratosis)	Dog bite	Copenhagen strain-based rabies vaccine	Technically, not a lab accident, but unintentional exposure to a recombinant virus via a dog bite	Reportedly, no prior smallpox vaccination (born 1971)	3 days after exposure, developed blisters on her forearm; 8 days after bite, hospitalized for progressive pain, erythema, and swelling of left forearm; 10 days after bite, swelling and erythema worsened, left axillary LN	Antibiotics and went to OR for incision and drainage of the forearm	30 days; developed antibodies to the recombinant VACV-expressed protein; no pregnancy complications and delivered a healthy baby	
4	EID, 2003 (14)	26	Needlestick during virus purification step	WR (~10^8 pfu)	Needlestick into the left thumb	Previously vaccinated in childhood (>20 years earlier)	Developed erythema and pain 3 days after inoculation; additional pustules on fourth and fifth fingers developed on days 5 and 6; day 6 axillary LN; day 8 necrotic areas a round lesion and a large erythematous lesion on left forearm	Day 9 began on antibiotics because of concern of bacterial superinfection; went to OR for surgical excision of necrotic tissue	Improved and lesions healed over ~3 weeks; evidence of increased anti-VACV antibodies	Fig. 5
5	J Invest Dermatol, 2003 (15)	40	Contact exposure through broken skin	TK-minus WR strain (10^9 pfu/ml)	Working with high titer in tissue culture with evidence of small erosions on both hands (from working in cold temperatures)	Prior vaccinations 28 and 39 years prior to exposure	Middle inner side of right second finger; second lesion developed (large nodule with central necrosis) on third finger of left hand 2 days later; no LN	Unsuccessful surgical incision followed by topical disinfectants (e.g., polyvidone iodine)	Two weeks and then healed; evidence of increased anti-VACV antibodies	Fig. 1

(continued)

Table 1 (continued)

Case no.	Journal, year (reference)	Age (years) or state (year) and underlying medical conditions	Exposure activity	Virus	Site and cause of infection	Prior vaccination status	Illness	Antibiotics/ surgery/antivirals	Resolution and follow-up	Figure
6	Can Commun Dis Rep, 2003 (16)	48 (history of eczema)	Contact exposure through broken skin	TK-minus virus	Chronic eczema on both hands and a cut on her finger, usually did NOT wear gloves when working with the virus	Vaccinated as a child	First developed pain and redness over dorsal aspect of her index finger; 5 days later admitted with blistering lesions on right index finger; also noted to have swollen axillary LN	Did not respond to antibiotics; treated with occlusive dressing	Spontaneously resolved	
7	J Clin Virol, 2004 (17)	25	Contact exposure through broken skin		Cut on finger with secondary spread by contact to another site	Never vaccinated	Developed a pustule at the site of a cut on the finger. Squeezed pus that squirted on to her face. 2 days later, a lesion formed on her chin; axillary and submental LN, malaise, fever; on day 20, four other lesions were noted on her palms, back of the knee, and upper back and felt to be generalized vaccinia	Did not respond to antibiotics	By day 28, lesions were fading, but continued to have fatigability; by day 36, just a scab on her finger but felt back to full strength; developed anti-VACV antibodies ~1 month after presentation	Fig. 2

8	EID, 2006 (18)	Graduate student PA (2004)	Unknown mechanism of infection	Recombinant WR strain	Unknown, but question of hand to eye or microscope eyepiece to eye or aerosol exposure	Never vaccinated	Painful eye infection (no keratitis or orbital cellulitis) requiring hospitalization	Antibiotics and then antiviral eye drops; VIG	Improvement 24 h after starting VIG; no sequelae, but recovery took a few weeks. Developed anti-VACV antibodies ~2 month after presentation; no secondary VACV infections of contacts were identified	Fig. 4
9	Military Medicine, 2007 (19)	28	Splash into eye		Sprayed ~1 ml of fluid containing virus in eye; washed eye for 2 min	Never vaccinated	Developed eye burning several hours after exposure		No infection occurred	
10	J Viral Hepatitis, 2007 (20)	30	Needlestick injury	Recombinant non-TK-minus WR strain (10^8 pfu/ml)	Needlestick into the left thumb	Never vaccinated	8 days after needle-stick, developed pain and erythema of the thumb and axillary LN; painful swelling of thumb worsened	15 days after injury, necrosis at the injection site was surgically removed	Developed antibodies and T-cell responses to the recombinant VACV-expressed protein	Fig. 6
11	MMWR, 2008 (21)	CT (2005)	Injecting mice	TK-minus WR strain	Needlestick in finger	Vaccinated as a child and ~10 years before the accident	3 days after accident, developed fever, LN, and bulla at the inoculation site	Hospitalized for 1 day	Symptoms improved rapidly	

(continued)

Table 1 (continued)

Case no.	Journal, year (reference)	Age (years) or state (year) and underlying medical conditions	Exposure activity	Virus	Site and cause of infection	Prior vaccination status	Illness	Antibiotics/ surgery/antivirals	Resolution and follow-up	Figure
12	MMWR, 2008 (21)	PA (2006)	Injecting mice	TK-minus WR strain	Needlestick in thumb	Never vaccinated	6 days after accident, sought medical attention for a lesion at the site of inoculation and a secondary lesion near the nail; 9 days after accident had malaise, fever, and LN	Finger surgically debrided 14 days after the accident	Began feeling better 13 days after the accident	
13	MMWR, 2008 (21)	IA (2007)	Needlestick	TK-minus WR strain (3×10^6 pfu)	Needlestick to finger while unsheathing a sterile needle	Never vaccinated	11 days after injury, developed fever, chills, and lesions with swelling at the inoculation site		Recovered fully	
14	MMWR, 2008 (21)	MD (2007)	Injecting animals	TK-minus WR strain (10^4 pfu in 5 μl)	Needlestick to finger	Unsuccessful immunization ~6 years prior to the accident	No infection	After accident, put finger into disinfectant containing hypochlorite and then was vaccinated on the day of accident	No infection	
15	MMWR, 2008 (21)	NH (2007)	Needle scratch while working with mice	WR strain (5×10^4 pfu/ml)	Needle scratch to finger	Never vaccinated	7 days after the accident, developed a pustule; afebrile	Hospitalized with streaking up arm	Recovered	

16	MMWR, 2009 (22)	20s	Unknown mechanism of infection	WR strain (a contaminating virus in a stock of recombinant virus the lab usually works with)	Ear and eye, with additional lesions on chest, shoulder, arm, and leg	Never vaccinated	Pain and swelling of right earlobe and cervical LN and fevers developed 4–6 days after working with vaccinia virus; 4 days after the onset of symptoms, pustular lesions were on right ear, left eye, chest, shoulder, left arm, and right leg	Symptoms worsened on antibiotics and steroids; hospitalized; acyclovir given	Full recovery and returned to work ~1 month after infection; no secondary VACV infections of contacts were identified	Fig. 8
17	MJA, 2009 (23)	26	Injecting mice	WR strain	Needlestick into the left second finger	Vaccinated within 5 years of accident	2 days after injury, developed a cloudy vesicle; 5 days after injury, finger became inflamed with streaking up the arm and axillary LN		All symptoms resolved after 10 days	Fig. 7
18	MMWR, 2009 (24)	35 (taking immuno-suppressive medication for inflammatory bowel disease)	Skin abrasion	Copenhagen strain-based rabies vaccine	Technically, not a lab accident, but exposed to recombinant VACV while handling raccoon rabies vaccine bait	Never vaccinated	4 days after exposure, developed some red papules that then increased in number; day 9 had 26 lesions on her arm with edema; afebrile	Hospitalized and treated with VIG on day 6 after exposure and then a repeat dose on day 12. Started on investigational antiviral agent, ST-246 × 14 days	Discharged on day 19 after exposure. By day 28, all lesion scabs had separated	Fig. 3
19	Unpublished report appearing in Med-scape News[a]	IL (2010)	Cut on finger	Cowpox virus contaminating a stock of nonpatho-genic poxvirus		Never vaccinated	Painful, ulcerated lesion on a finger that lasted 3 months	No further information available	No further information available	

Abbreviations: *LN* lymph node, *OR* operating room, *pfu* plaque-forming units, *TK* thymidine kinase, *VACV* vaccinia virus, *VIG* vaccinia immune globulin, *WR* Western Reserve

[a]Medscape News, February 8, 2011 (http://www.medscape.com/viewarticle/737030)

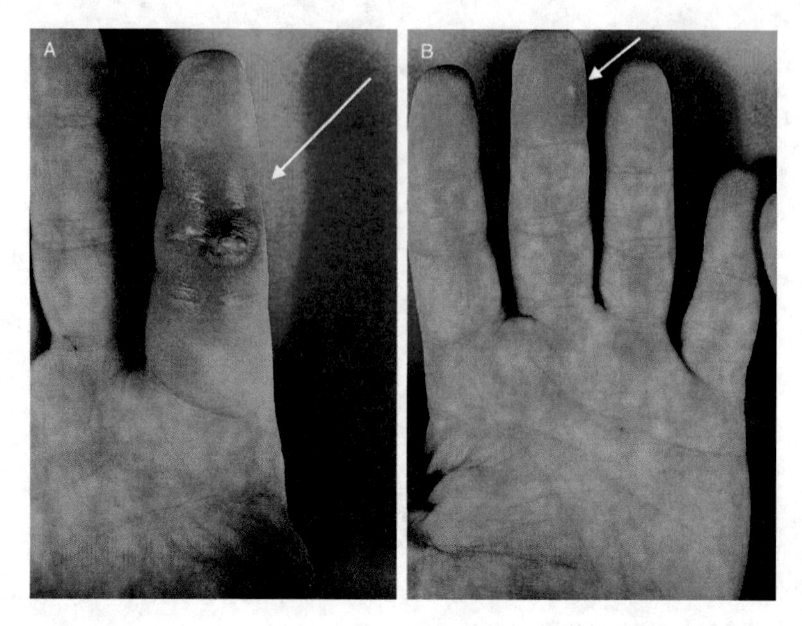

Fig. 1. Photograph of non-needlestick infection of fingers ~5 to 7 days after the onset of symptoms. (**a**) Right hand and (**b**) left hand. Reprinted by permission from Macmillan Publishers Ltd.: The Journal of Investigative Dermatology, see ref. 15, copyright © 2003.

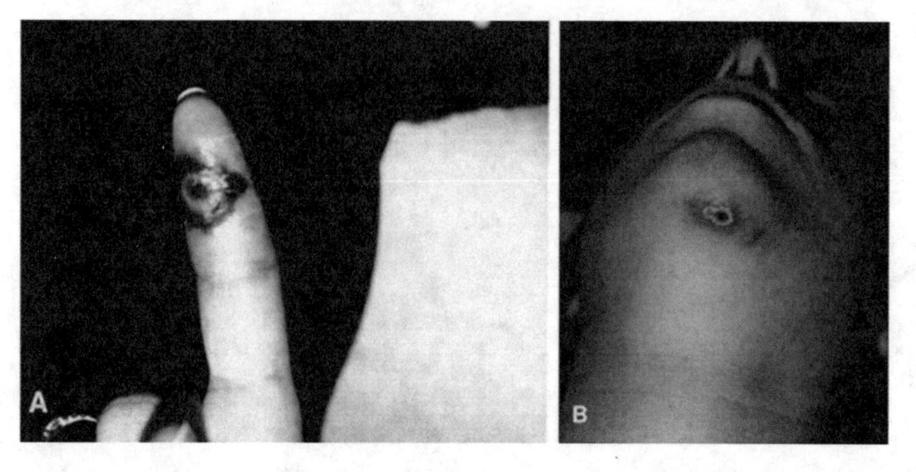

Fig. 2. Photographs of primary and secondary lesions 18 days after the onset of symptoms. (**a**) Primary lesion on finger at the site of a prior cut and (**b**) lesion on chin developing a few days after the finger lesion. Reprinted from the Journal of Clinical Virology, see ref. 17, copyright © 2004, with permission from Elsevier.

an outer garment can be quickly removed and decontaminated. Furthermore, a lab coat will prevent accidentally carrying the virus out of the laboratory environment. Since the virus can be stable in the environment, after protective equipment is removed, good hand washing with soap and water is important (28). Cases 1, 5–9, 16, 18, and 19 in Table 1 (see Figs. 1–4) represent potentially preventable accidents if proper biosafety practices were followed.

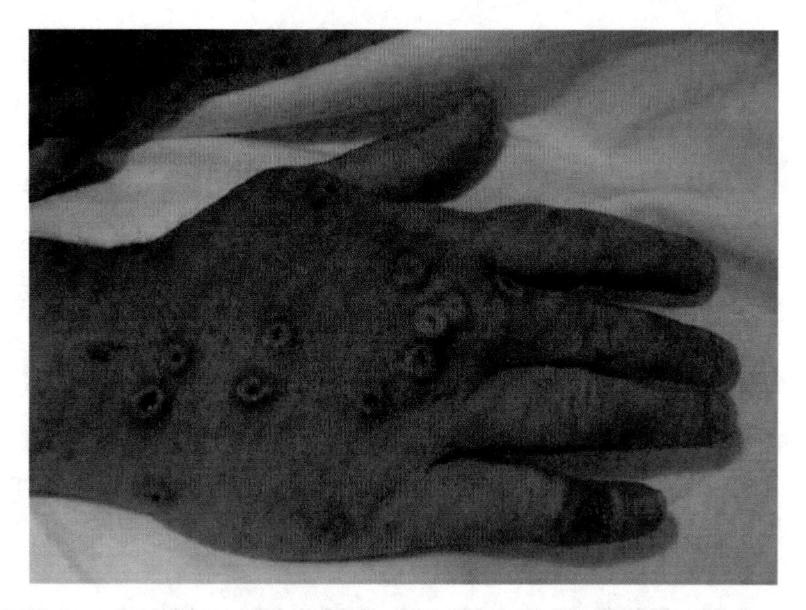

Fig. 3. Photograph of the right hand of a woman 11 days after contact with a raccoon rabies vaccine bait in Pennsylvania in 2009.

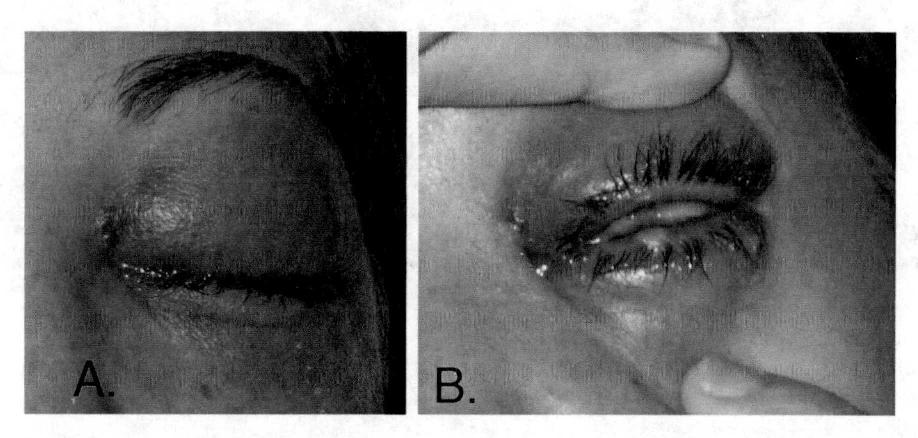

Fig. 4. Vaccinia virus infection of eye. (**a**) Left eye 5 days after the onset of symptoms. The primary pox lesion is located at the inner canthus. (**b**) Satellite lesion on lower conjunctiva developing 7 days after the onset of symptoms. Photographs by E. Claire Newbern. Figures and legend reproduced from ref. 18. These published materials are in the public domain.

Also note that in some published accidents, prompt interventions may have prevented potential infections. In Case 9, prompt flushing of the eye with water after a splash exposure may have prevented infection. In Case 14, disinfecting the site of inoculation, as well as active smallpox vaccination on the day of the accident may have prevented the infection.

3.2. Laboratory Safety

In addition to VACV being handled at Biosafety Level-2 (29), as with all biohazardous agents, routine good laboratory safety practices need to be fully implemented. This includes such things as no eating or drinking in the laboratory. To decrease the chance of accidental infections, the use of any sharps or glass should be minimized while working with VACV (see Note 6). Syringes and needles still need to be used when performing some experiments in animals, and thus personnel who need to be working with needles to inject animals with the virus would be considered to be performing a higher risk procedure (see Note 7 and Subheading 3.3). If sharps or disposable glassware is necessary, the proper leakproof, puncture-resistant sharps disposal container needs to be conveniently located close to the work area to prevent accidents during disposal of needles and glass (see Note 8). Multiple cases of laboratory accidents due to needlesticks have been reported. Cases 2, 4, 10–15, and 17 are examples of needlestick accidents while working with VACV (see Table 1 and Figs. 5–7).

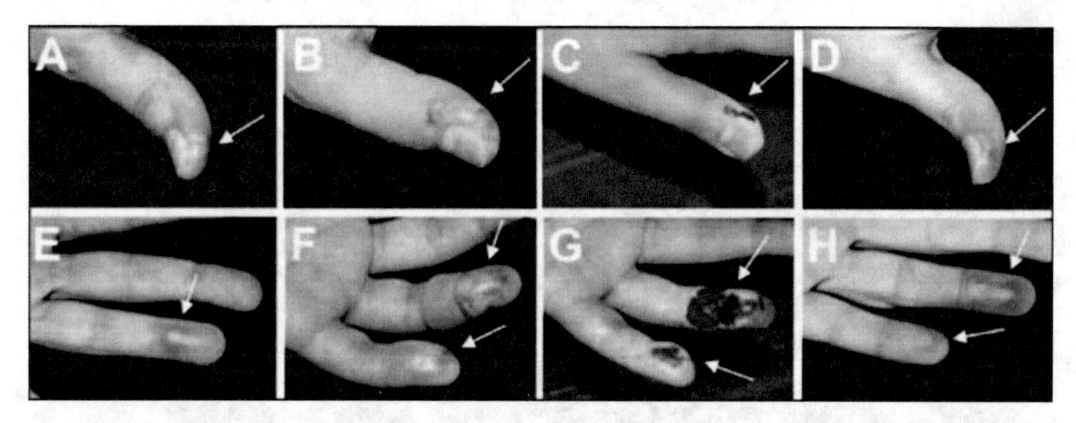

Fig. 5. Progression of the local reaction on the left hand after accidental needlestick inoculation with vaccinia virus: thumb, (**a**) day 4, (**b**) day 11, (**c**) day 12, and (**d**) day 20; fourth and fifth fingers, (**e**) day 7, (**f**) day 11, (**g**) day 12, and (**h**) day 20. Lesions were surgically excised to remove necrotic tissue on day 11. *Arrows* indicate the lesion areas. Figure and legend reproduced from ref. 14. These published materials are in the public domain.

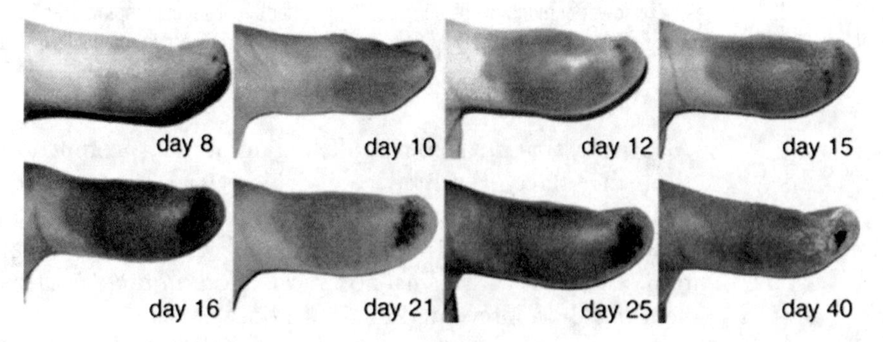

Fig. 6. Progression of thumb infection after needlestick injury with VACV. Time after accident is indicated. Surgery was performed on day 15. Reprinted from the Journal of Viral Hepatitis, see ref. 20, copyright © 2007, with permission from John Wiley & Sons.

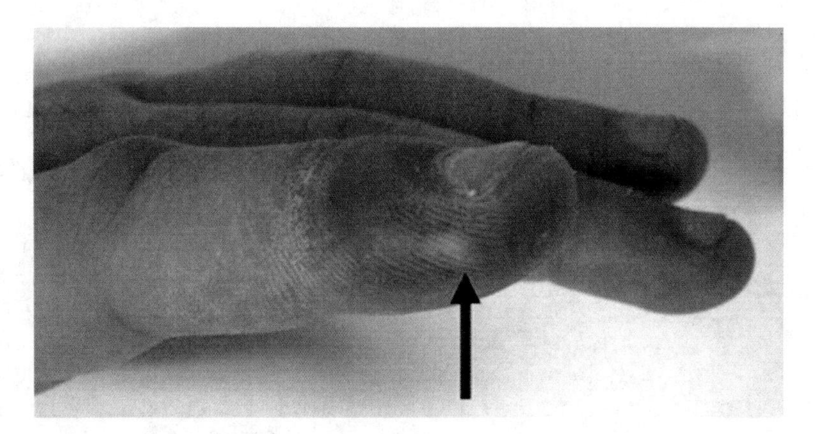

Fig. 7. Needlestick injury. Photograph of a finger 2 days after a needlestick injury. *Arrow* points to cloudy vesicle forming at the site of inoculation. Figure reproduced with permission from ref. 23: Senanayake SN. Needlestick injury with smallpox vaccine. MJA 2009 (11/12):657. Copyright ©2009. *The Medical Journal of Australia.*

As discussed earlier, a BSC is a requirement when working with VACV. As a virus that has an outer membrane envelope, VACV is susceptible to inactivation by a variety of detergents and disinfectants (30). Thus, after working with the virus in the BSC, surfaces should be wiped down with freshly prepared 1% bleach. Since experiments with VACV in tissue culture frequently involve aspirating and discarding virus-contaminated growth medium, one must properly inactivate the virus in the media prior to disposal. For aspirating off media from infected cells using the lab vacuum system, one must include a trap to collect the aspirated media and a vacu-guard to prevent contamination of the house vacuum system with the virus (see Note 9). The virus contained in the liquid must be inactivated by the addition of disinfectant (see Note 10). The virus is also susceptible to heat, and thus autoclaving of contaminated instruments, dryware, animal cages, and bedding exposed to the virus is also required. Properly packaged disposable plasticware used for culturing the virus should be sterilized in a humidified autoclave. While small samples containing virus have been shown to be inactivated in autoclaves in as little as 15 min (31), infectious waste should be autoclaved at 121°C for at least 60 min at 15 pounds/in^2 and then disposed of according to the institutional guidelines (see Note 10). When centrifuging large volumes of media containing virus, it is best to use centrifuge buckets that have safety lids to contain a spill and contain potential aerosolization of the virus if a tube should leak during centrifugation. Sonication of infected cells, which is frequently done in poxvirus protocols as a means of releasing virus from cells and breaking up virus clumps, can also cause aerosolization of the virus (see Note 11). Therefore, sonication should be performed in a cup sonicator with the virus or virus-infected cells remaining in a closed tube. Larger virus preparations may need to be sonicated with a probe sonicator, which should only be performed if the sonicator device is contained in a proper BSC that will properly filter the air and remove any potential aerosolized virus (see Note 4).

3.3. Occupational Smallpox Vaccination

In the USA, the currently licensed smallpox vaccine (ACAM 2000 (32)) is recommended for people who work with poxviruses (8) (see Note 12). While there is no debate that this policy should be followed for researchers who work with such poxviruses as variola virus, camelpoxvirus, and monkeypox virus (see Note 13), there is considerable debate whether smallpox vaccination should be carried out in all people who have contact with VACV (33–38). The ACIP and the CDC recommend vaccination every 10 years of all people who have contact with VACV (8). As mentioned previously, this recommendation does not apply to those working with highly attenuated strains of VACV (i.e., MVA, NYVAC, ALVAC, and TROVAC). These viruses are considered to be extremely safe because they do not replicate in mammalian cells and are avirulent in normal and immunosuppressed animal models (see Note 14). However, ACIP recommendations about smallpox vaccination of laboratory workers handling replication-competent strains of VACV remain unchanged from earlier decades (6, 7). The issue is that these older recommendations were at a time when essentially all adults working with VACV had previously been vaccinated at least once in childhood. Since routine vaccination in the USA ended in the early 1970s, there is now a growing population of workers who have never been vaccinated and thus the recommendation to vaccinate such workers would represent primary vaccinations. The rate of complications from primary vaccination is 10 to 20 times greater than the rate of complications in those who had previously been vaccinated (39, 40) (see Table 2 and Note 15). Thus, the risk–benefit ratio for routine vaccination of all lab workers who handle VACV in the USA has significantly changed from the 1980s to now. Also, a recent survey of lab workers revealed that previously unvaccinated people experience more postvaccination symptoms than previously vaccinated people (43). In addition, since vaccinated individuals can accidentally transmit the virus to close contacts, the potential infection of unvaccinated contacts is more problematic now than in previous times when most of the population had been vaccinated (see Note 16). Recent cases of accidental transmission of the vaccine have been documented (44–49) and at least one of these accidental transmissions resulted in significant morbidity (45). In contrast to the ACIP recommendation to vaccinate lab workers, advisory committees in other countries have reached different conclusions than the ACIP and do not recommend routine vaccination (e.g., 50). These committees concluded that the risk of vaccination (i.e., knowingly infecting an individual with VACV) outweighs the potential benefit of protecting them from an accidental exposure that results in an infection.

Table 2
Rates of reported VACV vaccine complications in adults (number of cases/million vaccinations)

Vaccination group	Accidental transfer	Generalized vaccinia	Eczema vaccinatum	Progressive vaccinia	Postvaccinial encephalitis	Total^
Primary*	606	212	30	–[a]	–[a]	1,515
Secondary*	25	9	4	7	4	114
The US Military (2003)#	153	80[b]	0	0	2[c]	320
Civilian (2003)&	510	77	0	0	26	1,224

*Data from ref. 40 for individuals ≥20 years of age. Receiving the vaccine for the first time (primary); receiving the vaccine in those who had previously been vaccinated (secondary)
#Data from ref. 41 reporting on 450,293 vaccinations (70% were primary vaccinations)
&Data from ref. 42 reporting on 39,213 civilian health-care and public health workers vaccinated between January 24 and December 31, 2003
^Includes other complications, such as severe reactions, bacterial super infection, and erythema multiforme
[a]No report of this complication was included in the 1968 10-state survey (40)
[b]All were in primary vaccinees
[c]One case in a primary vaccinee and another case in a secondary vaccinee

Some feel strongly that vaccination should be mandatory (51, 52). Another approach that has been implemented at some institutions in the USA is to offer mandatory counseling by Occupational Medicine regarding vaccination, but to then allow each individual to make an informed personal decision whether or not to be vaccinated (33–38). The types of procedures being done with VACV by the worker should enter into this decision-making process (see Notes 7 and 17). Since accidental infections have occurred in workers with distant past vaccination (see Cases 1, 4–6; Table 1, Figs. 1 and 5) or recent vaccination (see Cases 2, 11, and 17; Table 1 and Fig. 7), it is unclear the role prior vaccination may have played. With that said, the majority of published cases are in workers who have never been vaccinated (see Cases 3, 7–10, 12–16, 18, and 19; Table 1, Figs. 1, 2, 4, 6, and 8). Some would argue that prior vaccination prevented the accidental infection from becoming more serious. Others would argue that it is unclear if the cases of accidental infection in the setting of no prior vaccination were any different than those who had prior vaccination. However, it has been pointed out that more workers with accidental infections who had never been vaccinated were hospitalized (see Cases 3, 8, 15, 16, and 18; Table 1, Figs. 3, 4, and 8) than those who had prior vaccination (see Case 11; Table 1 and ref. 52).But the problem with such data is that not all lab accidents are reported. Furthermore, it is more likely that serious infections resulting in hospitalization would come to the attention of the CDC and thus further skew the data.

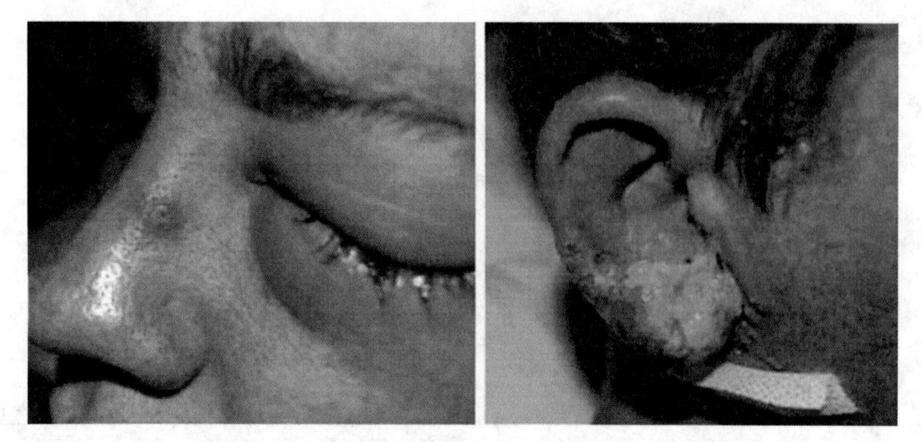

Fig. 8. Left eye and right ear of a man with laboratory-acquired VACV infection from Virginia in 2008. Photographs were taken ~4 days after the onset of symptoms when pustular lesions at similar stages of development developed on his right ear and left eye. Figures and legend reproduced from ref. 22. These published materials are in the public domain.

Another important issue related to the counseling session is that the worker should have a complete medical assessment to determine whether the smallpox vaccine can safely be given. Thus, workers may be identified who have medical conditions that would preclude them from vaccination. Such conditions include an immunodeficiency, a history of a skin condition like atopic dermatitis/eczema (see Note 18), pregnancy, and cardiac disease (see Note 15). Importantly, some feel that the presence of these conditions should preclude someone from working with VACV. Others think that for a properly trained researcher, working with VACV while pregnant or with mild eczema can be allowed, as long as the individual is not performing high-risk procedures with the virus (see examples of higher risk procedures in Note 7). Laboratory directors and workers need to be made aware of the risks and preventative measures available in order to make educated decisions. Each institution should develop a policy that addresses these concerns.

3.4. Future Considerations

Over the past decade, the concerns about smallpox bioterrorism led to the development of a number of new products, including a newly licensed vaccine grown in tissue culture (53), next-generation smallpox vaccines with better safety profiles (54), and a number of promising new anti-poxvirus therapeutics (55, 56). As these new therapeutics and safer next-generation vaccines move forward, they will certainly remedy some safety concerns of working with the non-attenuated laboratory strains of VACV. For example, the future availability of a smallpox vaccine that is safer than what is currently FDA approved in the USA will likely end the controversy of whether or not laboratory workers handling VACV should be vaccinated. A vaccine with few to no side effects (e.g., MVA) would clearly shift the risk–benefit ratio of vaccination in favor of vaccination to prevent infections resulting from a laboratory accident. In addition, the development of anti-poxvirus therapeutics (e.g., 55, 56) will be beneficial for treatment of serious infections after a lab accident.

4. Notes

1. VACV strain WR is the most commonly used laboratory strain of VACV and was selected for its neurovirulence in mice.
2. The smallpox vaccine is administered by scarification with a bifurcated needle. When the needle is dipped into the vaccine stock (~10^8 pfu/ml), approximately 2.5 μl of solution is taken up and thus approximately 2.5×10^5 pfu is delivered.
3. Institutions can obtain the smallpox vaccine from the CDC, Drug Services, National Center for Infectious Diseases (404-639-3670).
4. If infectious aerosol-generating procedures are to be performed outside of a BSC and cannot be contained, respiratory protection is required. Individuals requiring respiratory protection should be enrolled in the institution's respiratory protection program.
5. Nitrile or powder-free latex gloves should be used when working with virus.
6. Plasticware that is now commonly used in the modern-day lab has significantly decreased the need for glassware while working with the virus. For example, instead of using sterilized glass Pasteur pipets to aspirate medium while culturing the virus in tissue culture plates, one can use plastic pipet tips attached to vacuum tubing. A simple system used by many labs is to have a P-1,000 pipet tip attached to the vacuum tubing. One can then use P-200 pipet tips over the P-1,000 tip for direct contact with various culture wells. This allows easy changing of the tip to prevent any cross contamination of specimens.
7. Workers performing higher risk procedures like vaccinating animals, direct handling of infected animals (see Note 17), or preparing and purifying high-titer virus are individuals who might benefit the most from prophylactic smallpox vaccination. Also included in the group for whom vaccination should be strongly considered are those generating or working with

recombinant VACVs that express potentially toxic proteins, proteins that might enhance virulence of the recombinant VACV, or proteins to which seroconversion may be problematic (e.g., HIV, dengue). However, to successfully vaccinate such individuals, workers need additional education about the potential benefits of vaccination to overcome the fear of adverse effects of smallpox vaccination (57).

8. Sharps containers (when 2/3 full) should be autoclaved prior to disposal in accordance with the institution's infectious waste policy.
9. Labs frequently use a large-volume Erlenmeyer flask placed underneath the BSL-2 cabinet. This flask can be filled with detergent or bleach to inactivate the virus. Bleach should be at a final concentration of ~10%. The level of fluid in the flask should be monitored to prevent overfilling. It is also recommended that a vent/gas filter with a hydrophobic membrane be placed in-line between the Erlenmeyer flask and the vacuum outlet as an additional safety measure to prevent contamination of the vacuum system if the trap accidentally becomes overfilled. Decontaminated liquid waste should be disposed in accordance with the institution's regulatory requirements.
10. Special consideration must be made when working with virus and radioactivity. The waste stream for all radioactivity must be accounted for and disposed of following institutional guidelines. Thus, virus-contaminated radioactive waste must first be inactivated. Radioactive liquid waste should be treated with disinfectant and then disposed of with the normal radioactive liquid waste. Since one should not autoclave radioactive waste, radioactive contaminated disposable plasticware should be washed down with disinfectant solution to inactivate virus prior to disposing it as radioactive solid waste.
11. During sonication, use of proper ear protection is recommended.
12. Since the publication of the first edition of VACV methods (58), a relatively large number of civilians and an even larger number of people in the US military have been vaccinated with VACV. With careful screening of potential vaccine recipients, the experience with the US military found that the rate of the most serious adverse events was less than that previously reported (see Table 2). However, there has been a recent case of progressive vaccinia in a military recruit (59) and the data for the vaccination of the US civilians revealed that the rate of the most serious adverse events was similar to that previously reported (see Table 2).
13. For those working with these highly virulent poxviruses (e.g., variola virus, monkeypox virus), vaccination every 3 years is recommended (8).

14. In fact, these highly attenuated strains of VACV can now be handled in the laboratory at Biosafety Level-1 (25).
15. The potential serious complications from smallpox vaccine include eczema vaccinatum, postvaccinial encephalitis, and progressive vaccinia in immunocompromised hosts who were inadvertently vaccinated or were accidentally infected by exposure to a vaccinated individual (see Table 2). Photographs of these types of complications are available (8, 60). In the course of the pre-event smallpox vaccination programs during the first decade of the twenty-first century, myocarditis/pericarditis has also been reidentified as a potential complication from the vaccine (10, 61, 62).
16. Because of potential spread of the virus from the vaccination site, close contact with individuals who have the conditions listed in Note 15 are considered to be a contraindication to elective vaccination. Another potential contraindication for vaccination would be close contact with infants <1 year old. However, a worker who is not vaccinated because of these contraindications should still be able to work with VACV following guidelines of personal protection equipment and good hand washing. Alternatively, an individual who still wants to be vaccinated can choose to isolate himself or herself from contact with such at-risk people until the vaccination site scab has fallen off (typically anywhere from 2 to 4 weeks). These issues affect the risk–benefit ratio decision.
17. The ACIP recommends that workers who care for VACV-infected animals also receive smallpox vaccination (8). This is likely based on the finding that within a cage, VACV can spread from infected mice to naïve mice (63–65). But here too, the decision to vaccinate such workers could be a personal one based on what type of contact the worker has with the infected animals. If, for example, a worker has no direct contact (e.g., use of tongs to transfer animals from cage to cage), then the potential for exposure would be quite small. This type of conclusion is supported by studies that showed no transmission of virus between subcutaneously vaccinated and unvaccinated guinea pigs (66), and a recent study (65) that looked for VACV transmission to sentinel mice by the soiled bedding from cages that housed mice infected by the subcutaneous or intrarectal route with VACV. None of the mice exposed to the soiled bedding seroconverted.
18. There is increased likelihood of serious complications in those with skin conditions like atopic dermatitis or eczema whether the skin disease is active or non-active at the time of vaccination. While it is not known why such individuals have increased complications, it is theorized that there is some type of cutaneous immune abnormality (67).

Acknowledgments

I thank Edward Alexander for helpful discussions. The author is supported by the Philadelphia Veterans Affairs Medical Center, NIH grants U01 AI077913 and U01 AI066333, and the Middle Atlantic Regional Center of Excellence in Biodefense and Emerging Infectious Diseases (U54 AI057168).

References

1. Moss B (2007) Poxviridae: the viruses and their replication. In: Knipe DM, Howley PM (eds) Fields' virology. Wolters Kluwer Health/ Lippincott Williams & Wilkins, Philadelphia, Chapter 74
2. Seet BT, Johnston JB, Brunetti CR, Barrett JW, Everett H, Cameron C, Sypula J, Nazarian SH, Lucas A, McFadden G (2003) Poxviruses and immune evasion. Annu Rev Immunol 21:377–423
3. Haga IR, Bowie AG (2005) Evasion of innate immunity by vaccinia virus. Parasitology 130:S11–S25
4. Moss B (1996) Genetically engineered poxviruses for recombinant gene expression, vaccination, and safety. Proc Natl Acad Sci USA 93: 11341–11348
5. Carroll MW, Moss B (1997) Poxviruses as expression vectors. Curr Opin Biotechnol 8: 573–577
6. CDC (1985) Recommendations of the Immunization Practices Advisory Committee smallpox vaccine. Morb Mortal Wkly Rep 34: 341–342
7. CDC (1991) Vaccinia (smallpox) vaccine. Recommendations of the Immunization Practices Advisory Committee (ACIP). Morb Mortal Wkly Rep 40:1–10
8. CDC (2001) Vaccinia (smallpox) vaccine recommendations of the Advisory Committee on Immunization Practices (ACIP), 2001. Morb Mortal Wkly Rep 50:1–25
9. CDC (2003) Recommendations for using smallpox vaccine in a pre-event vaccination program. Supplemental recommendations of the Advisory Committee on Immunization Practices (ACIP) and the Healthcare Infection Control Practices Advisory Committee (HICPAC). MMWR Recomm Rep 52:1–16
10. CDC (2003) Supplemental recommendations on adverse events following smallpox vaccine in the pre-event vaccination program: recommendations of the Advisory Committee on Immunization Practices. Morb Mortal Wkly Rep 52:282–284
11. Jones L, Ristow S, Yilma T, Moss B (1986) Accidental human vaccination with vaccinia virus expressing nucleoprotein gene. Nature 319:543
12. Openshaw PJ, Alwan WH, Cherrie AH, Record FM (1991) Accidental infection of laboratory worker with recombinant vaccinia virus. Lancet 338:459
13. Rupprecht CE, Blass L, Smith K, Orciari LA, Niezgoda M, Whitfield SG, Gibbons RV, Guerra M, Hanlon CA (2001) Human infection due to recombinant vaccinia-rabies glycoprotein virus. N Engl J Med 345:582–586
14. Moussatche N, Tuyama M, Kato SE, Castro AP, Njaine B, Peralta RH, Peralta JM, Damaso CR, Barroso PF (2003) Accidental infection of laboratory worker with vaccinia virus. Emerg Infect Dis 9:724–726
15. Mempel M, Isa G, Klugbauer N, Meyer H, Wildi G, Ring J, Hofmann F, Hofmann H (2003) Laboratory acquired infection with recombinant vaccinia virus containing an immunomodulating construct. J Invest Dermatol 120:356–358
16. Loeb M, Zando I, Orvidas MC, Bialachowski A, Groves D, Mahoney J (2003) Laboratory-acquired vaccinia infection. Can Commun Dis Rep 29:134–136
17. Wlodaver CG, Palumbo GJ, Waner JL (2004) Laboratory-acquired vaccinia infection. J Clin Virol 29:167–170
18. Lewis FM, Chernak E, Goldman E, Li Y, Karem K, Damon IK, Henkel R, Newbern EC, Ross P, Johnson CC (2006) Ocular vaccinia infection in laboratory worker, Philadelphia, 2004. Emerg Infect Dis 12:134–137
19. Peate WF (2007) Prevention of vaccinia infection in a laboratory worker. Mil Med 172: 1117–1118
20. Eisenbach C, Neumann-Haefelin C, Freyse A, Korsukewitz T, Hoyler B, Stremmel W, Thimme R, Encke J (2007) Immune responses against HCV-NS3 after accidental infection with HCV-NS3 recombinant vaccinia virus. J Viral Hepat 14:817–819

21. CDC (2008) Laboratory-acquired vaccinia exposures and infections—United States, 2005–2007. Morb Mortal Wkly Rep 57: 401–404
22. CDC (2009) Laboratory-acquired vaccinia virus infection—Virginia, 2008. Morb Mortal Wkly Rep 58:797–800
23. Senanayake SN (2009) Needlestick injury with smallpox vaccine. Med J Aust 191:657
24. CDC (2009) Human vaccinia infection after contact with a raccoon rabies vaccine bait—Pennsylvania, 2009. Morb Mortal Wkly Rep 58:1204–1207
25. NIH (1996) Modifications to NIH vaccinia immunization policy. U.S. Department of Health and Human Services, Bethesda, MD
26. Niederkorn JY (2002) Immune privilege in the anterior chamber of the eye. Crit Rev Immunol 22:13–46
27. Ruben FL, Lane JM (1970) Ocular vaccinia. An epidemiologic analysis of 348 cases. Arch Ophthal 84:45–48
28. Jonczy EA, Daly J, Kotwal GJ (2000) A novel approach using an attenuated recombinant vaccinia virus to test the antipoxviral effects of handsoaps. Antiviral Res 45:149–153
29. Richmond JY, McKinney RW (1999). In: HHS publication; no. (CDC) 93-8395. U.S. Department of Health and Human Services, PHS, CDC, NIH, Washington, DC
30. Block SS (2001) Disinfection, sterilization, and preservation. Lippincott Williams & Wilkins, Philadelphia
31. Espy MJ, Uhl JR, Sloan LM, Rosenblatt JE, Cockerill FR 3rd, Smith TF (2002) Detection of vaccinia virus, herpes simplex virus, varicella-zoster virus, and Bacillus anthracis DNA by LightCycler polymerase chain reaction after autoclaving: implications for biosafety of bioterrorism agents. Mayo Clin Proc 77: 624–628
32. CDC (2008) Notice to readers: newly licensed smallpox vaccine to replace old smallpox vaccine. Morb Mortal Wkly Rep 57:207–208
33. Baxby D (1989) Smallpox vaccination for investigators. Lancet 2:919
34. Wenzel RP, Nettleman MD (1989) Smallpox vaccination for investigators using vaccinia recombinants. Lancet 2:630–631
35. Perry GF (1992) Occupational medicine forum. J Occup Med 34:757
36. Baxby D (1993) Indications for smallpox vaccination: policies still differ. Vaccine 11: 395–396
37. Williams NR, Cooper BM (1993) Counselling of workers handling vaccinia virus. Occup Med (Oxf) 43:125–127
38. Isaacs SN (2002) Critical evaluation of smallpox vaccination for laboratory workers. Occup Environ Med 59:573–574
39. Lane JM, Ruben FL, Neff JM, Millar JD (1969) Complications of smallpox vaccination, 1968. National surveillance in the United States. New Engl J Med 281:1201–1208
40. Lane JM, Ruben FL, Neff JM, Millar JD (1970) Complications of smallpox vaccination, 1968: results of ten statewide surveys. J Infect Dis 122:303–309
41. Grabenstein JD, Winkenwerder W Jr (2003) US military smallpox vaccination program experience. JAMA 289:3278–3282
42. CDC (2004) Update: adverse events following civilian smallpox vaccination—United States, 2003. Morb Mortal Wkly Rep 53: 106–107
43. Baggs J, Chen RT, Damon IK, Rotz L, Allen C, Fullerton KE, Casey C, Nordenberg D, Mootrey G (2005) Safety profile of smallpox vaccine: insights from the laboratory worker smallpox vaccination program. Clin Infect Dis 40:1133–1140
44. CDC (2004) Secondary and tertiary transfer of vaccinia virus among U.S. military personnel—United States and worldwide, 2002–2004. Morb Mortal Wkly Rep 53:103–105
45. CDC (2007) Household transmission of vaccinia virus from contact with a military smallpox vaccinee—Illinois and Indiana, 2007. Morb Mortal Wkly Rep 56:478–481
46. CDC (2007) Vulvar vaccinia infection after sexual contact with a military smallpox vaccinee–Alaska, 2006. Morb Mortal Wkly Rep 56: 417–419
47. CDC (2010) Vaccinia virus infection after sexual contact with a military smallpox vaccinee—Washington, 2010. Morb Mortal Wkly Rep 59:773–775
48. Young GE, Hidalgo CM, Sullivan-Frohm A, Schult C, Davis S, Kelly-Cirino C, Egan C, Wilkins K, Emerson GL, Noyes K, Blog D (2011) Secondary and tertiary transmission of vaccinia virus from US military service member. Emerg Infect Dis 17:718–721
49. Hughes CM, Blythe D, Li Y, Reddy R, Jordan C, Edwards C, Adams C, Conners H, Rasa C, Wilby S, Russell J, Russo KS, Somsel P, Wiedbrauk DL, Dougherty C, Allen C, Frace M, Emerson G, Olson VA, Smith SK, Braden Z, Abel J, Davidson W, Reynolds M, Damon IK (2011) Vaccinia virus infections in martial arts gym, Maryland, USA, 2008. Emerg Infect Dis 17:730–733
50. Advisory Committee on Dangerous Pathogens and Advisory Committee on Genetic

Modifications (1990) HMSO Publications Center, London, pp 1–16
51. Fulginiti VA (2003) The risks of vaccinia in laboratory workers. J Invest Dermatol 120:viii
52. MacNeil A, Reynolds MG, Damon IK (2009) Risks associated with vaccinia virus in the laboratory. Virology 385:1–4
53. Greenberg RN, Kennedy JS (2008) ACAM2000: a newly licensed cell culture-based live vaccinia smallpox vaccine. Expert Opin Investig Drugs 17:555–664
54. Kennedy JS, Greenberg RN (2009) IMVAMUNE: modified vaccinia Ankara strain as an attenuated smallpox vaccine. Expert Rev Vaccines 8:13–24
55. Jordan R, Goff A, Frimm A, Corrado ML, Hensley LE, Byrd CM, Mucker E, Shamblin J, Bolken TC, Wlazlowski C, Johnson W, Chapman J, Twenhafel N, Tyavanagimatt S, Amantana A, Chinsangaram J, Hruby DE, Huggins J (2009) ST-246 antiviral efficacy in a nonhuman primate monkeypox model: determination of the minimal effective dose and human dose justification. Antimicrob Agents Chemother 53:1817–1822
56. Lanier R, Trost L, Tippin T, Lampert B, Robertson A, Foster S, Rose M, Painter W, O'Mahony R, Almond M, Painter G (2010) Development of CMX001 for the treatment of poxvirus infections. Viruses 1: 2740–2762
57. Benzekri N, Goldman E, Lewis F, Johnson CC, Reynolds SM, Reynolds MG, Damon IK (2010) Laboratory worker knowledge, attitudes and practices towards smallpox vaccine. Occup Med (Lond) 60:75–77
58. Isaacs SN (2004) Vaccinia virus and poxvirology: methods and protocols. Humana Press, Totowa, NJ
59. CDC (2009) Progressive vaccinia in a military smallpox vaccinee—United States, 2009. Morb Mortal Wkly Rep 58:532–536
60. Cono J, Casey CG, Bell DM (2003) Smallpox vaccination and adverse reactions. Guidance for clinicians. MMWR Recomm Rep 52:1–28
61. CDC (2003) Update: cardiac-related events during the civilian smallpox vaccination program—United States, 2003. Morb Mortal Wkly Rep 52:492–496
62. Halsell JS, Riddle JR, Atwood JE, Gardner P, Shope R, Poland GA, Gray GC, Ostroff S, Eckart RE, Hospenthal DR, Gibson RL, Grabenstein JD, Arness MK, Tornberg DN (2003) Myopericarditis following smallpox vaccination among vaccinia-naive US military personnel. JAMA 289:3283–3289
63. Briody BA (1959) Response of mice to ectromelia and vaccinia viruses. Bacteriol Rev 23:61–95
64. Lee SL, Roos JM, McGuigan LC, Smith KA, Cormier N, Cohen LK, Roberts BE, Payne LG (1992) Molecular attenuation of vaccinia virus: mutant generation and animal characterization. J Virol 66:2617–2630
65. Gaertner DJ, Batchelder M, Herbst LH, Kaufman HL (2003) Administration of vaccinia virus to mice may cause contact or bedding sentinel mice to test positive for orthopoxvirus antibodies: case report and follow-up investigation. Comp Med 53:85–88
66. Holt RK, Walker BK, Ruff AJ (2002) Horizontal transmission of recombinant vaccinia virus in strain 13 guinea pigs. Contemp Top Lab Anim Sci 41:57–60
67. Engler RJ, Kenner J, Leung DY (2002) Smallpox vaccination: risk considerations for patients with atopic dermatitis. J Allergy Clin Immunol 110:357–365

Chapter 2

In-Fusion® Cloning with Vaccinia Virus DNA Polymerase

Chad R. Irwin, Andrew Farmer, David O. Willer, and David H. Evans

Abstract

Vaccinia virus DNA polymerase (VVpol) encodes a 3′-to-5′ proofreading exonuclease that can degrade the ends of duplex DNA and expose single-stranded DNA tails. The reaction plays a critical role in promoting virus recombination in vivo because single-strand annealing reactions can then fuse molecules sharing complementary tails into recombinant precursors called joint molecules. We have shown that this reaction can also occur in vitro, providing a simple method for the directional cloning of PCR products into any vector of interest. A commercial form of this recombineering technology called In-Fusion® that facilitates high-throughput directional cloning of PCR products has been commercialized by Clontech. To effect the in vitro cloning reaction, PCR products are prepared using primers that add 16–18 bp of sequence to each end of the PCR amplicon that are homologous to the two ends of a linearized vector. The linearized vector and PCR products are coincubated with VVpol, which exposes the complementary ends and promotes joint molecule formation. Vaccinia virus single-stranded DNA binding protein can be added to enhance this reaction, although it is not an essential component. The resulting joint molecules are used to transform *E. coli*, which convert these noncovalently joined molecules into stable recombinants. We illustrate how this technology works by using, as an example, the cloning of the vaccinia N2L gene into the vector pETBlue-2.

Key words: Vaccinia virus, Recombineering, In-Fusion® cloning, DNA polymerase, PCR cloning

1. Introduction

By an odd coincidence, vaccinia virus-encoded enzymes have been utilized in two distinctly different methods for rapid cloning of PCR products, both of which have been commercialized. The first method was invented by Stuart Shuman and is sold as Topo® cloning kits by Invitrogen (1, 2). The Shuman method exploits the reversible association between vaccinia virus topoisomerase I and its

Stuart N. Isaacs (ed.), *Vaccinia Virus and Poxvirology: Methods and Protocols*, Methods in Molecular Biology, vol. 890,
DOI 10.1007/978-1-61779-876-4_2, © Springer Science+Business Media, LLC 2012

DNA cleavage target, and permits rapid cloning of PCR amplicons into vectors that are supplied bearing molecules of topoisomerase covalently attached to the vector ends. The advantages of this elegant method have been reviewed and compared with other high-throughput cloning technologies (3).

We have described an alternate method for ligase-independent cloning of PCR products, which instead uses vaccinia virus DNA polymerase (VVpol). The principles of the method are shown in Fig. 1. These studies have shown that VVpol can fuse linear DNA molecules into concatemers, in vitro, if the ends of these DNAs share ten or more nucleotides of sequence identity (4–8). The joining reaction depends upon the 3′-to-5′ proofreading exonuclease

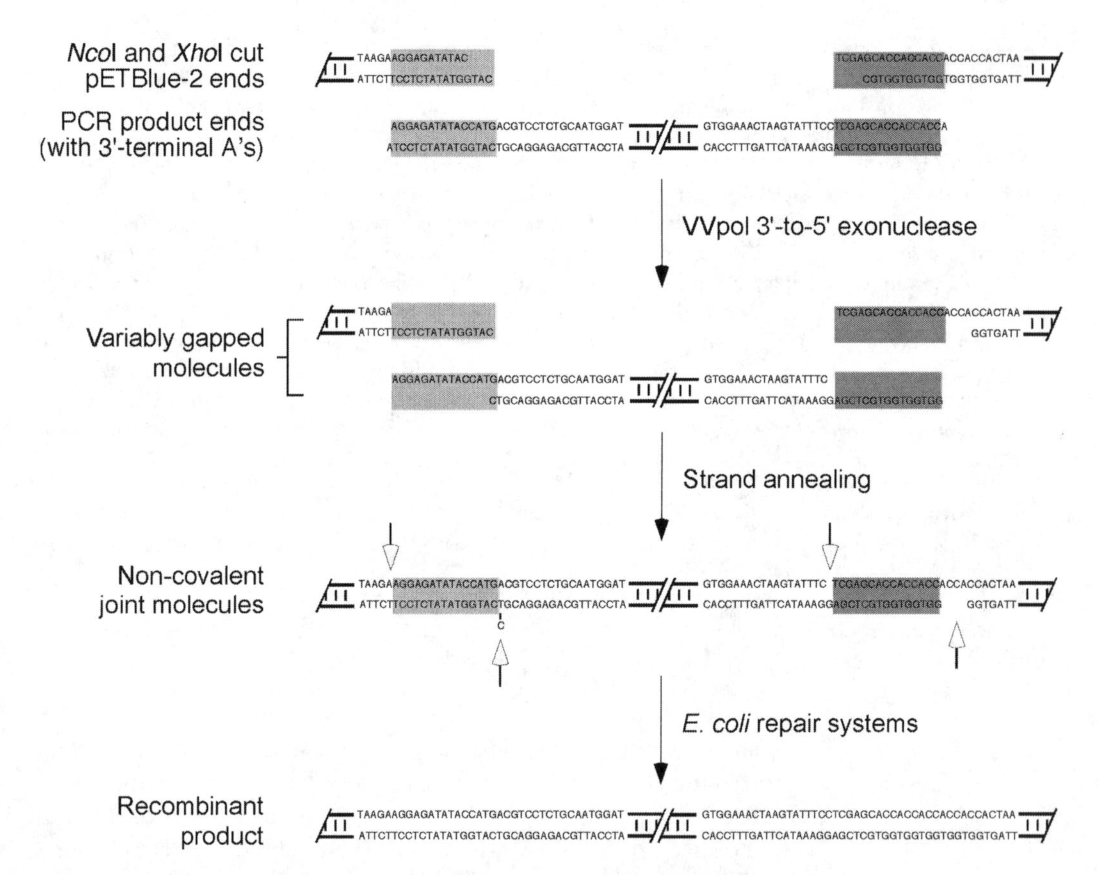

Fig. 1. Cloning DNA with vaccinia virus DNA polymerase. The target vector is digested with restriction enzyme(s) and the PCR products are prepared using primers that encode homology to the vector ends (*dark boxes*, the recommended minimal length of the homology is 16–18 bp). These products are then mixed together and incubated with VVpol. The 3′-to-5′ exonuclease activity degrades the ends of the DNAs, exposing areas of complementary sequence. These ends can then anneal (in a cloning reaction that is stimulated by vaccinia virus I3 single-strand DNA binding protein) and create noncovalently linked joint molecules that are sufficiently stable to survive transfection of *E. coli*. *E. coli* DNA repair systems convert the joint molecules into stable recombinants. Note that the joint molecules will contain a mix of nicks, gaps, and extra nucleotides (*white arrows*) due to variation in the extent of exonuclease attack. We illustrate the technique using a specific combination of *NcoI* and *XhoI* cut vector (pETBlue-2) and PCR amplicon (the N2L gene), but the method can be used with any combination of homologous ends.

function of the polymerase and is enhanced by adding vaccinia virus single-strand DNA binding protein (the I3 protein) (4, 5, 9). The proofreading exonuclease can attack a variety of duplex ends and the joining reaction depends upon it exposing 5′-overhangs of single-stranded DNA. Such single-stranded DNAs can anneal spontaneously if they encode complementary sequences and will form joint molecules that can be readily converted into covalently joined recombinants after transfection of *E. coli*. One of the properties of VVpol that may help promote these reactions is that it has a reduced capacity to attack joint molecules, once formed, and this may help stabilize such products against continued exonucleolytic degradation (10). These in vitro systems exhibit many of the same properties that characterize in vivo vaccinia virus recombination reactions (i.e., they have the same exonuclease and homology requirements, and the products show evidence of attack by a 3′-to-5′ exonuclease), suggesting that vaccinia virus also uses the DNA polymerase as a recombinase in infected cells (4, 7, 8).

This approach has the advantage that it requires no modification of existing vectors beyond linearizing them, which can be accomplished either by digestion with restriction enzymes(s) or by PCR amplification of the entire vector. This second approach thus permits the cloning of fragments at any location in a vector, even if there are no convenient cloning sites. Significantly, the VVpol 3′ to 5′ exonuclease activity is able to process DNA irrespective of the type of duplex end (5′-overhang, blunt, or 3′-overhang). Thus, the method permits directional cloning of PCR products irrespective of the form of the linear ends of the vector. The method does require extending the PCR primers to add sequences that duplicate the 16–18 bp of sequences flanking the vector cut site, but this costs very little nowadays and is often done anyway to produce clones encoding precisely modified junction sequences (e.g., peptide epitope tags and altered reading frames). The benefit, however, is that the desired sequence is cloned without the addition of any extra bases, since the primer extensions are simply homologous to the ends of the target vector itself. Moreover, there appears to be no significant restriction on the sequence of homology used, permitting cloning at any desired location. In this chapter, we outline this approach and illustrate the method by showing how it was used to clone the vaccinia N2L gene into the protein expression vector pETBlue-2.

2. Materials

2.1. DNA Substrate Preparation

1. 60 mm dish of BSC-40 cells.

2. Media: minimal-essential media supplemented with 5% FBS and 1% of each of the following: nonessential amino acids, L-glutamine and antibiotic–antimycotic.
3. Cell lysis buffer: 1.2% (w/v) SDS, 50 mM Tris–HCl, pH 8.0, 4 mM EDTA, 4 mM $CaCl_2$, 0.2 mg/mL proteinase K.
4. Buffer saturated phenol.
5. Cold 95% and 70% ethanol.
6. 3 M sodium acetate, pH 5.2.
7. Forward primer: 5′-AGG AGA TAT ACC ATG ACG TCC TCT GCA ATG GAT-3′ (see Notes 1 and 2).
8. Reverse primer: 5′-GGT GGT GGT GCT CGA GGA AAT ACT TAG TTT CCA C-3′ (see Notes 1 and 2).
9. *Taq* DNA polymerase (e.g., Fermentas) (see Note 3).
10. pETBlue-2 plasmid (Novagen).
11. Restriction enzymes.
12. 0.8% Agarose gel in 1× Tris–acetate buffer (TAE).
13. Qiagen gel extraction kit.
14. NanoDrop spectrophotometer.

2.2. Cloning by the Strand Joining Reaction and Plasmid Analysis

1. 10× Reaction buffer: 300 mM Tris–HCl, pH 7.9, 50 mM $MgCl_2$, 700 mM NaCl, 18 mM dithiothreitol, 0.1 mg/mL acetylated BSA.
2. Polymerase dilution buffer: 25 mM potassium phosphate pH 7.4, 5 mM β-mercaptoethanol, 1 mM EDTA, 10% (v/v) glycerol, 0.1 mg/mL acetylated BSA.
3. VVpol, 25 ng/μL in polymerase dilution buffer (see Note 4).
4. I3 single-strand DNA binding protein, 0.5 mg/mL in PBS (see Note 5).
5. 37°C water bath.
6. 55°C heating block.
7. Electro- and chemically competent *E. coli*, strain DH10B (see Note 6).
8. Bio-Rad Gene Pulser and electroporation cuvettes.
9. SOC medium.
10. LB/X-gal/amp plates: LB agar supplemented with 40 μg/mL X-gal, 100 μg/mL ampicillin (see Note 7).
11. LB/amp media: LB supplemented with 100 μg/mL ampicillin.
12. Mini-prep kit (e.g., Fermentas).
13. Ethidium bromide solution: 0.1 μg/mL ethidium bromide in 1× TAE.
14. Kodak Gel Logic 200L photodocumentation system.

15. Reaction stopping solution: 2.4 μL 0.5 M EDTA, 0.6 μL 20 mg/mL proteinase K, 0.2 μL 10% (w/v) SDS. Add 3.2 μL per 20 μL reaction.

3. Methods

3.1. DNA Substrate Preparation

Any DNA can be used, but as an example we used a 553-bp amplicon encoding the N2L gene. This DNA was prepared by PCR using purified vaccinia virus genomic DNA (strain Western Reserve) as a template, *Taq* DNA polymerase and standard PCR reaction and cycling conditions.

3.1.1. Isolation of Vaccinia Virus Genomic DNA

1. Infect a 60-mm dish of BSC-40 cells with vaccinia virus (strain Western Reserve) at an M.O.I. of five.
2. Twenty-four hour after infection, remove the media and add 1 mL of cell lysis buffer.
3. Incubate at 37°C for 3–4 h.
4. Transfer the mixture to a 1.5-mL tube and add 0.5 mL of buffer saturated phenol. Mix well and then centrifuge at room-temperature for 10 min at 18,000 × *g*.
5. Transfer 0.3 mL of the aqueous layer to a new tube and add 0.05 mL of 3 M sodium acetate and 1.25 mL 95% ethanol.
6. Allow the DNA to precipitate for 15 min at −80°C before centrifuging at 18,000 × *g* for 15 min.
7. Remove the supernatant and wash pellet with 70% ethanol.
8. Allow pellet to air-dry for 20–30 min at room temperature before resuspending in water.
9. Check OD and dilute DNA to 25 ng/μL.

3.1.2. PCR Amplification of VACV Gene and Plasmid DNA Preparation

1. To PCR amplify the N2L gene two 50 μL PCR reactions were assembled as recommended by the supplier of *Taq* polymerase, using 25 ng of vaccinia virus DNA as a template. The cycling parameters were as follows: 2 min initial denaturation at 94°C followed by 30 cycles of 94°C for 30 s, 50°C for 30 s, and 72°C for 60 s. This was followed by a final elongation step of 72°C for 7 min.
2. Digest pETBlue-2 plasmid DNA with *Nco*I and *Xho*I.
3. Gel-purify both the PCR product and plasmid digestion on a 0.8% agarose gel (see Note 8).
4. Bands of the correct size were excised and purified using a Qiagen gel extraction kit.
5. A NanoDrop spectrophotometer was used to determine the yield and purity of the DNA.

3.2. Cloning Reaction and Plasmid Confirmation

In parallel to the "home-made" reactions outlined below, we also assembled a cloning reaction using In-Fusion® enzyme and incubated the mix as directed by the supplier (Clontech). The commercial reaction is simpler and more efficient, as discussed in Subheading 3.4.

1. Cloning reactions were prepared as outlined in Table 1 using a 3:1 molar ratio of insert to vector.
2. The cloning reactions were started by adding 25 ng of VVpol last and then incubating for 20 min at 37°C.
3. The cloning reactions were stopped by heating to 55°C for 15 min.
4. After heat inactivating each strand-joining cloning reaction, 1 μL of each product from cloning reactions 1–5 was used to transform electrocompetent DH10B *E. coli* cells using a Bio-Rad Gene Pulser (see Subheading 3.3 for what to do with remaining material in each cloning reaction tube).
5. We added 0.25 mL of SOC medium to each electroporation cuvette, and the bacteria were incubated for 1 h at 37°C before being plated in duplicate on LB/X-gal/amp plates (see Note 9).
6. The plates were scored next day for the yield of blue or white colonies (see Table 2 for an example of the results and discussion).
7. Five white colonies and one blue colony were selected from the transformants produced in cloning reaction 5, Table 1, and inoculated into 3 mL of LB/amp media and cultured overnight at 37°C.

Table 1
Cloning reaction composition

	Cloning reaction number				
Component	**1**	**2**	**3**	**4**	**5**
10× Reaction buffer	2 μL	2 μL	2 μL	2 μL	2 μL
Vector (50 ng/μL)	2 μL	2 μL	2 μL	2 μL	2 μL
Insert (25 ng/μL)	–	–	2 μL	2 μL	2 μL
VVpol (25 ng/μL)	–	1 μL	–	1 μL	1 μL
I3 (0.5 mg/mL)	–	–	–	–	1 μL
Water	16 μL	15 μL	14 μL	13 μL	12 μL
Total	20 μL	20 μL	20 μL	20 μL	20 μL

Table 2
Cloning efficiency

Cloning reaction # (from Table 1)	1	2	3	4	5
Cloning reaction composition					
Vector	✓	✓	✓	✓	✓
PCR amplicon		✓	✓	✓	✓
VVpol				✓	✓
I3					✓
Cloning reaction yield					
Blue colonies	25	14	9	19	2
White colonies	9	22	17	35	304
Percent white	26%	61%	65%	65%	99%

The addition of VVpol and I3 greatly increased the number of transformants as well as the proportion of white clones. DNA isolated from In-Fusion® cloning reactions were transfo✓rmed in parallel (see Subheading 3.4)

8. The recombinant plasmids were purified using a mini-prep kit.
9. The plasmid concentrations determined by spectrophotometry, and 0.5 μg of each DNA cut with restriction enzymes. In the example here, the plasmid DNA was digested with either *Bam*HI or *Nco*I or *Xho*I.
10. The resulting cloning reaction products were size fractionated by electrophoresis using a 0.8% agarose gel, stained with ethidium bromide, and imaged using a Kodak Gel Logic 200L photo documentation system (see Fig. 2 and Note 10).

3.3. Detection of End-Joining by Agarose Gel Electrophoresis

To detect the production of joint molecules, one can use the remainder of the cloning reactions which were not used for bacterial transformation, to determine the success of the end-joining step. To do this (detailed below) the DNA is incubated with SDS and proteinase K to remove the VVpol and I3 proteins. The material is fractionated by electrophoresis on a 1.2% agarose gel and visualized with ethidium bromide. The exonucleolytic activity of VVpol is readily detected in many reactions as a blurring of the bands and a shift to smaller sizes (e.g., see Fig. 3, cloning reaction 2). When both insert and vector are present, the formation of slower migrating bands, indicative of the presence of joint molecules, is also seen. The greatest yield of joint molecules is seen in reactions containing VVpol plus I3 protein (see Fig. 3, cloning reaction 5). Several different products can be seen in these cloning reactions, which appear to comprise a mixture of concatemers and circles.

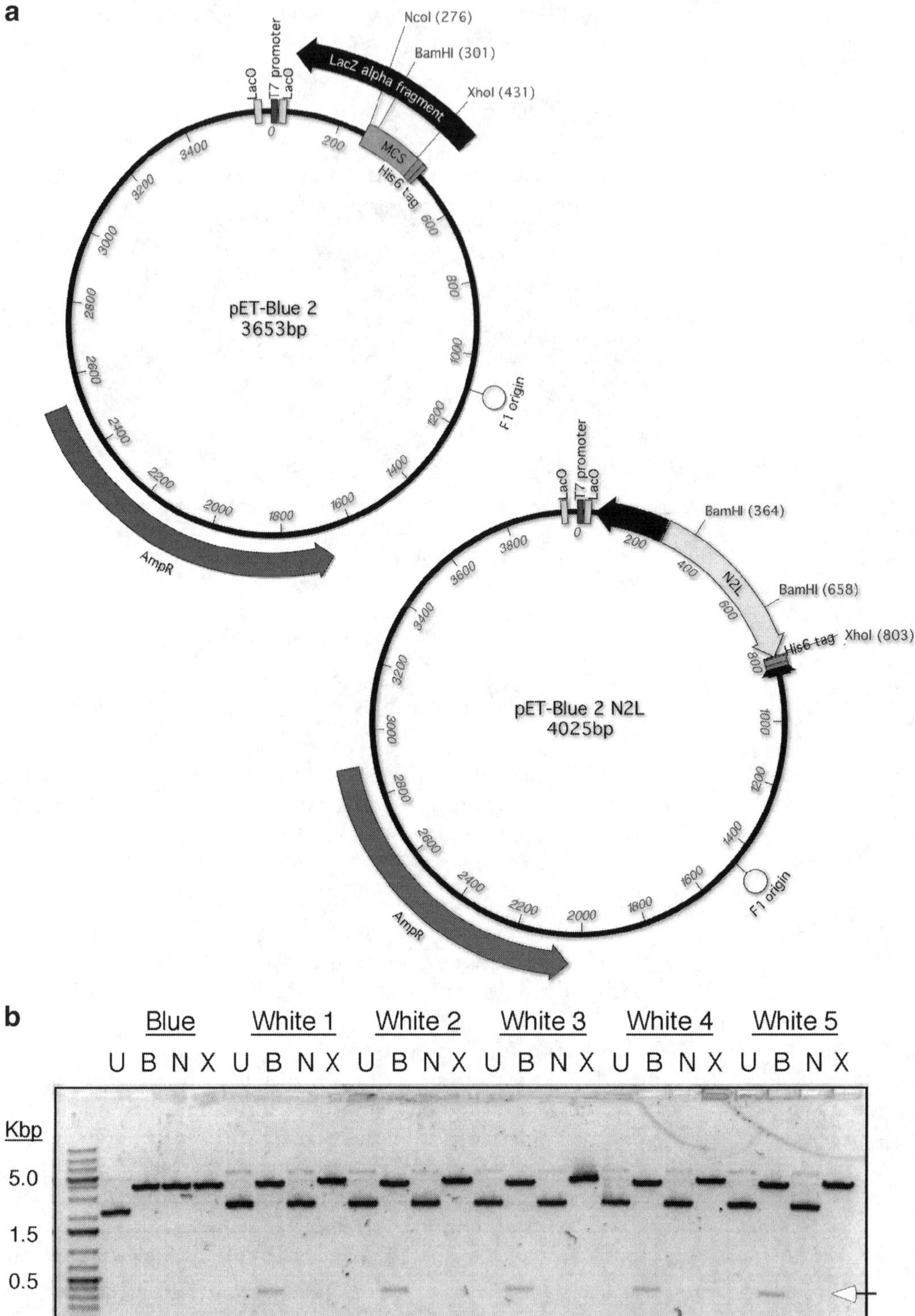

Fig. 2. Restriction analysis recombinant plasmids. (**a**) Plasmid maps showing the original vector (pETBlue-2) and predicted recombinant (pETBlue-2 N2L). (**b**) Restriction analysis of recombinant clones. Plasmid DNA was extracted from six colonies (5 white and 1 blue) and digested with *Bam*HI (B), *Nco*I (N), or *Xho*I (X). The products were separated on a 0.8% agarose gel along with uncut (U) DNA and stained with ethidium bromide. All five clones from the white colonies encoded N2L inserts (*arrow*). The blue colony appeared to be the original vector.

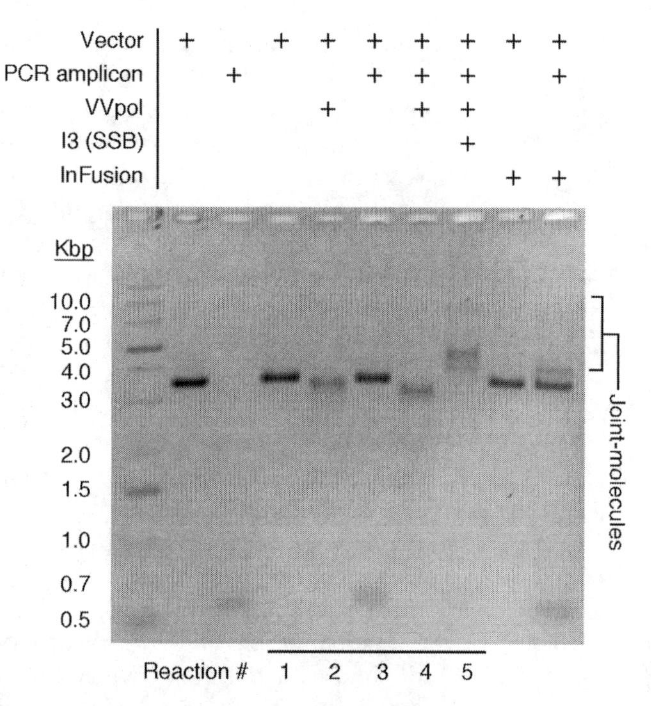

Fig. 3. Agarose gel analysis of cloning reaction products. The cloning reaction products were deproteinized and then separated on a 1.2% agarose gel. Under these conditions, the exonuclease activity causes a reduction in apparent mass and blurring of the bands. The greatest yield of joint molecules is seen in cloning reactions containing both VVpol and I3 single-strand DNA binding protein (cloning reaction #5). Although the commercial In-Fusion® cloning reactions produce lesser numbers of joint molecules (*last lane*), they actually produce a far greater yield of recombinants.

The In-Fusion® cloning reaction mix produces primarily a single new joint molecule in a manner dependent upon the presence of both vector and PCR amplicons (see Fig. 3).

1. To measure the efficiency of strand joining, 3.2 μL of the reaction stopping solution was added to the remainder of each of cloning reactions 1–5.
2. Tubes were incubated for 30 min at 37°C.
3. The cloning reaction products were then separated by electrophoresis for 4 h at 40 V/cm in a 1.2% agarose gel.
4. The DNA was stained with ethidium bromide solution for 30 min and imaged using a Kodak Gel Logic 200L photodocumentation system (see Fig. 3 for an example of results).

3.4. Conclusions

We have shown that VVpol can be used in conjunction with vaccinia virus single-strand DNA binding protein to catalyze a ligase-independent method for cloning PCR products. This approach is adaptable to any vector, allows for the rapid directional cloning of DNAs, and readily facilitates the engineering of flanking regulatory and other sequence elements. While we illustrated the joining of two pieces of DNA, this approach has also been used to clone multiple pieces of DNA together at once (11, 12).

Clontech has optimized this method to generate a more efficient reaction. The commercial method produces many more (>50-fold) transformants than do our original methods. For example, the reactions we assembled including vector, insert, VVPol, and I3 yielded 304 white and 2 blue transformants from 1/20th of the DNA (see Table 2, reaction 5), whereas the vector plus insert In-Fusion® cloning reaction yielded 410 white and 3 blue colonies from 1/800th of the DNA. The commercial method yielded just a single white colony when supplied with only vector DNA. Both methods produce about the same proportion of white clones (>99%) when cloning reactions contain both vector and insert DNA.

This method produces very few aberrant clones, as illustrated by the observation that all five of the plasmids isolated from white colonies encoded an insert. Clontech's internal quality control data indicates that >90% of constructs produced in their proprietary approach contain inserts, even in the absence of blue–white or other selection methods. Furthermore, when Clontech analyzed one side of the junction region in a much larger library, only 43 out of 3,650 clones (1.1%) were found to contain mutations in the region of homologous sequence common to the vector and PCR product. Assuming the error rate is the same for both junctions, the likely percentage of clones that have at least one mutation in either end is just 2.3%. This highlights the accuracy of this process. We have noted some very rare examples of codons being lost within repetitive sequences (e.g., CAT repeats associated with His_6 tags), presumably by a process resembling the Streisinger frameshift error mechanism (13). None of these error frequencies are high enough to routinely inconvenience the investigator, but illustrate why one should still always sequence any new clone as a matter of good lab practice.

Although few research laboratories would have the technical capacity (or desire) to prepare VVpol in practice, the availability of a commercial version of the method makes it readily accessible to interested researchers, and the utility of the method has been documented in a number of large-scale cloning exercises (14–16).

4. Notes

1. Although one could use many different vectors, in this chapter we choose pETBlue-2 because it permits blue/white screening of inserts and regulated expression of a His_6-tagged protein from an IPTG-inducible T7 RNA polymerase promoter. The two primers that were used to amplify the N2L gene are bipartite in nature. The 3′-ends of these primers encode the N- and

C-terminal ends of the N2 protein and were designed using standard PCR design principles. The 5′-ends of these primers encode sequences that are homologous to the ends of *Nco*I- and *Xho*I-digested pETBlue-2 DNA. The forward primer encodes promoter sequences found upstream of the *Nco*I cut site and the reverse primer encodes the His_6-tag found downstream of the *Xho*I site. The underlined nucleotides in the primer denote nucleotides that are identical to the pETBlue-2 sequence. It should be noted that the forward primer was designed to delete the *Nco*I site and place the N2L gene in frame with the vector start codon, while the *Xho*I site was retained in the reverse primer. Vector maps were designed using MacVector 11.0.

2. Clontech provides a program that can be used to design primers for In-Fusion® cloning. It is available at http://bioinfo.clontech.com/infusion/convertPcrPrimersInit.do.
3. While *Taq* DNA polymerase was used here, high-fidelity polymerases can also be used to prepare the PCR amplicons. The method is not affected by the presence or absence of 3′-A residues.
4. Vaccinia DNA polymerase can be purified to homogeneity using methods developed by Traktman and her colleagues (17). The purity was ascertained by SDS-PAGE and the concentration determined using a Bradford assay (Bio-Rad). A working stock of polymerase was prepared by diluting it to a concentration of 25 ng/μL in polymerase dilution buffer. This reagent is part of the In-Fusion® cloning kit.
5. A C-terminal His_6-tagged form of recombinant vaccinia single-strand DNA binding protein (I3) can be expressed in *E. coli* strain BL21 and purified using nickel affinity columns as described by Tseng et al. (18). The purity was confirmed by SDS-PAGE, and the concentration determined using a Bradford assay. To produce a working stock, the I3 DNA binding protein was diluted to a concentration of 0.5 mg/mL in PBS.
6. We recommend using *E. coli* with a competency of at least 1×10^8 cfu/μg in order to increase the number of clones obtained.
7. LB plates containing X-gal permits blue–white screening of colonies if beta-galactosidase cassette included in the plasmid.
8. We recommend purifying the PCR products because the dNTPs in the PCR reaction can inhibit the 3′-to-5′ exonuclease and thus inhibit joint molecule formation (9). Although this step is not strictly necessary, gel purification also minimizes the later recovery of vector plasmid or cloning of undesirable PCR products.
9. The In-Fusion® cloning reaction (10 μL) was diluted to 100 μL with 10 mM Tris–HCl, pH 8 and 1 mM EDTA, and 2.5 μL

was used to transform chemically competent *E. coli*, and 1/20th of the mix plated in a similar fashion.

10. The plasmid DNA was digested with either *Bam*HI, *Nco*I or *Xho*I. These enzymes each cut pETBlue-2 once, although the *Bam*HI site is deleted during a *Nco*I and *Xho*I double digest. There are no *Nco*I and *Xho*I sites in the N2L gene and *BamHI* is found twice (see Fig. 2, panel a). The plasmids recovered from the five white colonies all exhibited the restriction patterns expected if they encoded an N2L gene insert and the presence of this insert, and its proper orientation, was later confirmed by DNA sequencing (data not shown). The DNA extracted from a blue colony had a restriction pattern identical to pETBlue-2, suggesting that it was derived from an uncut plasmid.

5. Acknowledgments and Disclosures

We would like to thank Dr. James Lin and Ms. Nicole Favis for their assistance with the purification of vaccinia DNA polymerase. Chad Irwin is a recipient of an Alberta Cancer Research Institute Graduate Studentship. These studies were originally supported by an operating grant to D.E. from the Canadian Institutes of Health Research. Research in D.E.'s laboratory is currently supported by the Canadian Institutes of Health Research and Natural Sciences and Engineering Research Council. Clontech Laboratories, Inc. holds an exclusive licence to use the In-Fusion® technology from the University of Guelph, Guelph, Ontario. As inventors and former employees of the University of Guelph, D.E. and D.W. receive a portion of the royalties paid under the terms of the licence agreement.

References

1. Shuman S (1992) Two classes of DNA end-joining reactions catalyzed by vaccinia topoisomerase I. J Biol Chem 267:16755–16758
2. Shuman S (1992) DNA strand transfer reactions catalyzed by vaccinia topoisom*erase I.* J Biol Chem 267:8620–8627
3. Marsischky G, LaBaer J (2004) Many paths to many clones: a comparative look at high-throughput cloning methods. Genome Res 14:2020–2028
4. Gammon DB, Evans DH (2009) The 3′-to-5′ exonuclease activity of vaccinia virus DNA polymerase is essential and plays a role in promoting virus genetic recombination. J Virol 83:4236–4250
5. Willer DO et al (1999) Vaccinia virus DNA polymerase promotes DNA pairing and strand-transfer reactions. Virology 257:511–523
6. Willer DO et al (2000) In vitro concatemer formation catalyzed by vaccinia virus DNA polymerase. Virology 278:562–569
7. Yao XD, Evans DH (2003) Characterization of the recombinant joints formed by single-strand annealing reactions in vaccinia virus-infected cells. Virology 308:147–156
8. Yao XD, Evans DH (2001) Effects of DNA structure and homology length on vaccinia virus recombination. J Virol 75:6923–6932
9. Hamilton MD et al (2007) Duplex strand joining reactions catalyzed by vaccinia virus

DNA polymerase. Nucleic Acids Res 35: 143–151
10. Hamilton MD, Evans DH (2005) Enzymatic processing of replication and recombination intermediates by the vaccinia virus DNA polymerase. Nucleic Acids Res 33:2259–2268
11. Sleight SC et al (2010) In-Fusion BioBrick assembly and re-engineering. Nucleic Acids Res 38:2624–2636
12. Zhu B et al (2007) In-fusion assembly: seamless engineering of multidomain fusion proteins, modular vectors, and mutations. BioTechniques 43:354–359
13. Streisinger G et al (1966) Frameshift mutations and the genetic code. This paper is dedicated to Professor Theodosius Dobzhansky on the occasion of his 66th birthday. Cold Spring Harb Symp Quant Biol 31:77–84
14. Benoit RM et al (2006) An improved method for fast, robust, and seamless integration of DNA fragments into multiple plasmids. Protein Exp Purif 45:66–71
15. Park J et al (2005) Building a human kinase gene repository: bioinformatics, molecular cloning, and functional validation. Proc Natl Acad Sci USA 102:8114–8119
16. Berrow NS et al (2007) A versatile ligation-independent cloning method suitable for high-throughput expression screening applications. Nucleic Acids Res 35:e45
17. McDonald WF, Traktman P (1994) Overexpression and purification of the vaccinia virus DNA polymerase. Protein Exp Purif 5:409–421
18. Tseng M et al (1999) DNA binding and aggregation properties of the vaccinia virus I3L gene product. J Biol Chem 274:21637–21644

Chapter 3

Genetic Manipulation of Poxviruses Using Bacterial Artificial Chromosome Recombineering

Matthew G. Cottingham

Abstract

Traditional methods for genetic manipulation of poxviruses rely on low-frequency natural recombination in virus-infected cells. Although these powerful systems represent the technical foundation of current knowledge and applications of poxviruses, they require long (≥500 bp) flanking sequences for homologous recombination, an efficient viral selection method, and burdensome, time-consuming plaque purification. The beginning of the twenty-first century has seen the application of bacterial artificial chromosome (BAC) technology to poxviruses as an alternative method for their genetic manipulation, following the invention of a long-sought-after method for deriving a BAC clone of vaccinia virus (VAC-BAC) by Arban Domi and Bernard Moss.

The key advantages of the BAC system are the ease and versatility of performing genetic manipulation using bacteriophage λ Red recombination (recombineering), which requires only ~50 bp homology arms that can be easily created by PCR, and which allows seamless mutations lacking any marker gene without having to perform transient-dominant selection. On the other hand, there are disadvantages, including the significant setup time, the risk of contamination of the cloned genome with bacterial insertion sequences, and the nontrivial issue of removal of the BAC cassette from derived viruses. These must be carefully weighed to decide whether the use of BACs will be advantageous for a particular application, making pox-BAC systems likely to complement, rather than supplant, traditional methods in most laboratories.

Key words: Poxvirus, Vaccinia, Bacterial artificial chromosome, Recombineering, Red, Fowlpox

1. Introduction

Bacterial artificial chromosome (BACs) are low-copy number F based plasmids into which at least 300 kb of foreign DNA can be inserted, making them eminently suitable for cloning of large DNA viruses such as the herpesviruses, as first described in 1997 (1). Use of BACs for poxviruses requires the production of circular DNA

Stuart N. Isaacs (ed.), *Vaccinia Virus and Poxvirology: Methods and Protocols*, Methods in Molecular Biology, vol. 890, DOI 10.1007/978-1-61779-876-4_3, © Springer Science+Business Media, LLC 2012

molecules from the normally linear viral genome, and it was not until 2002 that a method was developed for isolating and cloning such species (2) by using pharmacological block of viral late gene expression to promote formation of head-to-head concatemeric genomic DNA, which is then able to self-recombine to form head-to-tail concatemers and hence circular DNA molecules. These are able to propagate in *E. coli* by virtue of a mini-F BAC plasmid cassette previously inserted into the viral genome using traditional poxviral recombination and plaque-purification. Once a BAC is derived, the power of *RecA*-independent, bacteriophage λ encoded Red recombination can be brought to bear. This involves transient induction of the phage *Exo*, *Bet*, and *Gam* functions, allowing recombination by single strand invasion (rather than via a Holliday junction) (3), requiring only short (50 bp) homologous flanking sequences, which can be conveniently synthesized as oligonucleotides. Various manifestations of this system allow facile seamless modifications, including insertions, deletions, and substitutions (4).

Since the original description of "VAC-BAC", based on the vaccinia virus Western Reserve (WR) strain, by Domi and Moss (2, 5), other research groups, both academic and commercial, have described the application of the technique to other poxviruses, first the attenuated strain modified vaccinia virus Ankara (MVA) (6), and then its replication competent parental strain chorioallantois vaccinia virus Ankara (CVA), together with an independently derived MVA-BAC (7), and most recently a BAC clone of cowpox virus (8).

The advantages of a pox-BAC system are the versatility and facility of recombineering, especially for "markerless" mutants, and the ability to perform multiple manipulations in parallel. Following recombineering, the modified BAC should be clonal, obviating the necessity for plaque-purification, and allowing generation of mutants with impaired growth that may be impossible to isolate using selection at the viral level (9). The defined nature of a BAC clone may also be valuable for vaccine clinical manufacturing applications, where traceability is paramount, since poxviral strains are typically heterogeneous (10).

Despite its convenience, there are also significant limitations of the pox-BAC system, which need to be taken into account when deciding whether to use BAC methods or a traditional-style approach. Since poxviral DNA is noninfectious, a helper virus is required to initiate "rescue" or reconstitution of the BAC clone to a replicating viral genome. This requires some replication block or host range restriction of the helper virus (typically fowlpox virus) to prevent its growth at the expense of the BAC-derived virus (see Subheading 3.3). Although very efficient, there are several unknowns associated with helper virus mediated BAC rescue – for example, is there potential for recombination with the helper virus (though it has been reported that fowlpox virus and vaccinia virus do not recombine (11)); and exactly how is the fused terminus in the BAC converted back to a hairpin with extruded nucleotides in the virus? Although VAC-BAC is remarkably genetically stable in DH10B *E. coli*, the repeat regions of the inverted terminal repeats are subject to expansion and contraction (2). Bacterial insertion sequences can jump into the BAC (see Subheading 3.5) (7), necessitating careful PCR-based quality control of derived viruses, especially in the case of an apparently non-rescuable (i.e., potentially lethal) mutation. Even when such an event has been ruled out, it is not necessarily safe to assume clonality of the input BAC, and rescued viruses must be extensively checked ensure they are pure and correct, since even an infinitesimal carryover of wild-type BAC can contaminate the rescue. Furthermore, BAC-derived poxviruses still carry the plasmid in the viral genome, and although this does not affect viral growth (12), its removal may be required for some applications (e.g., clinical vaccines) and is not necessarily straightforward, even when a "self-excising" BAC is utilized (12, 13). Despite these difficulties, the BAC recombineering method has remarkable power, as has recently been beautifully demonstrated by the introduction of the 6 large MVA deletions into CVA using 6 rounds or selection–counterselection (12 Red recombineering steps) in a VAC-BAC clone (7).

2. Materials

2.1. Isolation of BAC Clones from pre-BAC Poxviral DNA

1. IβT (isatin-β-thiosemicarbazone): 5 mg/mL in acetone, stored at –20°C. Immediately prior to use, dilute this stock to 1 mg/mL in 0.25 M NaOH (see Note 1).
2. Polypropylene cell lifter.
3. TBSE: 20 mM Tris–HCl, pH 8.0, 10 mM EDTA, 150 mM NaCl.
4. LSEB: 20 mM Tris–HCl, pH 8.0, 10 mM EDTA, 0.75% (w/v) SDS. Add 0.65 mg/mL proteinase K just before use.
5. PhaseLok Gel (see Note 2).
6. Tris-saturated phenol (see Note 3).
7. Tris-saturated phenol–chloroform–isoamyl alcohol (25:24:1) (see Note 3).
8. 3 M sodium acetate, pH 5.0 (e.g., ready-made solution from Sigma).
9. TE: 10 mM Tris–HCl, pH 8.0, 1 mM EDTA.
10. ElectroMAX maximum efficiency DH10B (e.g., Invitrogen) or home-made electrocompetent cells (see Subheading 3.4, steps 9 and 10).

11. Bio-Rad Gene Pulser electroporator and cuvettes (1 mm electrode gap).
12. SOC medium: Super Optimal Broth with glucose.
13. 14 mL polypropylene snap-cap tubes (see Note 4).
14. Chloramphenicol: 12.5 mg/mL (1,000×) stock in methanol.

2.2. BAC Miniprep

1. STET: 8% sucrose; 5% Triton X-100; 50 mM EDTA; 50 mM Tris–HCl, pH 8.0.
2. Alkaline SDS: 1% SDS in 0.2 M NaOH (see Note 5).
3. 7.5 M ammonium acetate.
4. Refrigerated microcentrifuge (e.g., Eppendorf 5415R).
5. DNAse-free RNAse A.
6. Restriction enzymes.

2.3. BAC Rescue

1. Baffled flasks (e.g., Wheaton) (see Note 6).
2. QIAGEN Plasmid Maxi kit (see Note 7).
3. Fowlpox virus (see Note 8).
4. Lipofectamine 2000 (see Note 9).
5. OptiMEM (see Note 10).

2.4. BAC Recombineering Using GalK

1. DNA oligonucleotide primers (see Note 11).
2. Finnzymes F-540: Phusion Hot Start polymerase (see Note 12).
3. p*GalK* plasmid and recombineering strains (obtained from NCI-Frederick (http://web.ncifcrf.gov/research/brb/recombineeringInformation.aspx) (14)).
4. QIAGEN Minelute gel purification kit or similar.
5. M9 minimal salts, 5× (e.g., BD Difco) solution prepared according to manufacturer's instructions and autoclaved.
6. 5× M63 minimal medium: 24 g KH_2PO_4 (anhydrous), 56 g K_2HPO_4 (anhydrous), 10 g $(NH_4)_2SO_4$, 1 mL 2.5 mg/mL $FeSO_4$ in 1 L; check pH=7 and autoclave.
7. M63 chloramphenicol/galactose agar plates (for positive selection): autoclave 7.5 g agar in 400 mL water and add 100 mL 5× M63 to molten agar. Equilibrate to 50°C. Add 0.5 mL of 1 M $MgSO_4$ (autoclaved); 2.5 mL of 0.2 mg/mL D-biotin (sterile-filtered); 2.5 mL of 9 mg/mL L-leucine (sterile-filtered); 500 μL of 12.5 mg/mL chloramphenicol in MeOH; and 5 mL of 20% D-galactose (autoclaved). Pour up to 20 Petri dishes. (All reagents can be obtained from Sigma.)
8. M63 chloramphenicol/deoxygalactose/glycerol agar plates (for negative selection): autoclave 7.5 g agar in 400 mL water and add 100 mL 5× M63 to molten agar. Equilibrate to 50°C.

Add 0.5 mL of 1 M $MgSO_4$ (autoclaved); 2.5 mL of 0.2 mg/mL D-biotin (sterile-filtered); 2.5 mL of 9 mg/mL L-leucine (sterile-filtered); 500 μL of 12.5 mg/mL chloramphenicol in MeOH; and 5 mL of 20% glycerol (autoclaved); and 5 mL of 20% deoxygalactose (sterile filtered). Pour up to 20 Petri dishes.

9. MacConkey galactose indicator plates: MacConkey agar base (e.g., BD Difco) prepared with 1% galactose according to manufacturer's instructions and autoclaved.
10. Phire polymerase (Finnzymes) (see Note 13).

2.5. Bacterial Insertion Sequence PCR

1. QIAGEN DNA blood mini kit or similar.
2. DNA oligonucleotide primers (see Table 6).
3. Native Taq polymerase kit (Fermentas) or similar (see Note 14).
4. 10× $(NH_4)_2SO_4$ Taq buffer: 750 mM Tris–HCl, pH 8.8 at 25°C, 200 mM $(NH_4)_2SO_4$, 0.1% Tween 20.

3. Methods

A full discussion of the options for design of a poxviral BAC system and subsequent strategies for genetic manipulation is outside the scope of this chapter. Brief considerations of some of the main points are given in the introductory paragraphs of the protocols provided here for obtaining BAC clones from a pre-BAC virus (see Subheading 3.1), miniprepping BAC DNA to screen candidate clones by PCR or restriction digest (see Subheading 3.2), rescuing an orthopoxvirus BAC clone in mammalian cells using an avipoxvirus helper (see Subheading 3.3), BAC recombineering using *GalK* selection in the SW102 strain, using deletion of *B15R* from MVA as an example (see Subheading 3.4), and checking that bacterial insertion sequences are absent from rescued viruses (see Subheading 3.5).

3.1. Isolation of BAC Clones from pre-BAC Poxviral DNA

The first stage in generation of a BAC clone of a poxvirus is to insert the BAC sequences required for propagation in *E. coli* into the viral genome using traditional endogenous recombination in infected cells. To accomplish this, a shuttle vector is constructed based on the selected mini-F BAC plasmid (e.g., pMBO131 (7) or pBELO-BAC11 (6, 8)) and containing the following features: (i) GFP under a poxviral promoter (or other convenient means for selection of pre-BAC recombinant virus and/or visualization of rescued BAC-derived viruses); (ii) flanking sequences (~500 bp) for recombination with the desired nonessential insertion locus in the viral genome (e.g., thymidine kinase (2, 8), the vaccinia virus *I3L-I4L* intergenic region (7), or the deletion 3 locus of MVA (6)); and (iii) a unique restriction site between the recombination flanks to allow linearization of the plasmid prior to transfection into infected cells and isolation of recombinants by plaque purification. The method as originally described (2) also added *loxP* sites to promote circularization, but this is not necessary (6–8). With the pure pre-BAC virus in hand, clones are isolated using the following protocol, based on Domi and Moss' original paper (2).

1. Prepare a 6-well plate containing confluent monolayers of the cell line normally utilized for propagation of the poxvirus of choice using the standard culture conditions.
2. Infect two wells with 5 pfu/cell of pre-BAC recombinant virus in the presence and absence (control) of 45 μM IβT in a volume of 1 mL/well. Typically, a reduced concentration of serum (2% instead of 10%) in the medium is used for this step.
3. After incubation for 1–2 h, replace the inoculum with 2 mL/well of reduced serum medium containing or lacking (control) 45 μM IβT and incubate overnight (see Note 15).
4. Scrape off the cell monolayer and pellet cells for 30 s at maximum speed in a microfuge. Wash the cells twice with PBS, pelleting after each wash.
5. Resuspend pellet in 50 μL TBSE by vortexing, then add 250 μL LSEB, mix by gentle inversion, and incubate at 37°C for 5 h to overnight.
6. Add 300 μL Tris-saturated phenol, mix well, and pour (do not pipette) into a prepared PhaseLok tube, then spin at maximum speed in a microfuge for 5 min. Decant (do not pipette) the supernatant into a fresh PhaseLok tube.
7. Repeat step 6.
8. Repeat step 6 using Tris-saturated phenol–chloroform–isoamyl alcohol (25:24:1).
9. Repeat step 6 using chloroform–isoamyl alcohol (24:1), but decant the supernatant into a 1.5-mL tube and precipitate DNA by adding 30 μL 3 M sodium acetate and 750 μL ethanol. Optionally incubate at –20°C for a few hours or overnight to improve precipitation.
10. Pellet DNA for 15 min at maximum speed in a microfuge (preferably refrigerated), wash pellet with 250 μL 70% EtOH, respin, air-dry, and resuspend in 20 μL of TE without pipetting or vortexing (see Note 16). The solution should appear viscous and contain >1 μg/mL DNA.
11. Electroporate DNA into electrocompetent *E. coli* by pipetting 1–3 μL of DNA (pipetted gently with a wide-bore or cut-off

pipette tip) into 50 μL electrocompetent DH10B *E. coli* in a cold 1 mm cuvette. Mix by gentle shaking and pulse at 1.8 kV/25 μF/200 Ω (see Note 17).

12. After electroporation, immediately add 1 mL SOC medium to the cuvette and transfer contents to a 14 mL snap-cap tube, and incubate for 1 h at 37°C with shaking.
13. Spread 1 μL, 10 μL (both diluted in 100 μL SOC), 100 μL and the remainder onto LB agar plates containing the correct antibiotic and incubate overnight at 37°C. The appropriate antibiotic is determined by the BAC cassette inserted into the virus. Typically, BeloBAC11 (or a similar mini-F) is used, in which case 12.5 μg/mL chloramphenicol is added to LB for selection.
14. Screen bacterial colonies by PCR and/or restriction digest (see Subheading 3.2) and send away one clone for complete sequencing (see Note 18).

3.2. BAC Miniprep

This protocol, based on one kindly supplied by Dr Richard Wade-Martins, Department of Anatomy and Human Genetics, Oxford University, UK, is useful for characterizing colonies by PCR or restriction digest. Direct PCR of a colony can also be used but is not always as reliable as miniprepping the DNA prior to PCR.

1. Inoculate colonies into cultures of 2 mL LB plus appropriate antibiotic (e.g., 12.5 μg/mL chloramphenicol) and incubate overnight with shaking at 37°C (or 32°C for recombineering strains).
2. Transfer 1.5 mL to microfuge tubes and pellet cells for 30 s at 16,100 × *g*. Aspirate the supernatant and resuspend the bacterial pellet in 70 μL STET by vortexing. Add STET to 4–6 tubes at a time: do not leave the unresuspended pellet in STET for a prolonged period.
3. Lyse by adding 200 μL alkaline SDS while vortexing. Immediately neutralize the solution by adding 150 μL of 7.5 M ammonium acetate while maintaining vortexing and then continue to vortex for another few seconds.
4. Place the neutralized lysate on ice for 5 min and then centrifuge at 16,100 × *g* for 20 min at 4°C.
5. Pour (do not pipette) the supernatant into a fresh 1.5-mL microfuge tube. Add 250 μL room-temperature isopropanol, mix by repeated inversion, and then pellet the DNA by centrifuging at 9,300 × *g* for 8 min. Orientate tube in centrifuge so that predicted location of pellet is known.
6. Aspirate supernatant (pellet may not be visible) and wash pellet with 200 μL room temperature 70% ethanol.

Table1
Primers for multiplex MVA-BAC integrity PCR

C19L/B25R[a]	acgggatcgcagtctttatg	ccggagacgtcatctgttct	232 bp
F10L	tgccggataaaagtgggata	caaaattgggctccatcagt	525 bp
J6R	tacggttttggggtgacatt	cgaccaccatatcctccatc	592 bp
A32L	ttcaccttcacaaaatacggagt	ttgctgtcgcacaaaatcat	359 bp
BAC identity[b]	atagaacttacgcaaatatta gcaaaaat	tggaaagcgggcagtga	444 bp

[a]ITR is highly variable: check these sequences in the strain of interest
[b]Specific for BAC insertion, with one primer in the viral flank (*A51R*) and the other in the BAC—this is for the author's MVA-BAC construct (6). A similar pair would need designing for other VAC-BAC constructs, and an analogous approach could be taken for other viruses

7. Aspirate supernatant and air-dry the pellet for 5–10 min. Resuspend the DNA in 50 μL of TE containing 5 μg/mL RNAse A (see Note 16).
8. For a restriction digest, use 8–10 μL in a 15 μL reaction and load the whole reaction into a single lane of an agarose electrophoresis gel (1% or less) (see Note 19).
9. For PCR, dilute the DNA 1:100 and use 1–5 μL per reaction. Table 1 shows a suggested multiplexed PCR screening assay for integrity of vaccinia virus BAC clones, which yields a ladder of five bands if all targets are present and Table 2 shows the PCR cocktail and reaction conditions.

3.3. BAC Rescue

In order to obtain virus from a genomic clone, a helper virus is required to supply the molecular machinery to initiate poxviral transcription from the transfected BAC. All poxvirus BACs created to date (2, 6–8) are orthopoxviruses so can be conveniently rescued using fowlpox virus in a mammalian cell line in which the BAC-cloned virus is able to propagate and the helper avipoxvirus is unable to replicate (11). Other options for helper viruses include the following: shope fibroma virus, a leporipoxvirus which does not replicate in mammalian cells (7, 15); a temperature-sensitive mutant helper virus (16); or potentially a UV-psoralen inactivated virus (17), since only early viral gene expression is required for rescue.

1. Grow up a 500 mL LB overnight culture of *E. coli* containing the poxviral BAC clone.
2. Purify the BAC for transfection using a standard QIAGEN maxiprep kit with the following modifications: (i) perform lysis in double volumes (20 mL) of each of Buffers P1, P2, and P3.

Table 2
PCR cocktail and reaction conditions for screening the integrity of vaccinia virus BAC clones

		Volume (μL)
C19L/B25R primers from Table 1 (10 μM stock)	Forward	0.4
	Reverse	0.4
F10L primers from Table 1 (10 μM stock)	Forward	0.4
	Reverse	0.4
J6R primers from Table 1 (10 μM stock)	Forward	0.4
	Reverse	0.4
A32L primers from Table 1 (of 10 μM stock)	Forward	0.4
	Reverse	0.4
BAC identity primers from Table 1 (10 μM stock)	Forward	0.4
	Reverse	0.4
dNTP mix (8 mM stock)		0.4
Titanium buffer		2.0
1:100 BAC miniprep DNA		2.0
Titanium Taq		0.4
Water		11.2
Total		20.0

Program: 95°C for 60 s then 35 cycles of 95°C for 30 s, 54°C for 10 s, 72°C for 60 s; final elongation 72°C for 3 min. Analyze on a 2% agarose gel

It may be necessary to gently break up viscous aggregates with a pipette tip after P3 addition (use the optional LyseBlue reagent). (ii) Heat Buffer QF to 65°C before eluting DNA from the Tip-500 column. (iii) Observe precautions in Note 16 when resuspending in ~100 μL TE. This procedure should yield up to 50 μg BAC DNA.

3. Prepare T25 flasks containing ~70–80% confluent monolayers of the cell line normally utilized for propagation of the poxvirus using the standard culture conditions.
4. Infect with fowlpox virus at 1 pfu/cell using medium with reduced serum (typically 2% instead of 10%) medium or OptiMEM in a volume of 2 mL and incubate for 1–2 h (see Note 20).

5. Meanwhile, prepare transfection mix: add 20 μL Lipofectamine 2000 to 600 μL OptiMEM and incubate at room temperature for 5 min; dilute 8 μg BAC DNA in 600 μL OptiMEM; add diluted Lipofectamine to diluted BAC-DNA, mix gently by inversion, and incubate at room temperature for 20–40 min. Aspirate the inoculum and replace with the transfection mixture, using a cut-off pipette tip. Rock the flask to distribute liquid evenly. If possible, rock flask gently every hour or so (see Note 21).
6. Incubate for 5 h (optimal) or overnight, then add 1.2 mL prewarmed OptiMEM.
7. After 24–48 h, examine by epifluorescence microscopy (if the poxvirus-BAC has a poxviral fluorescent reporter). The signal will be quite dim, and positive cells indicate only transfection, not necessarily rescue.
8. After 5 days, harvest the culture and freeze–thaw cells up to three times.
9. To assess rescue efficiency, make tenfold serial dilutions from undiluted rescue lysate down to 10^{-5} and replate onto fresh cell monolayers in 6-well plates, by infecting for 1–2 h then replacing medium. Inspect for viral CPE and marker gene fluorescence after 2–5 days. An efficient rescue by this method contains $>10^4$ pfu/mL in the lysate. Proceed to amplify and check the rescued virus by PCR (see Subheading 3.5).

3.4. BAC Recombineering

There are an increasing number of techniques and systems for genetic manipulation of BACs utilizing Red recombineering, which employs transient induction of the phage *exo*, *bet*, and *gam* genes to promote homologous recombination with very short (≥50 bp) flanking sequences (4). For poxviruses, the *GalK*-based system (14) has been utilized (6) as well as antibiotic selection (5, 9), the *rpsL-neo* selection–counterselection cassette (7), and "*en passant*" recombineering (8, 12). Other selection methods that may also be suitable have been described (18, 19). The protocol here is an example of deletion of the *B15R* open reading frame (encoding an IL-1β binding protein) from MVA-BAC using *GalK* selection in the SW102 strain and is partly based on Søren Warming's recombineering protocol (14) available at < http://web.ncifcrf.gov/research/brb/protocol/>. *GalK* or other bacterial selectable markers can be left in the virus (see Note 22) or removed in a second targeting reaction by counterselection (with optional additional mutations) to enable a subsequent round of insertion using the same marker (see Note 23). For introduction of long sequences such as transgenic expression cassettes, the tandem insertion/*en passant* removal method (8, 12, 20) is recommended instead of counterselection, which is rather inefficient, at least in MVA-BAC (6).

Table 3
Primers used for *B15R* deletion in MVA by *GalK* recombineering

Left homology arm	ctcttctccctttcccagaaacaaactttttttacccactataaaataaa
Right homology arm	tgaatgtatgttgttacatttccatgtcaattgagtttataagaatttt
Left *GalK* primer	ctcttctccctttcccagaaacaaactttttttacccactataaaataaaCCTGTTGACAATTAATCATCGGCA
Right *GalK* primer[a]	aaaaattcttataaactcaattgacatggaaatgtaacaacatacattcaTCAGCACTGTCCTGCTCCTT

[a]Right homology arm is reverse-complemented. *GalK* primer sequences (capitalized) are from Warming et al. (14)

1. The BAC of interest is first electroporated into the SW102 recombineering strain following the electroporation procedure described in Subheading 3.1, step 11, using electrocompetent SW102 prepared as described below in step 10.
2. Check the BAC from a single colony by multiplex PCR, restriction mapping (see Subheading 3.2), and rescue (see Subheading 3.3).
3. Design homology arms to delete the ORF of interest, and order gel purified, mass spectrometry checked primers for amplification of *GalK* with addition of the flanking sequences. As an example, we show the deletion of the *B15R* gene (encoding an IL-1β binding protein) from MVA (see Table 3 for primer sequences used for the deletion of *B15R* from MVA).
4. Amplify homology arms containing *GalK* using the PCR cocktail and reaction conditions shown in Table 4.
5. When PCR reaction is complete, add 1–2 μL *Dpn*I per reaction and incubate at 37°C for 30–60 min to remove template DNA (which is methylated, unlike the PCR product).
6. Run out the whole reaction on a 1% agarose gel, excise the band (expected size: 1.2 kb), and gel-purify the product. Check DNA concentration by A_{260}/A_{280}.
7. Grow up an overnight culture (32°C) of SW102 *E. coli* carrying MVA-BAC and dilute 1:50 the following morning into 100 mL LB with 12.5 μg/mL chloramphenicol in a 500 mL baffled flask. Grow to mid-log phase (3–5 h; OD_{600} 0.5–0.7) in a shaking water bath at 32°C.
8. Switch on second water bath to 42°C and prepare ice-cold water and prechilled centrifuge.
9. Take half the culture (50 mL) into a new baffled flask and heat shock for *exactly* 15 min in a shaking water bath at 42°C (and keep the other half of the culture at 32°C). Immediately immerse both the heat-shocked culture flask and control culture flask that was kept at 32°C in an ice-water slurry (see Notes 24 and 25).

Table 4
PCR cocktail and reaction conditions for the deletion of *B15R* from MVA

	Volume (μL)
Phusion HF buffer	10
Phusion	0.5
dNTP mix (8 mM)	1.25
Left *GalK* homology arm primer[a] (1 μM stock)	2.5
Right *GalK* homology arm primer[a] (1 μM stock)	2.5
p*GalK* template plasmid (4 ng/μL)	2.5
DMSO	1.5
Water	29.25
Total	50

[a]Primer concentrations are 1/10th standard. Program: 98°C for 30 s then 35 cycles of 98°C for 5 s, 68.4°C for 10 s, 72°C for 30 s; final elongation 72°C for 5 min

10. Make the cells electrocompetent: pour into 50 mL Falcon tubes and pellet at 6,000×*g* for 5 min at 0°C; resuspend in 5 mL ice-cold water by gentle shaking (not vortexing or pipetting), then add water to 50 mL. Repeat this washing step twice. Ensure that everything is absolutely ice-cold (in ice-water slurry, not merely on ice) throughout. Bacterial pellets will become very mobile when well washed, so exercise caution to avoid losing the pellets. Resuspend final pellets in a minimal volume of water (250–500 μL).
11. Electroporate the heat-induced *GalK*-positive BAC-containing SW102 cells and the nonheat induced control cells with 1–5 μL of PCR product: pipette 50 μL of bacteria into a 1-mm cuvette, add DNA and mix by gentle agitation, pulse at 1.8 kV/25 μF/200 Ω, immediately add 1 mL LB, transfer to a 14-mL snap-cap tube, and then incubate for 1 h at 32°C with shaking (see Notes 17 and 26).
12. Wash the cells in M9 salts to remove LB medium prior to galactose selection by transferring cells to a 1.5-mL tube, pelleting at maximum speed for 30 s in a microfuge, resuspending in 1 mL M9 salts by vortexing, and repeating twice. The nonheat induced pellet may have long sticky strands (this is normal).
13. Plate out 1, 10 μL (diluted in 100 μL M9 salts), 100 μL, and the remainder of the induced cells plus 100 μL un-induced as a control onto M63 chloramphenicol/galactose agar plates. Incubate at 32°C for 3–5 days.

Table 5
Primers for quality control of *B15R* deletion in MVA-BAC by *GalK* recombineering

Forward identity primer	tcccttttcccagaaacaaac	*B14R* (upstream of deletion)
Reverse identity primer	acgcgaactttacggtcatc	In *GalK* (generic primer)
Forward purity primer	ttctgaacccgacacaatca	*B15R* (in deleted region)
Reverse purity primer	ttgtgggaggtctcaacgat	*B15R* (in deleted region)
Forward sequencing primer	tgcgctggacaattgtattc	Upstream of deletion
Reverse sequencing primer	gtatcgcattccaccctttc	Downstream of deletion

Identity PCR should be positive and negative on wild-type. Purity PCR should be negative on deletion mutant and positive on wild-type. Forward and reverse sequencing primers are used to obtain a sequence read through the homology arm, using BAC DNA or a Phusion PCR product made with these primers as the template

14. If a good ratio of colonies is seen for the induced compared to un-induced cells, streak out three colonies onto MacConkey galactose plates containing 12.5 μg/mL chloramphenicol then incubate at 32°C for 2 days. Pick a single, well-isolated, bright red colony. If *GalK* is to remain in the construct for viral rescue, restreak at least twice more to ensure removal of wild-type hitchhikers. If the clone is destined for counterselection (to remove *GalK*), a single MacConkey streak is sufficient.
15. Check constructs by multiplex PCR (see Subheading 3.2, step 9), an identity PCR across the insertion locus, and a purity PCR to verify absence of wild-type BAC (if a *GalK*-containing virus is to be made). Check sequences of the homology arms are correct either by sequencing the BAC or a PCR product. Examples of primers for *B15R* are shown in Table 5.
16. Proceed to virus rescue (see Subheading 3.2) or counterselection (below), remembering to grow SW102 cells at 32°C. Restriction mapping is not usually necessary unless rescue fails. Perform the same PCR characterization on the derived virus (see Note 27 and Subheading 3.5).
17. For counterselection (to remove *GalK*), generate a linear targeting DNA by annealing of complementary long synthetic oligos (see Note 28), or PCR amplification of a conventionally cloned or synthetic sequence. For the *B15R* example here, this sequence consisted simply of the adjacent left and right homology arms.
18. Repeat steps 6–10 to electroporate this DNA into heat-induced and control *GalK*-positive BAC-containing SW102 cells (see Note 29), omitting *Dpn*I digestion and band excision (see Note 30).

Table 6
***E. coli* insertion sequence (IS) primers**

PCR	Size (bp)	Forward primer	Reverse primer	Sequencing primer
IS1	448	atgggcgttggcctcaac	atgactttgtcatgcagctcc	gatatacgcagcgaattgagc
IS2	1,165	attggagaacagatgattga	attcccgtggcgagcgataa	gaatggcatccgcatagtg
IS3	238	gtaaaaaaccccgtaaacag	aggatagccagctcttcatc	ctgcgaaattcaggcgaatg
IS4	468	tcgttgccttgccgaatcag	tcgctgttcttcatcgtgcc	ccgagagtggtaaaggagag
IS5	810	tcagcagtaagcgccgt	gccgaactgtcgcttga	aagccagatacaaggggttg
IS10	360	aagcgaactgttgagagtac	gacacggactcattgtcac	tcaagtaaggcgtggcaag
IS30	366	ctcacctgacactgtctgag	cacggctacgaaagtacagc	gctcagctaaacaacagac
IS150	414	ttcaagatcctcaatgcgtc	cgccttgaagtcgtgaatcac	tgccgagatgatcctgtaac
IS186	414	gactggtttggcatacttgc	gttcttaccgcaatccagc	agcggctcaaaagtttgctg

These oligos cover all known IS elements of *E. coli* K12 DH10B based on the published sequence (GenBank NC_010473). Size refers to expected product size using forward and reverse primers. The sequencing primer is provided to confirm identity of product. Oligo sequences kindly provided by Michaela Späth, Kay Brinkmann, and Jürgen Hausman from Bavarian-Nordic GmbH, Martinsreid, Germany

19. Plate recovered cells onto M63 chloramphenicol/deoxygalactose/glycerol plates (see Note 31).
20. Restreaking is unnecessary, but be prepared to screen at least 12 colonies for the correct deletion (see Note 23).

3.5. PCR to Verify Absence of Bacterial Insertion Sequences in Rescued BAC Derived Viruses

This protocol was kindly supplied by Michaela Späth, Kay Brinkmann, and Jürgen Hausman from Bavarian-Nordic GmbH, Martinsreid, Germany. Bacterial transposons or insertion sequences (IS) can jump into a BAC and be propagated in the genome of the resulting virus following rescue. If such a transposition occurs, it is likely to affect only a small proportion of the BAC DNA in the prerescue maxiprep, so will normally be lost upon serial passage of the virus, but in some circumstances it may also become fixed in the viral population. It is pointless to directly analyze the BAC DNA preps by PCR, since these are always contaminated with trace amounts of bacterial chromosome. So the rescued virus must be passaged before PCR analysis using the *E. coli* insertion sequence (IS) primers provided in Table 6.

1. Use a QIAGEN DNA blood mini kit or similar kit to purify DNA from rescued virus after at least two passages in the producer cell line, using 200 μL of virus containing >10^8 pfu/mL. In the final elution step, elute DNA into 50–100 μL water.

Table 7
PCR cocktail and reaction conditions to verify absence of bacterial insertion sequences in rescued BAC derived viruses

	Volume (μL)
DNA	1.0
10× $(NH_4)_2SO_4$ Taq buffer	2.0
$MgCl_2$ (25 mM stock)	1.6
BSA (20 mg/mL)	0.1
IS forward primer from Table 6 (10 μM stock)	0.8
IS reverse primer from Table 6 (10 μM stock)	0.8
dNTP mix (10 mM)	0.4
Fermentas native Taq	0.1
Water	13.2
Total	20.0

Program 94°C for 5 min, then 30 cycles of 94°C for 30 s, 55°C for 30 s, 72°C for 1 min; final elongation at 72°C for 4 min, then 4°C hold

2. Set up PCR and reaction conditions as in Table 7. As a positive control, use DNA extracted from 200 μL log-phase DH10B culture. As a negative control, use water or DNA extracted from MDS42 bacteria (see Note 32).
3. Analyze products on 1% agarose gel (see Note 33).

4. Notes

1. The author is not aware of a commercial supplier of small quantities of isatin-β-thiosemicarbazone (IβT; 1H-indole-2,3-dione-3-thiosemicarbazone). If unable to obtain from a friendly poxvirologist, one option is to find a tame organic chemist to perform a very simple synthesis using widely available reagents (condensation of isatin and thiosemicarbazide in the presence of glacial acetic acid) (21). Exercise caution with IβT as the toxicity of the substance is not fully characterized.
2. PhaseLok Gel (PLG) reduces the hazards of phenol use and enables the aqueous phase to be simply poured off. We use PLG Heavy 2 mL tubes (5-PRIME; distributed by VWR). Note that the gel must be briefly pelleted before use.

3. Reagents for phenol extraction are available as ready-to-use solutions from Sigma. Store at 4°C and check the phenol solutions before use to make sure they are colorless. Perform in a fume cabinet with appropriate safety precautions. Attempting to substitute this phenol prep with a commercial silica-based kit (e.g., QIAGEN) is not recommended.
4. The use of 14 mL polypropylene snap-cap round-bottom tubes (e.g., Falcon tubes #352059 or similar) for recovery is recommended. These allow good shaking and aeration, promoting growth and protein expression prior to selection, unlike, for example, conical tubes or microtubes.
5. This is identical to QIAGEN buffer P2. Check for SDS precipitation before use and keep lid closed to prevent acidification.
6. Baffled flasks are recommended for BAC growth. Ensure flask is large enough for excellent aeration (i.e., need five- to tenfold larger than culture volume). Shaking water baths are highly recommended for this protocol.
7. QIAGEN Plasmid Maxi kit (Cat. No. 12162) is recommended. Do not use Plus or QIAfilter kits; the Large Construct (exonuclease) kit is also not recommended. An alternative is the Macherey-Nagel NucleoBond BAC 100 kit.
8. Any vaccine strain of fowlpox virus, available from most veterinary vaccine manufacturers, is suitable. The attenuated FP9 or HP-438 Munich strain (22) is recommended and can be obtained from Dr Michael A. Skinner, Imperial College London (email: m.skinner@imperial.ox.ac.uk). Other avipoxvirus vaccine strains should perform similarly (e.g., canarypox virus). It is advisable to use a recombinant virus carrying a readily detectable marker gene to facilitate verification of absence of helper in the rescued virus if desired.
9. Lipofectamine 2000 protocol is slightly modified from the manufacturer's instructions. Use of Fugene HD (Roche) has also been reported (7).
10. OptiMEM serum-free medium is recommended, but other media can also be used, with potentially reduced transfection efficiency. If the cells used do not tolerate OptiMEM for long periods, the transfection mix can be supplemented with a different medium (mixed with the OptiMEM already in the well).
11. Order primers from a trusted supplier. Mass spectrometry cannot detect swapped bases, so the homology arm regions must always be sequenced in the final construct.
12. Phusion Hot Start from Finnzymes is highly recommended, as it has the lowest error rate of any commercially available thermophilic polymerase.

13. Phire from Finnzymes gives a very rapid PCR screen result and has excellent sensitivity.
14. Use of native rather than recombinant Taq polymerase avoids possible false-positives from contaminating *E. coli* DNA.
15. Normally the "hairpin resolvase" (A22) enzyme cuts the dimeric poxviral genome at the fused terminus after DNA replication. IβT is used to inhibit late gene expression: it binds to the viral RNA polymerase, impairing postreplicative transcriptional termination and leading to an accumulation of dsRNA due to excess read-through, with resulting activation of 2′–5′ oligoadenylate synthetase and RNAse L, all of which has the effect of stalling viral growth at the DNA replication stage (23). Under these conditions, *A22R* (and the other viral late genes) are not expressed, and thus the concatemers that are required for BAC production are generated.
16. When working with the viral genomic BAC DNA, it pays to be paranoid about shearing the DNA. Always handle gently, never vortex, and never pipette unless necessary, in which case cut the ends off pipette tips to increase the bore size. Store DNA at 4°C or at −20°C as an ethanol precipitate. Although repeated freeze–thaw should be avoided, a single cycle can help dissolve a stubborn pellet without inflicting too much shearing.
17. For efficient electroporation (response time >4.5 ms) it is crucial to keep all reagents and cuvettes absolutely ice-cold. Arcing results from bubbles, high temperature, or excess salt. Pipette bacteria gently and handle as little as possible.
18. Obtaining the complete sequence of the clone is recommended: the adoption of so-called "next-generation" sequencing technology continues to reduce costs.
19. We have found that *Hin*dIII works well, but some restriction enzymes do not cut BAC DNA efficiently, possibly due to supercoiling of the DNA.
20. This rescue protocol employs a higher MOI of the helper virus than originally described (2). The inventors of VAC-BAC used a low MOI of fowlpox virus so that surrounding cells would be uninfected and thus available to the BAC-derived virus, resulting in a plaque around the original infected-transfected cell during rescue. Although this may be rather exciting, it means that most of the tricky BAC transfection is wasted, since the cells are unlikely to be fowlpox-infected. Therefore, in the method described here, most of the cells are infected with fowlpox virus to maximize the frequency of rescue, with the disadvantage of reduced growth of the rescued virus beyond the initial infected-transfected cells. Thus, in the final step of Subheading 3.3, the rescue is harvested and replated onto fresh cells to reveal the reconstituted MVA.

21. BAC DNA transfects poorly compared to smaller plasmids, and some transfection reagents are ineffective for BAC DNA. This transfection protocol works well and should not require optimization for most continuous mammalian cell lines.
22. For most in vitro studies, it is usually acceptable to leave the bacterial selectable marker in the construct, and therefore in the rescued virus, which in any case contains the entire BAC cassette as well (unless a system for its removal has been set up (12, 13)). It should be noted that although no explicit poxviral promoter is present, the nonspecific nature of late poxviral transcriptional termination might allow some expression of the gene. These issues need to be considered if the BAC derived viruses will be used for animal studies.
23. Insertion by counterselection (replacement of *GalK* or another selectable marker with an exogenous sequence) is possible, but inefficient for poxviral BACs (6), especially for long sequences. The reason for this is that marker removal can occur via a large deletion instead of the desired recombination. It is nevertheless useful for removal of the marker, allowing "marker recycling" (a subsequent insertion of the same marker at a different locus), as well as for seamless generation of point mutants and small insertions. The advantage of *GalK* over tandem positive–negative markers is that the same enzymatic activity is required for both positive and negative selection, but the principal disadvantage (other than making the minimal medium) is the need for restreaking on indicator plates to ensure purity, since wild-type "hitch-hikers" may grow within *GalK*+ colonies.
24. At this step, cells can be left in an ice-water slurry over a lunch break. Some operators claim this increases electrocompetence.
25. For the non-bacteriologist, it may be worth noting that this heat shock step has nothing to do with the subsequent transformation (unlike for chemically competent bacteria), but instead induces the expression of the phage recombination proteins via temporary inactivation of the temperature-sensitive repressor. The uninduced bacteria are a negative control: any resulting colonies arise from transformation (by electroporation) with the *GalK* plasmid (used as the template in the PCR) rather than from integration of *GalK* at the desired site in the BAC. The gel extraction and *Dpn*I digest serves to reduce this background as much as possible (*Dpn*I cuts only the methylated plasmid DNA and not the unmethylated PCR product).
26. The more DNA electroporated, the better: 100 ng as a minimum; preferably up to 1 μg.
27. Even with antibiotic selection of BAC mutants, and more so with *GalK* selection where restreaking is required, it is not necessarily safe to assume that the rescued virus is pure, even if

identity and purity PCRs are correct at the BAC stage, and especially when multiple mutants are being handled in parallel or when the mutant has a growth disadvantage or is lethal. It is therefore advisable to conduct a similar level of quality control on BAC-derived viruses as would be used for conventional recombinants (the same PCR assays can generally be used before and after rescue). If a minor contamination is suspected, it is sometimes more convenient to perform a couple of rounds of viral plaque-purification than to repeat the BAC isolation and rescue. The recombineering system can be modified to insert a poxviral marker in tandem with the bacterial selection cassette.

28. Anneal by mixing equimolar quantities, heating to 95°C for 1 min, then place at room temperature until cool, then on ice. The advantage of long oligos is that a high concentration for electroporation can be easily achieved, but they carry a possibly greater risk of sequence errors compared to PCR amplification of targeting DNA with Phusion.
29. It is even more crucial to electroporate as much targeting DNA as possible at this stage to drive the desired recombination and prevent large deletions (see Note 23). At least 1 μg is required, preferably even more, so it may be necessary to ethanol-precipitate the DNA to concentrate it and remove salt (see Subheading 3.1, step 9).
30. During positive selection, the cells can become *GalK* positive either via recombination of *GalK* into the desired locus, or simply by transformation with the plasmid (used as the PCR template). But during removal, the only way the cells can become *GalK* negative is to get rid of the *GalK* from the BAC: transformation with the plasmid makes no difference. Furthermore, only linear DNA is a substrate for Red-recombination.
31. During negative selection, the deoxygalactose is toxic to *GalK* positive clones. Since there is no carbon source in the base M63 minimal medium, the glycerol is needed to fuel the *GalK* negative bacteria.
32. Optionally, use DNA extracted from MDS42 bacteria (Scarabgenomics) as a negative control. This DH10B-derived strain was genetically engineered to delete all IS elements (24). The IS4 primers (see Table 6) often produce a nonspecific PCR product from MDS42 DNA that is derived from bacterial helicase, so if MDS42 DNA is to be used as negative control, different IS4 primers should be designed. As a positive control for the presence of this DNA, the following primers produce a 433 bp product specific to bacterial RNA polymerase: forward primer 5′-gctggactgctgatgtcgaag-3′; reverse primer 5′-ttcgccagaataccgcgatcg-3′.

33. If the virus is unluckily positive for IS, it may still be suitable for the envisaged experiment depending on the nature of the insertion locus (identified by sequencing); but if an isolate free of IS must be derived, the options are either to analyze viral plaques, to repeat the rescue with an alternative BAC clone (it is advisable to make glycerol stocks to cover this eventuality), or to repeat the recombineering.

Acknowledgments

The author would like to thank Dr Richard Wade-Martins, Department of Anatomy and Human Genetics, University of Oxford, UK for supplying the protocol upon which that in Subheading 3.2 is based; Michaela Späth, Kay Brinkmann, and Jürgen Hausman from Bavarian-Nordic GmbH, Martinsreid, Germany for the protocol in Subheading 3.4; Dr Michael Skinner, Imperial College London, UK for agreeing to supply FP9; and principal investigators Prof. Adrian V. S. Hill and Dr Sarah C. Gilbert, Jenner Institute, University of Oxford, UK.

References

1. Messerle M, Crnkovic I, Hammerschmidt W, Ziegler H, Koszinowski UH (1997) Cloning and mutagenesis of a herpesvirus genome as an infectious bacterial artificial chromosome. Proc Natl Acad Sci USA 94:14759–14763
2. Domi A, Moss B (2002) Cloning the vaccinia virus genome as a bacterial artificial chromosome in *Escherichia coli* and recovery of infectious virus in mammalian cells. Proc Natl Acad Sci USA 99:12415–12420
3. Court DL, Sawitzke JA, Thomason LC (2002) Genetic engineering using homologous recombination. Annu Rev Genet 36:361–388
4. Sharan SK, Thomason LC, Kuznetsov SG, Court DL (2009) Recombineering: a homologous recombination-based method of genetic engineering. Nat Protoc 4:206–223
5. Domi A, Moss B (2005) Engineering of a vaccinia virus bacterial artificial chromosome in *Escherichia coli* by bacteriophage lambda-based recombination. Nat Methods 2:95–97
6. Cottingham MG, Andersen RF, Spencer AJ, Saurya S, Furze J, Hill AV, Gilbert SC (2008) Recombination-mediated genetic engineering of a bacterial artificial chromosome clone of modified vaccinia virus Ankara (MVA). PLoS One 3:e1638
7. Meisinger-Henschel C, Spath M, Lukassen S, Wolferstatter M, Kachelriess H, Baur K, Dirmeier U, Wagner M, Chaplin P, Suter M, Hausmann J (2010) Introduction of the six major genomic deletions of Modified Vaccinia Virus Ankara (MVA) into the parental vaccinia virus is not sufficient to reproduce an MVA-like phenotype in cell culture and in mice. J Virol 84:9907–9919
8. Roth SJ, Hoper D, Beer M, Feineis S, Tischer BK, Osterrieder N (2011) Recovery of infectious virus from full-length cowpox virus (CPXV) DNA cloned as a bacterial artificial chromosome (BAC). Vet Res 42:3
9. Domi A, Weisberg AS, Moss B (2008) Vaccinia virus E2L null mutants exhibit a major reduction in extracellular virion formation and virus spread. J Virol 82:4215–4226
10. Osborne JD, Da Silva M, Frace AM, Sammons SA, Olsen-Rasmussen M, Upton C, Buller RM, Chen N, Feng Z, Roper RL, Liu J, Pougatcheva S, Chen W, Wohlhueter RM, Esposito JJ (2007) Genomic differences of vaccinia virus clones from Dryvax smallpox vaccine: the Dryvax-like ACAM2000 and the mouse neurovirulent Clone-3. Vaccine 25:8807–8832
11. Scheiflinger F, Dorner F, Falkner FG (1992) Construction of chimeric vaccinia viruses by molecular cloning and packaging. Proc Natl Acad Sci USA 89:9977–9981

12. Cottingham MG, Gilbert SC (2010) Rapid generation of markerless recombinant MVA vaccines by en passant recombineering of a self-excising bacterial artificial chromosome. J Virol Methods 168:233–236
13. Tischer BK, Kaufer BB, Sommer M, Wussow F, Arvin AM, Osterrieder N (2007) A self-excisable infectious bacterial artificial chromosome clone of varicella-zoster virus allows analysis of the essential tegument protein encoded by ORF9. J Virol 81:13200–13208
14. Warming S, Costantino N, Court DL, Jenkins NA, Copeland NG (2005) Simple and highly efficient BAC recombineering using galK selection. Nucleic Acids Res 33:e36
15. Yao XD, Evans DH (2004) Construction of recombinant vaccinia viruses using leporipoxvirus-catalyzed recombination and reactivation of orthopoxvirus DNA. Methods Mol Biol 269:51–64
16. Merchlinsky M, Moss B (1992) Introduction of foreign DNA into the vaccinia virus genome by in vitro ligation: recombination-independent selectable cloning vectors. Virology 190:522–526
17. Tsung K, Yim JH, Marti W, Buller RM, Norton JA (1996) Gene expression and cytopathic effect of vaccinia virus inactivated by psoralen and long-wave UV light. J Virol 70:165–171
18. Wong QN, Ng VC, Lin MC, Kung HF, Chan D, Huang JD (2005) Efficient and seamless DNA recombineering using a thymidylate synthase A selection system in *Escherichia coli*. Nucleic Acids Res 33:e59
19. DeVito JA (2008) Recombineering with tolC as a selectable/counter-selectable marker: remodeling the rRNA operons of *Escherichia coli*. Nucleic Acids Res 36:e4
20. Tischer BK, Smith GA, Osterrieder N (2011) En passant mutagenesis: a two step markerless red recombination system. Methods Mol Biol 634:421–430
21. Bal TR, Anand B, Yogeeswari P, Sriram D (2005) Synthesis and evaluation of anti-HIV activity of isatin beta-thiosemicarbazone derivatives. Bioorg Med Chem Lett 15:4451–4455
22. Mayr A, Malicki K (1966) Attenuation of virulent fowl pox virus in tissue culture and characteristics of the attenuated virus. Zentralbl Veterinarmed B 13:1–13
23. Cresawn SG, Prins C, Latner DR, Condit RC (2007) Mapping and phenotypic analysis of spontaneous isatin-beta-thiosemicarbazone resistant mutants of vaccinia virus. Virology 363:319–332
24. Posfai G, Plunkett G 3rd, Feher T, Frisch D, Keil GM, Umenhoffer K, Kolisnychenko V, Stahl B, Sharma SS, de Arruda M, Burland V, Harcum SW, Blattner FR (2006) Emergent properties of reduced-genome *Escherichia coli*. Science 312:1044–1046

Chapter 4

Easy and Efficient Protocols for Working with Recombinant Vaccinia Virus MVA

Melanie Kremer, Asisa Volz, Joost H.C.M. Kreijtz, Robert Fux, Michael H. Lehmann, and Gerd Sutter

Abstract

Modified vaccinia virus Ankara (MVA) is a highly attenuated and replication-deficient strain of vaccinia virus that is increasingly used as vector for expression of recombinant genes in the research laboratory and in biomedicine for vaccine development. Major benefits of MVA include the clear safety advantage compared to conventional vaccinia viruses, the longstanding experience in the genetic engineering of the virus, and the availability of established procedures for virus production at an industrial scale. MVA vectors can be handled under biosafety level 1 conditions, and a multitude of recombinant MVA vaccines has proven to be immunogenic and protective when delivering various heterologous antigens in animals and humans. In this chapter we provide convenient state-of-the-art protocols for generation, amplification, and purification of recombinant MVA viruses. Importantly, we include methodology for rigid quality control to obtain best possible vector viruses for further investigations including clinical evaluation.

Key words: Viral vector, Vector vaccine, Quality control, Genetic stability, Large-scale production, Clinical testing

1. Introduction

Recombinant vaccinia viruses are excellent research tools that find a broad range of applications in different fields including cell biology, protein sciences, virology, immunology, and medicine (1). However, vaccinia virus can replicate in humans and has to be handled under biosafety level 2 laboratory conditions. Additionally, its imperfect safety record as a smallpox vaccine has been a concern for its use as a vector in clinical applications. Therefore, poxvirus vector systems based on highly attenuated vaccinia viruses such as modified vaccinia virus Ankara (MVA) or genetically modified vaccinia virus NYVAC have been established (2, 3). These viruses

Stuart N. Isaacs (ed.), *Vaccinia Virus and Poxvirology: Methods and Protocols*, Methods in Molecular Biology, vol. 890, DOI 10.1007/978-1-61779-876-4_4, © Springer Science+Business Media, LLC 2012

share replication deficiency in human and most mammalian cells, allowing for handling under biosafety 1 conditions, but have maintained the advantage of high-level gene expression and immunogenicity as recombinant vaccines. MVA vectors are currently evaluated as prophylactic or therapeutic candidate vaccines against various infectious diseases and various types of cancer (for reviews see (4–6)). In addition, nonrecombinant MVA is being developed as a safe third-generation smallpox vaccine also suitable for immunization of risk populations in which the use of replicating vaccinia viruses is contraindicated (e.g., patients diagnosed with atopic dermatitis or HIV-infected individuals) (7). Importantly, the experience from the recent use of MVA vaccines in humans is in line with the excellent safety record established when MVA was clinically tested for immunization against smallpox in more than 100,000 humans in Germany in the 1970s.

The most frequently practiced approach to generate recombinant MVA (rMVA) utilizes the well-established method of homologous DNA recombination in infected cells (8), which occurs frequently during vaccinia virus replication (0.1%). The event of recombination is typically directed by an MVA transfer plasmid containing the following features: (1) The actual expression unit including a virus-specific promoter usually followed by a multiple cloning site for insertion of foreign target gene sequences, (2) selectable or nonselectable marker genes to ease the clonal isolation of rMVA, (3) two stretches of genomic MVA sequences flanking the foreign and marker gene sequences (1 and 2) to direct the recombination to a desired locus in a nonessential region of the MVA genome. rMVA are readily generated by infection of cells with MVA and simultaneous transfection with this transfer plasmid DNA and viruses are clonally isolated by plaque purification in additional tissue culture passages. Alternatively, the entire MVA genome cloned into a bacterial artificial chromosome can be elegantly used in *E. coli* to produce rMVA genomes containing expression cassettes for foreign genes (9, 10). This rMVA DNA is transfected into cells, and upon activation through simultaneous coinfection with helper poxviruses rMVA viruses are formed. Importantly, independent from the technology used for MVA vector virus generation, the engineered MVA must be thoroughly controlled for genetic stability and fitness for large-scale amplification (e.g., for vaccine production). Expression of some foreign target gene products can exert suppressive effects on MVA replication which may prompt, sooner or later, the occurrence of non-expressing mutant viruses as it was exemplarily described for the construction of rMVA expressing HIV-1 antigens (11).

Here, we describe a collection of straightforward methodologies for generation and characterization of rMVA. The protocols include well-established techniques for isolation of cloned viruses, convenient procedures for virus amplification and titration, and methods for proper control of the quality of the isolated vector viruses.

2. Materials

2.1. Cell Culture

1. DF-1: Chicken fibroblast cell line UMNSAH/DF-1 (ATCC CRL-12203).
2. BHK-21: Baby hamster kidney cells (ATCC CCL-10).
3. HeLa cells (ATCC CCL-2).
4. RK-13: Rabbit kidney cells (ATCC CCL-37).
5. Tissue culture pates (e.g., 96-well tissue culture plates, six-well tissue culture plates, T75 and T175 tissue culture flasks).
6. Cell growth medium: MEM or DMEM supplemented with 10% Heat-inactivated fetal calf (FCS) and 1% antibiotics (penicillin, streptomycin)/antimycotics (amphothericin B) (AB/AM) (see Note 1).
7. VP-SFM: Growth media for serum-free conditions.
8. 0.05% Trypsin/EDTA: 0.05% Trypsin and 0.02% EDTA.

2.2. Preparation and Cultivation of Primary Chicken Embryo Fibroblasts

1. Ten 11-day-old embryonated chicken eggs.
2. Chicken Embryo Fibroblasts (CEF) growth medium: MEM supplemented with 10% FCS and 1% AB/AM.
3. 0.25% Trypsin/EDTA: 0.25% Trypsin and 0.02% EDTA.
4. EtOH abs: laboratory-grade absolute ethanol.
5. 50-ml Falcon tubes.
6. Sterile petri dishes, 10 cm diameter.
7. Sterile instruments for dissection (e.g., scissors, forceps).
8. 10-ml syringes.
9. Sterile Glassware: one Erlenmeyer flask, two beakers covered with two layers of gauze (attached by autoclave tape).
10. DPBS: Sterile Dulbecco's phosphate buffered saline.
11. Recombinant trypsin solution: 50 ml TrypLE Select in 150 ml DPBS.

2.3. Virus Growth and Purification

1. Virus growth medium: DMEM or MEM supplemented with 2% FCS and 1% AB/AM.
2. Virus Strain: Virus: Vaccinia virus strain MVA (e.g., clonal isolate F6 at 584th passage CEF (12)).
3. Confluent monolayers of CEF, DF-1 or BHK-21 cells grown on 60-mm^2 dishes, T75 and T175 tissue culture flasks.

4. Cell scraper.
5. 10 mM Tris–HCl, pH 9.0, autoclaved (stored 4°C).
6. 1 mM Tris–HCl, pH 9.0, autoclaved (store 4°C).
7. Cup sonicator and sonification needle (e.g., Sonopuls HD 200).
8. 36% sucrose: 35% sucrose (w/v) in 10 mM Tris–HCl, pH 9.0, sterile filtered (store 4°C).

2.4. Titration of MVA

1. Fixing solution: 1:1 mixture of acetone:methanol, stored at −20°C.
2. PBS: Phosphate buffered saline, pH 7.5.
3. Blocking buffer: PBS and 3% bovine serum albumin.
4. Primary antibody (primary Ab): Polyclonal rabbit anti-vaccinia virus (Lister strain) antibody (IgG fraction, Acris Antibodies GmbH (e.g., Cat. No. BP1076)) diluted 1:2,000 in blocking buffer.
5. Secondary antibody (secondary Ab): horseradish peroxidase-conjugated polyclonal goat anti-rabbit antibody (IgG (H + L)) diluted 1:5,000 in blocking buffer.
6. TrueBlue Peroxidase Substrate solution (1-Component).

2.5. Generation of Recombinant MVA

2.5.1. Molecular Cloning of Recombinant Gene Sequences, Transfection, Plaque Purifications

1. MVA targeting vector plasmids.
2. Restriction endonucleases.
3. DNA-modifying enzymes, e.g., Klenow DNA polymerase, T4 DNA ligase.
4. Serum-free medium: MEM supplemented with 1% AB/AM (store at 4°C).
5. FUGENE™ Transfection Reagent stored at 4°C.
6. 1.5-ml Eppendorf vials.
7. Air-displacement pipette and autoclaved tips.
8. Inverted research-grade fluorescence microscope (e.g., Zeiss Axiovert 200 M).
9. 2×Medium: 2× MEM or DMEM supplemented with 4% FCS, 2% AB/AM, stored at 4°C.
10. 2% LMP-agarose: 2% low-melting-point agarose in distilled sterile water stored at room temperature (see Note 2).

2.5.2. Characterization of rMVA Genomes by PCR, Extraction Viral Genomic DNA

1. 10× TEN buffer: 100 mM Tris–HCl, pH 7.4, 10 mM EDTA, 1 M NaCl.
2. Proteinase K: DNA grade proteinase K prepared at 1 mg/ml stock solution in 1 mM $CaCl_2$ stored at −20°C.
3. 20% SDS: 20% sodium dodecyl sulfate in distilled water, DNase free, sterile filtered.

4. Phenol–chloroform: 1:1 mixture, stored at 4°C.
5. 3 M NaAc: 3 M sodium acetate in distilled water.
6. 70% ETOH in distilled water.

2.6. Characterization of Clonal rMVA Isolates and Quality Control of rMVA

1. Oligonucleotide primers (see Table 1).
2. Viral genomic DNA prepared as described in Subheading 3.4.8.
3. Plasmid DNA diluted to a final concentration of 100 ng/μl.
4. PCR master kit (Roche, Mannheim, Germany) (store at –20°C).
5. Mastercyler pro gradient.
6. PCR Buffer: 10 mM Tris–HCl, pH 8.8, 50 mM KCl, 0.1% Triton X-100, 200 μM each of four dNTPs, 1.5 mM $MgCl_2$.

Table 1
Oligonucleotide primers

Primer	Sequence (5′ to 3′)
MVA-III-5′	GAATGCACATACATAAGTACCGGCATCTCTAGCAGT
MVA-III-3′	CACCAGCGTCTACATGACGAGCTTCCGAGTTCC
MVA-I8R/G1L-5′	ATTCTCGCTGAGGAGTTGG
MVA-I8R/G1L-3′	GTCGTGTCTACAAAAGGAG
K1L-int-1	TGATGACAAGGGAAACACCGC
K1L-int-2	GTCGACGTCAATTAGTCGAGC
MVA_Del1-F	CTTTCGCAGCATAAGTAGTATGTC
MVA_Del1-R	CATTACCGCTTCATTCTTATATTC
MVA_Del2-F	GGGTAAAATTGTAGCATCATATACC
MVA_Del2-R	AAAGCTTTCTCTCTAGCAAAGATG
MVA_Del3-F	GATGAGTGTAGATGCTGTTATTTTG
MVA_Del3-R	GCAGCTAAAAGAATAATGGAATTG
MVA_Del4-F	AGATAGTGGAAGATACAACTGTTACG
MVA_Del4-R	TCTCTATCGGTGAGATACAAATACC
MVA_Del5-F	CGTGTATAACATCTTTGATAGAATCAG
MVA_Del5-R	AACATAGCGGTGTACTAATTGATTT
MVA_Del6-F	CGTCATCGATAACTGTAGTCTTG
MVA_Del6-R	TACCCTTCGAATAAATAAAGACG

Oligos dissolved in sterile distilled water to a final concentration of 5 pmol/μl and stored at –20 C.

7. DyNAzyme™ II (Finnzymes).
8. Phage X174 DNA-*Hae*III digest and λDNA-*Hin*dIII digest.

2.7. Assay for Expression of Recombinant Gene by Immunocytochemistry and by Western Blotting

1. Saturated dianisidine solution: Add a small amount of o-dianisidine in 700 μl absolute EtOH, mix by vortexing for 2 min. Centrifuge for 30 s at top speed at room temperature (see Note 3).
2. Substrate solution for peroxidase staining: 240 μl of the supernatant only from the saturated dianisidine solution added to 12 ml PBS in a 15-ml conical centrifuge tube. Mix by vortexing (see Note 3). Add 12 μl hydrogen super oxide (H_2O_2, >30%), gently mix again, and use immediately (see Note 4).
3. Glycerine solution: 50% glycerine in water.
4. RIPA buffer: 25 mM Tris–HCl, pH 7.6, 150 mM NaCl, 1% NP-40, 1% sodium deoxycholate, 0.1% SDS.
5. Protease inhibitors: Complete protease inhibitor cocktail tablets (e.g., Roche Applied Sciences).

3. Methods

3.1. Cell Culture

RK-13, DF-1, BHK-21, and CEF cells are grown on 96-well pates, six-well plates, T75 or T175 flasks in cell growth medium at 37°C and 5% CO2. To expedite clinical use, the generation and amplification of rMVA can be performed in compliance with certain regulatory requirements. Most importantly, it is considered a safety advantage to exclusively use CEF cultures prepared from specific pathogen-free eggs with recombinant trypsin and maintained under serum-free medium conditions.

3.1.1. Preparation of Chicken Embryo Fibroblasts

1. Check for viability of the chicken embryo by candling the eggs. Will need about a half an embryo per T175.
2. Place eggs with blunt ends facing up so that the air space is facing upwards.
3. Wipe eggs with ethanol, crack eggshell with scissors, and cut it off, taking care not to damage the membrane.
4. Remove membrane and elevate the chick embryo legs with forceps. Using a second pair of forceps to help support the embryo. Transfer it into a Petri dish containing 10 ml DPBS.
5. Carefully remove head, wings, legs, and internal organs of the embryo with forceps.
6. Transfer remaining torso of the embryo to a second Petri dish containing 10 ml DPBS.

7. Homogenize the tissue by pressing five embryos at a time through a 10-ml syringe (without needle) into an Erlenmeyer flask.
8. Add 100 ml 0.25% trypsin/EDTA (prewarmed to 37°C) and trypsinize the tissue by stirring vigorously for 10 min at 37°C using a magnetic stirrer.
9. Place cell suspension through gauze into a beaker, taking care not to transfer remaining embryo clumps. Add 10 ml FBS to inactivate trypsin.
10. Place remaining embryo clumps in Erlenmeyer flask with another 100 ml 0.25% trypsin/EDTA (prewarmed to 37°C). Stir vigorously for 10 min at 37°C.
11. Pass this trypsinized material through the gauze of a second beaker and pool filtrates in a 250-ml centrifuge bottle.
12. Spin at 1,800 × *g* for 10 min at 4°C.
13. Discard supernatant, and resuspend pellet in no more than 10 ml of CEF growth media to allow breaking up of clumps by pipetting vigorously (10–15 times) to obtain a single cell suspension. Then add an additional 90 ml CEF growth medium to the cells.
14. Re-pellet cells by centrifuging for 10 min at 1,800 × *g*.
15. Prepare 20 T175 culture flasks containing 40 ml of CEF growth medium in each (calculate 1/2 embryo per flask).
16. Resuspend cell pellet in 5 ml CEF growth medium (pipetting 10–15 times) to obtain a single cell suspension, then transfer to 50-ml Falcon tube and fill up to a final volume of 20 ml.
17. Add 1 ml of cell suspension to each T175 flask.
18. Incubate for 3–4 days at 37°C until monolayers are confluent.
19. Split at ratio of 1:4 for six-well plates (2 ml/plate), 12-well plates (1 ml/plate), or T175 flasks.

3.1.2. Preparation of Chicken Embryo Fibroblasts Under Serum-Free Conditions

1. Check for viability of the chicken embryo by candling the eggs.
2. Place eggs with blunt ends facing up, in order to position the air space upwards.
3. Wipe eggs with ethanol, crack eggshell with scissors and cut it off, taking care not to damage the membrane.
4. Remove membrane and elevate the legs with forceps. Using a second pair of forceps to help support the embryo. Transfer it into a well of a six-well plate containing 3 ml of DPBS.
5. Remove head, wings, and legs and transfer embryo to the second well containing 3 ml DPBS.

6. Remove internal organs of the embryo carefully with forceps and transfer embryo to the third well containing 3 ml DPBS.
7. Repeat steps 4–6 for each embryo you need and use a fresh line-up of three wells with DPBS for each embryo.
8. Transfer the torso(s) of the embryo(s) into 10-ml syringe (without needle) and push them through into an Erlenmeyer flask to homogenize the tissue (can have up to five embryos per syringe).
9. Add 100 ml of the prewarmed recombinant trypsin solution and trypsinize the tissue by shaking for 20 min (150 rpm) at room temperature.
10. Prechill a beaker and attach four layers of sterile gauze fixed over the opening.
11. Pass the trypsinized cell suspension through the gauze into the prechilled beaker, while keeping the clumps in the Erlenmeyer.
12. Add 100 ml of the prewarmed recombinant trypsin solution and trypsinize the remaining tissue by shaking for 20 min (150 rpm) at room temperature.
13. Pass this trypsinized material through the gauze of a same beaker.
14. Divide the total of 200 ml cell suspension over four 50-ml tubes and pellet cells at 1,200 × *g* for 10 min at 4°C.
15. Discard the supernatant and carefully collect each pellet in 3 ml of DPBS. If possible leave the erythrocyte pellet in the tube.
16. Pool the four 3-ml cell suspensions and resuspend by pipetting to obtain a single cell suspension. Bring the volume up to 50 ml with DPBS and pellet cells at 1,200 × *g* for 10 min at 4°C.
17. Discard supernatant, collect pellet in 5 ml VP-SFM, transfer to a new 50-ml tube and resuspend by pipetting to obtain a single cell suspension.
18. Seed cells from one embryo in a total volume of 15 ml VP-SFM in a T75 flask (confluent in 1–2 days) or 30 ml VP-SFM in a T175 flask (confluent in 2–3 days).
19. Incubate for 1–3 days at 37°C until monolayers are confluent.
20. Wash cells on day 1 once with prewarmed DPBS and add fresh VP-SFM.
21. Once the cells are confluent split at ratio of 1:2 (based on culture surface). Cells will be confluent in 1–2 days. If split at a ratio of 1:4, cells will be confluent in 3–5 days.

3.2. Virus Growth

3.2.1. Amplification of MVA

This protocol can either be used to amplify parental MVA or to grow stocks of rMVA. If titered MVA starting material is available for amplification, use a multiplicity of infection (MOI) of 0.1 PFU/

cell for all infections. Either CEF, DF-1 or BHK-21 cells can be used for virus amplification, although growth in CEF cells yields higher virus output. CEF cells prepared under serum-free conditions should be used for amplification of virus for clinical use. Remove growth medium from cells before adding virus material to allow for efficient infection of cell monolayers.

1. Infect cell monolayers of ten to forty 175-cm^2 tissue culture flasks by inoculating each flask with 5 ml virus suspension (0.1 PFU/cell) in virus growth medium (see Notes 5 and 6).
2. Allow virus adsorption for 1 h at 37°C. Gently rock flasks in 20 min intervals (see Note 6).
3. Add 30 ml virus growth medium per flask.
4. Incubate at 37°C for 2 days or until cytopathic effect (CPE) is obvious.
5. Scrape cells and transfer to 250-ml centrifuge tubes.
6. Centrifuge at 38,000 × *g* for 90 min at 4°C.
7. Discard medium, resuspend and combine cell pellets in 10 mM Tris–HCl, pH 9.0. Use about 1 ml of this Tris buffer per five T175 culture flask.
8. Freeze–thaw virus material three times (3×) by sequentially putting tubes on dry ice and in a 37°C water bath. Vortex after each thaw.
9. Homogenize the material using a cup sonicator. Fill cup sonicator with ice water (50% ice), place tube containing virus material in ice water, and sonicate at maximal power for 1 min. Repeat 3×, taking care to avoid heating the sample by replenishing ice in cup. Store virus material at –80°C as crude material or until further purification (see Subheading 3.2.2) (see Note 7).

3.2.2. Purification of MVA

The use of purified virus preparations may be desirable for in vivo applications or for some in vitro experiments (e.g., the generation of clean viral DNA). Crude stock preparations of MVA can be semi-purified from cell debris and recombinant proteins by ultra centrifugation through a sucrose cushion.

1. Homogenize the virus material using a sonication needle. Place tube containing virus material in small beaker with ice water and plunge sterilized sonication needle into the virus suspension. Sonicate at maximal power for 15 s on ice. Repeat 10×, taking care to avoid heating the sample and allow suspension to cool in between sonications (see Note 8).
2. Transfer virus suspension to 50-ml Falcon tube and centrifuge 5 min at 500 × *g* and 4°C.

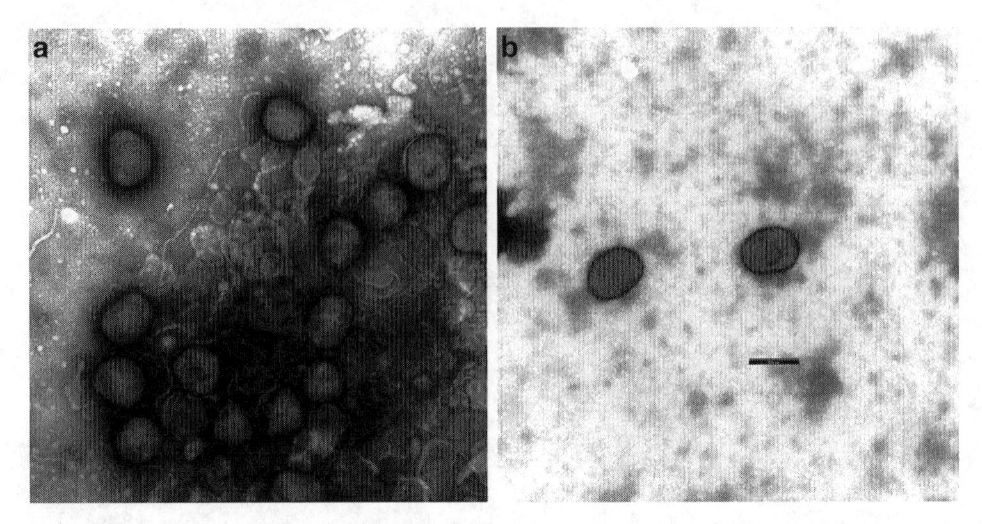

Fig. 1. Electron microscopy to control the quality of virus purification. Negative staining of (**a**) unpurified virus material or (**b**) after centrifugation through a sucrose cushion.

3. Collect supernatant in 50-ml Falcon tube, resuspend pellet in 10 ml of 10 mM Tris–HCl, pH 9.0 and repeat steps 1–3. Pool supernatants.
4. Prepare sterile 36% sucrose cushions by filling half volume of an ultracentrifuge tube (e.g., SW 28) with sucrose. Overlay with equal volume of virus suspension.
5. Spin 60 min at 30,000 × *g* and 4°C.
6. Discard supernatant (cell debris and sucrose) and resuspend pelleted virus material in 1 mM Tris–HCl, pH 9.0. Use about 1 ml per ten cell culture flasks. Virus is stored at –80°C (see Note 9).
7. If band purified virus is required, see Note 10.
8. Material may be quality controlled by electron microscopy (see Fig. 1 and Note 11).

3.3. Titration of MVA (See Note 12)

Plaque morphology of MVA differs in CEF and DF-1 cells (Fig. 2). MVA foci in CEF cell monolayers appear as large comet-shaped structures while in DF-1 cell monolayers MVA forms small round foci of infected cells. The small plaque phenotype of MVA in DF-1 cells makes counting of stained viral foci by eye difficult but is more suitable for using spot counting software (e.g., Eli.Analyse Software V6.0, A.EL.VIS GmbH).

3.3.1. Determining Amount of Plaque-Forming Units Per Ml (PFU/ml)

MVA does not form lytic plaques as known from conventional vaccinia virus, and plaques (representing foci of virus infected cells) are usually detected by staining procedures. Our preferred method to determine the PFU titer of MVA stock preparations is

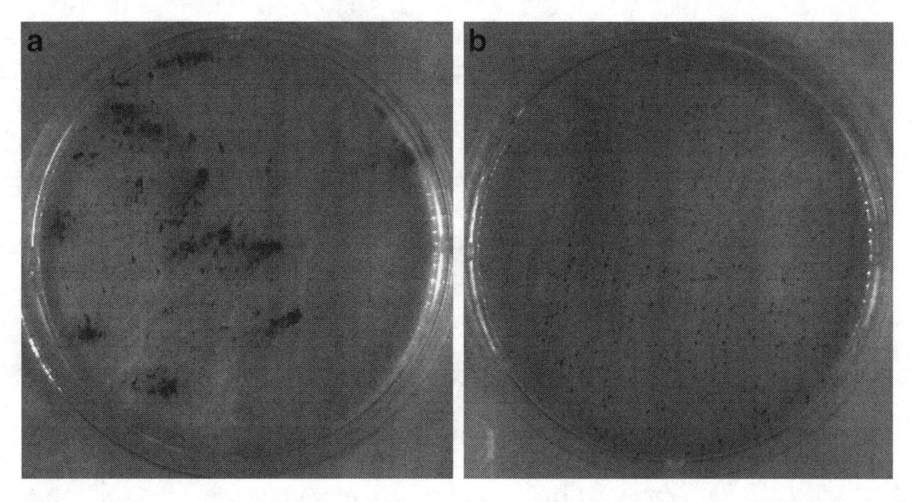

Fig. 2. Differences in plaque morphology of MVA in DF-1 or CEF monolayers. Immunostaining of MVA infected (**a**) CEF or (**b**) DF-1 cell monolayers 48 h post infection.

to visualize MVA plaques by antigen-specific immune-peroxidase staining of cells containing vaccinia virus antigens (see Note 12).

1. Thaw virus material and homogenize in cup sonicator as described in Subheading 3.2.1, step 9 (see Note 13).
2. Make tenfold serial dilutions (ranging from 10^{-1} to 10^{-9}) of virus material in 3 ml virus growth medium.
3. Plate on CEF, DF-1, or BHK-21 six-well plates 1 ml of virus suspension per well in duplicates, use 10^{-4}–10^{-9} dilutions.
4. Incubate for 2 h at 37°C.
5. Replace inoculum with virus growth medium.
6. Incubate for 48 h at 37°C.
7. Remove medium from infected tissue culture plates. Fix and permeabilize cells with 1 ml ice-cold fixing solution per well for 5 min at room temperature.
8. Remove fixing solution and air-dry fixed monolayers.
9. Add 1 ml Blocking solution per well and incubate for 30 min at room temperature or over night at 4°C.
10. Add 1 ml primary antibody solution per well and incubate for 1 h at room temperature with rocking.
11. Remove primary Ab and wash 3× with 1 ml PBS per well for 5 min each at room temperature with rocking.
12. Add 1 ml secondary antibody solution per well and incubate for 45–60 min at room temperature with rocking.
13. Remove Ab and wash 3× with 1 ml PBS per well for 5 min each at room temperature with rocking.

14. Add 0.5 ml TrueBlue Substrate solution per well and leave 15–30 min to clearly visualize stained viral foci at room temperature with rocking.
15. To determine the titer, count stained foci in a suitable dilution. Count in both wells of the dilution and calculate the mean. To express titer as PFU/ml multiply the counted number of foci by the dilution. Wells with 20–100 viral foci generate the most accurate results.

3.3.2. Determining the Tissue Culture Infectious Dose 50

To titrate the infectivity of MVA stock preparations, foci of MVA infected cells are visualized by specific immune-peroxidase staining of cells containing vaccinia viral antigen (see Note 12).

1. After thawing, homogenize MVA stock virus preparation by sonication as described in Subheading 3.2.1, step 9 (see Note 13).
2. Make tenfold serial dilutions (ranging from 10^{-1} to 10^{-11}) of virus material in 1 ml of virus growth medium.
3. Add 100 μl of each dilution in replicates of *eight* to subconfluent cell monolayers grown in 96-well plates using a multipipette and incubate at 37°C for 48 h.
4. Remove medium from infected tissue culture plates. Fix and permeabilize cells with 200 μl ice-cold fixing solution per well for 2 min at room temperature.
5. Add 200 μl Blocking solution per well and incubate for 30 min at room temperature or over night at 4°C.
6. Add 100 μl primary antibody solution per well and incubate for 1 h at room temperature with rocking.
7. Remove Ab and wash 3× with 200 μl PBS per well. For each washing step, allow to incubate with PBS for 10 min at room temperature with rocking.
8. Add 100 μl secondary antibody solution per well and incubate for 45–60 min at room temperature with rocking.
9. Remove Ab and wash 3× with 200 μl PBS per well.
10. Add 100 μl of TrueBlue Substrate solution per well and leave for 15–30 min at room temperature with rocking to clearly see stained viral foci.
11. Monitor 96-well plate under microscope and count all wells positive in which viral foci can be detected. Calculate titer according to the method of Kaerber (16) by determining the end-point dilution that will infect 50% of the wells inoculated calculating in the following way (see Note 14 for example of calculations):

    $\log_{10}$ 50% end-point dilution $= x - d/2 + (d \Sigma r/n)$

x = highest dilution in which all eight wells (8/8) are counted positive

d = the $\log_{10}$ of the dilution factor ($d = 1$ when serial tenfold dilutions are used)

r = number of positive wells per dilution

n = total number of wells per dilution ($n = 8$ when dilutions are plated out in replicates of eight).

3.4. Generation of rMVA

3.4.1. Molecular Cloning of Recombinant Gene Sequences

Subclone recombinant genes into MVA plasmid vectors such as those listed in Table 2. DNA-fragments containing the coding sequence of the gene of interest, including authentic start (ATG) and stop (TAA/TAG/TGA) codons, are cloned into the multiple cloning site of the respective vector under control of a VACV-specific promoter to generate the MVA transfer vector plasmid. For cloning it can be useful to obtain custom-made target gene sequences by gene synthesis (see Note 15). The correct orientation of the recombinant gene is determined by the direction of promoter-specific transcription. Prepare stocks of transfer plasmid DNA (see Note 16) for generation of recombinant MVA.

3.4.2. Transfection of MVA-Infected Cells with Vector Plasmids

Transfer vector plasmids are transfected into MVA-infected cells, and homologous recombination between MVA and plasmid DNA generates a recombinant virus.

1. Grow CEF or BHK-21 cell monolayers to 80% confluence in six-well tissue culture plates (see Note 17). Use one well per transfection.
2. Discard medium and overlay cells with virus growth medium containing MVA at an MOI of 0.01 (i.e., an inoculum of $\sim 1 \times 10^4$ PFU MVA in 1 ml medium for one well with 1×10^6 cells). Incubate for 90 min at 37°C.

Table 2
MVA transfer vector plasmids

Vector	Insertion site	Viral promoter	Selection	Screening	Reference
pIIIdHR-P7.5	Del III	P7.5, e/l	Transient HR	–	(2, 13)
pVIdHR-PH5	Del VI	PmH5, e/l	Transient HR	–	(14)
pIIIH5redK1L	Del III	PmH5, e/l	Transient HR	RFP	Unpublished
pIIIE3redK1L	Del III	PE3L, e	Transient HR	RFP	Unpublished
pIIIsynIIred	Del III	PsynII, l	–	RFP	Unpublished
pLW-73	I8R-G1L	PmH5, e/l	–	GFP	(11)

Promoter driving expression. *e/l* early/late, *e* early, *l* late, *HR* host range, *RFP/GFP* red/green fluorescent protein

3. 15–30 min post infection start preparing FUGENE/plasmid DNA-mix as described by the manufacturer in serum-free medium using 3 μg plasmid DNA.
4. Add FUGENE/plasmid DNA-mix directly to the medium. Incubate for 6–8 h at 37°C.
5. Replace supernatant with fresh virus growth medium and incubate for 48 h at 37°C.
6. Harvest cell monolayer with a cell scraper and transfer cells and medium into 1.5-ml microcentrifuge tubes. Store transfection harvest at –20 to –80°C.

3.4.3. Isolation of rMVA

For either selection technique, note that because of its highly host-restricted nature, MVA does not produce the rapid cytopathic effect accompanied by destruction of the cell monolayer seen with standard strains of vaccinia (e.g., Western Reserve, Copenhagen) in mammalian cells. Plaque formation is only observed in CEF, DF-1, or BHK-21 cells, or when providing K1L expression in RK-13 cells (13, 15). Also, when picking rMVA plaques, preferably choose well-separated viral foci from wells infected with highest dilutions. This will drastically reduce the number of plaque passages needed to isolate clonally pure rMVA.

3.4.4. Transient Host Range Selection

This technique relies on the inability of wild-type MVA to grow on rabbit kidney RK-13 cells, while a recombinant expressing K1L can grow (see Fig. 3 for flow chart of the procedure). Vector plasmids contain the vaccinia virus K1L gene (along with the cloned foreign gene of interest) flanked by segments of MVA-DNA that direct integration into the viral genome precisely at the site of a naturally disrupted MVA gene sequence (e.g., deletion III, (12)) rMVA expressing the recombinant antigen and expressing K1L coding sequences are isolated by consecutive rounds of plaque purification in RK-13 cell monolayers, selecting typical aggregates of infected RK-13 cells. The K1L expression cassette is designed to contain repetitive DNA sequences that allow for its deletion from the recombinant MVA genome when the virus is further plaque purified under nonselective growth conditions (i.e. by additional rounds of plaque purification on CEF or BHK-21 cells).

1. Freeze–thaw transfection harvest three times (3×) and homogenize in a cup sonicator (see Note 5).
2. Make four tenfold serial dilutions (10^{-1}–10^{-4}) of the virus suspension in virus growth medium. Start with diluting 200 μl of the transfection harvest (or plaque pick) in 1.8 ml infection medium.
3. Remove growth medium from subconfluent RK-13 cell monolayers grown in six-well plates and infect with 1 ml diluted virus suspension per well. Incubate at 37°C for 48–72 h.

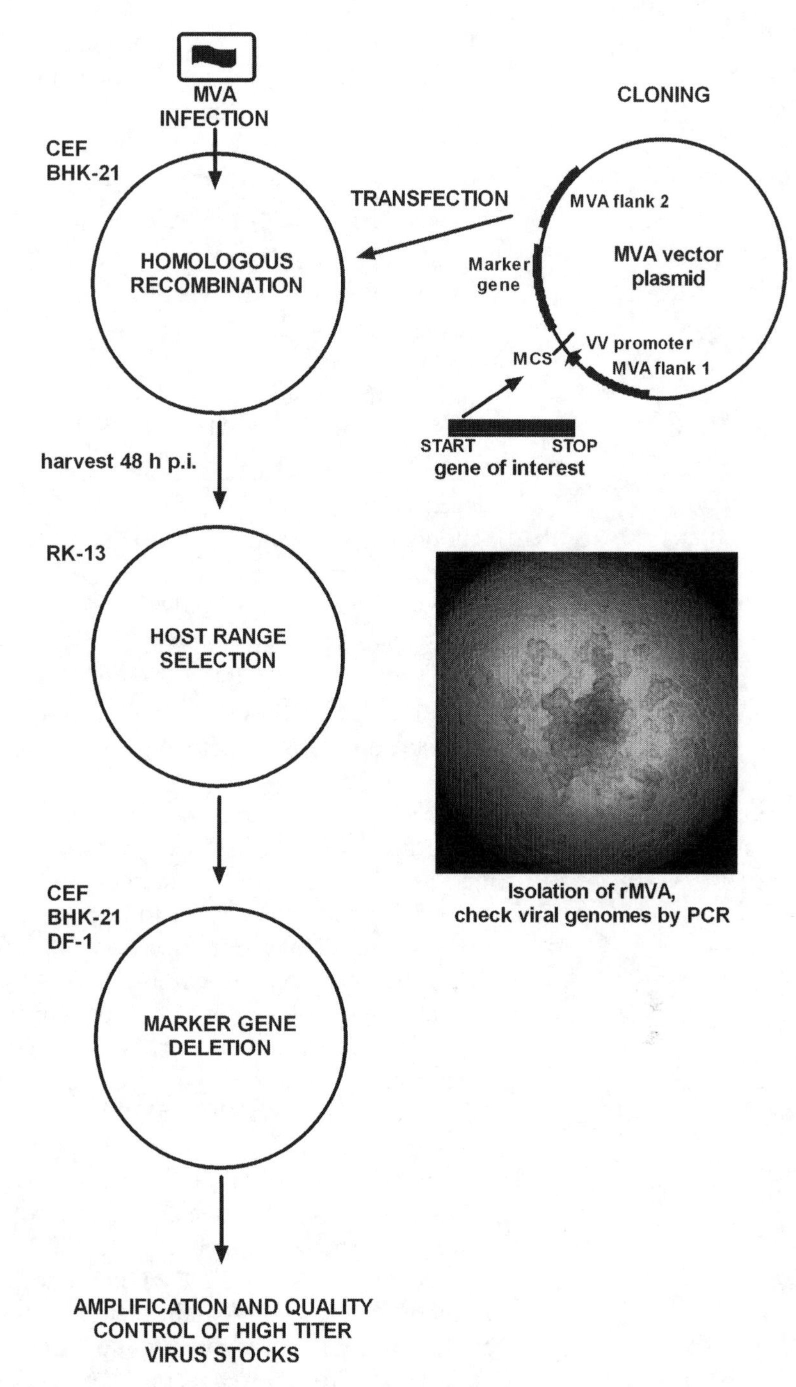

Fig. 3. Flow chart for generation of recombinant MVA using host-range selection.

4. Select typical cell aggregates of MVA/K1L infected RK-13 cells under a microscope. Mark foci with a permanent marker on the bottom of the culture well (see Note 18).
5. Add 0.5 ml virus growth medium to sterile microcentrifuge tubes.

6. Pick marked foci in a 20 μl volume by aspiration with an air-displacement pipette. Scrape and aspirate cells together with medium and transfer material to the tube containing 0.5 ml medium. Pick 5–15 foci, using new tips each time and placing aspirates in separate tubes.
7. Freeze–thaw, sonicate, and replate virus material obtained from plaque picks as described above (or store at −80°C until ready to continue plaque purification).
8. Repeat as described in steps 1–7 until clonally pure rMVA/K1L is obtained. This usually requires 2–4 rounds of plaque purification. Use PCR analysis of viral DNA to monitor for absence of wild-type MVA (see Subheading 3.4.7).
9. If there is no evidence of contaminating wild-type MVA, proceed to plaque purification on CEF, DF-1, or BHK-21 cell monolayers, selecting for rMVA-specific foci (see Note 19). Repeat steps as described in steps 1–7 in one of these cell lines. Use PCR analysis of viral DNA to monitor for absence of K1L selection cassette (Subheading 3.4.7). This usually requires 3–5 rounds of plaque purification.
10. Amplify isolated virus (Subheading 3.4.9) and analyze (Subheading 3.5) the cloned rMVA (see Note 20).

3.4.5. Screening for Transient Fluorescent Reporters

Isolation of rMVA using transient host-range selection is very effective but RK-13 cells have no record to generate rMVA for clinical use. Thus if this is a concern, rMVA may be easily generated in CEF cells which are prepared under serum-free medium conditions. Also the use of selecting agents and substrates may be undesirable. The most convenient method is the use of fluorescent reporters to screen for rMVA without requirement for other additives in the culture medium. rMVA expressing the recombinant antigen and transiently co-expressing green fluorescent (GFP) or red fluorescent protein (RFP) (rMVA/GFP, rMVA/RFP) are clonally isolated by consecutive rounds of plaque purification in CEF, DF-1, or BHK-21 cell monolayers. For screening green or red fluorescent cell foci are visualized by microscopy. The fluorescent cassette is designed to contain repetitive DNA sequences that allow for its deletion from the rMVA genome. Thus, to remove the reporter gene from rMVA additional rounds of plaque purification is carried out, screening for nonfluorescent foci of virus-infected cells. See Fig. 4 for flow chart of the procedure.

1. Freeze–thaw transfection harvest three times (3×) and homogenize in cup sonicator (see Note 5).
2. Make four tenfold serial dilutions (10^{-1}–10^{-4}) of the virus suspension in virus growth medium. Start with diluting 200 μl of the transfection harvest (or plaque pick) in 1.8 ml infection medium.

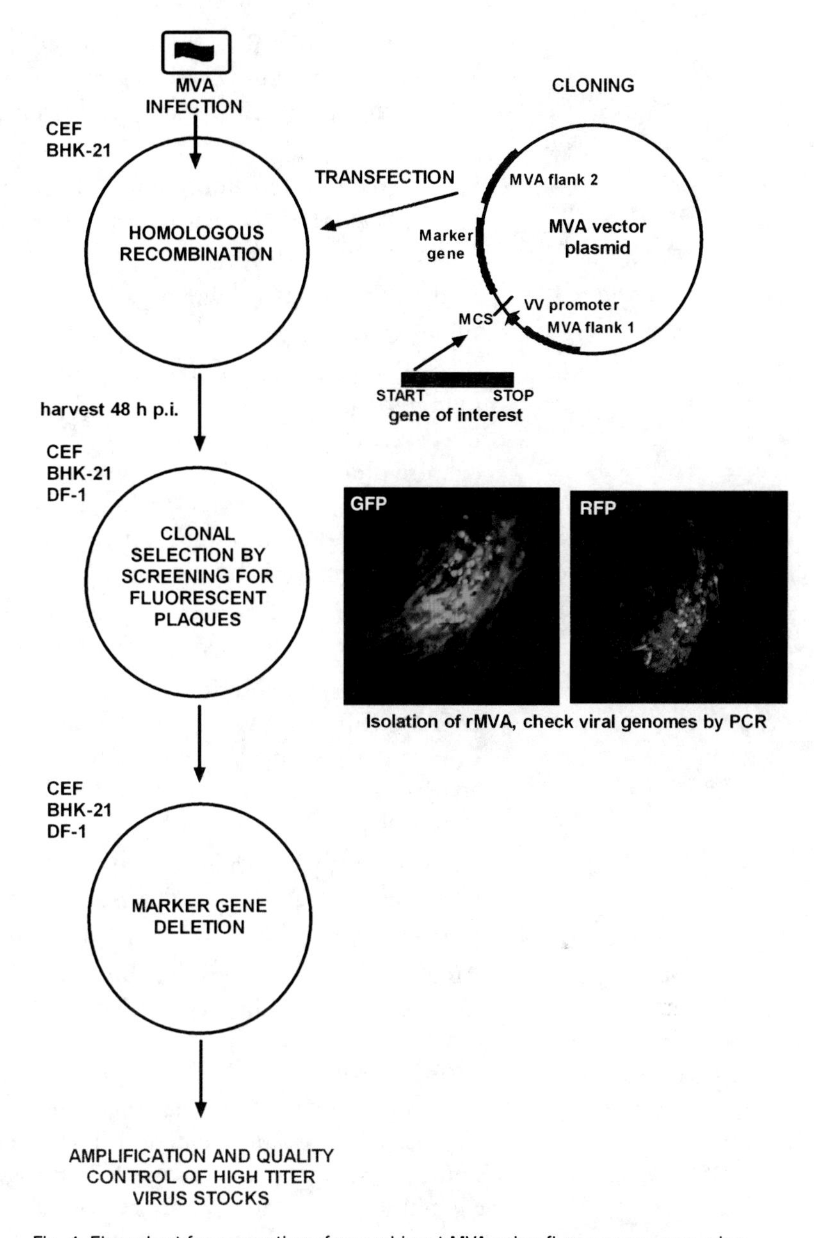

Fig. 4. Flow chart for generation of recombinant MVA using fluorescence screening.

3. Remove growth medium from confluent cell monolayers grown in six-well plates and infect with 1 ml diluted virus suspension per well. Incubate at 37°C for 2 h.
4. Melt 2% LMP-agarose, keep at 37°C until needed. Prewarm 2× medium and keep at 37°C until needed.
5. 2 h after infection of cell monolayers, mix equal amounts of 2% LMP-agarose and 2× medium.

6. Remove inoculum from cells and overlay cell monolayers with 1 ml of 2× medium/LMP-agarose mixture. Allow agar to solidify at room temperature, and then incubate for 48 h at 37°C.
7. Select green or red fluorescent viral foci under a fluorescent microscope. Mark foci with a permanent marker on the bottom of the culture well (see Note 21).
8. Add 0.5 ml virus growth medium to sterile microcentrifuge tubes.
9. Pick marked foci of cells by inserting the tip of a sterile cotton-plugged Pasteur pipette through the agarose into the marked fluorescent viral foci. Scrape and aspirate cells together with agarose plug into the Pasteur pipette tip, and transfer contents to the tube containing 0.5 ml medium by squeezing the rubber bulb on the Pasteur pipette. Pick 5–15 foci, using separate sterile pipettes and placing contents into separate tubes.
10. Freeze–thaw, sonicate, and replate virus material obtained from plaque picks as described in steps 1–6 or store at –80°C.
11. Repeat steps as described in steps 1–9 until clonally pure rMVA/GFP or rMVA/RFP is obtained. This usually requires 5–10 rounds of plaque purification (see Note 21). Use PCR analysis of viral DNA to monitor for absence of wild-type MVA (Subheading 3.4.7).
12. Continue plaque purification now selecting for *nonfluorescent* viral foci. Repeat steps as described in steps 1–9 until all viral isolates fail to produce any fluorescent foci.
13. Amplify isolated virus (Subheading 3.4.9) and analyze (Subheading 3.5) the cloned rMVA (see Note 20).

3.4.6. Combination of Host-Range Selection and Screening for Fluorescent Reporters

The combination of selection and screening methods is well established to allow for stringent and rapid isolation of rMVA (2, 16, 17). Transient host-range selection on RK-13 cells in presence of K1L and transient co-expression of RFP together with the target gene is a highly convenient combination of selection and screening without need for addition of antibiotics, chemotherapeutics, or substrates. Vector plasmids contain the RFP gene (along with the cloned foreign gene of interest) flanked by segments of MVA-DNA that direct integration into the viral genome precisely at the site of a naturally disrupted MVA gene sequence (e.g., deletion III (12)). Additionally, outside of the flanking MVA-DNA sequences the K1L expression cassette is located. During homologous recombination instable intermediates are formed that allow growth of rMVA on RK-13 cells. rMVA expressing the recombinant antigen, RFP, and K1L coding sequences are isolated by two consecutive rounds of plaque purification in RK-13 cell monolayers, selecting red fluorescent aggregates of infected RK-13 cells. This selection greatly reduces the amount of nonrecombinant MVA contained in the harvest after transfection. The K1L expression cassette as selectable marker is not stably integrated in the recombinant MVA genome and it is automatically lost upon plaque purification under nonselective growth conditions. Thus, rMVA expressing the recombinant antigen and transiently co-expressing RFP coding sequences (rMVA/RFP) are clonally isolated by three consecutive rounds of plaque purification in CEF, DF-1, or BHK-21 cells. Red fluorescent cell foci are visualized by microscopy. The RFP expression cassette is designed to contain repetitive DNA sequences that allow for its deletion from the rMVA genome. Thus, final marker-free rMVA are obtained upon additional two rounds of plaque purifications picking nonfluorescent foci of virus infected cells. See Fig. 5 for flow chart of the procedure.

1. Freeze–thaw transfection harvest three times (3×) and homogenize in a cup sonicator (see Note 5).
2. Make four tenfold serial dilutions (10^{-1}–10^{-4}) of the virus suspension in virus growth medium. Start with diluting 200 µl of the transfection harvest (or plaque pick) in 1.8 ml medium.
3. Remove growth medium from subconfluent RK-13 cell monolayers grown in six-well plates and infect with 1 ml diluted virus suspension per well. Incubate at 37°C for 48–72 h.
4. Select red fluorescent cell aggregates of rMVA/K1L infected RK-13 cells under a fluorescence microscope. Mark foci with a permanent marker on the bottom of the culture well (see Notes 18 and 21).
5. Add 0.5 ml virus growth medium to sterile microcentrifuge tubes.
6. Pick marked foci in a 20 µl volume by aspiration with an air-displacement pipette. Scrape and aspirate cells together with medium and transfer material to the tube containing 0.5 ml medium. Pick 5–15 foci, using new tips each time and placing aspirates in separate tube.
7. Freeze–thaw, sonicate, and replate virus material obtained from plaque picks as described above or store at −80°C.
8. Repeat as described in steps 1–6.
9. Make four tenfold serial dilutions (10^{-1}–10^{-4}) of the virus suspension in virus growth medium. Start with diluting 200 µl of the plaque pick in 1.8 ml medium.
10. Remove growth medium from confluent CEF, DF-1, or BHK-21 cell monolayers grown in six-well plates and infect with 1 ml diluted virus suspension per well. Incubate at 37°C for 2 h.
11. Melt 2% LMP-agarose, keep at 37°C until needed. Prewarm 2× medium and keep at 37°C until needed.

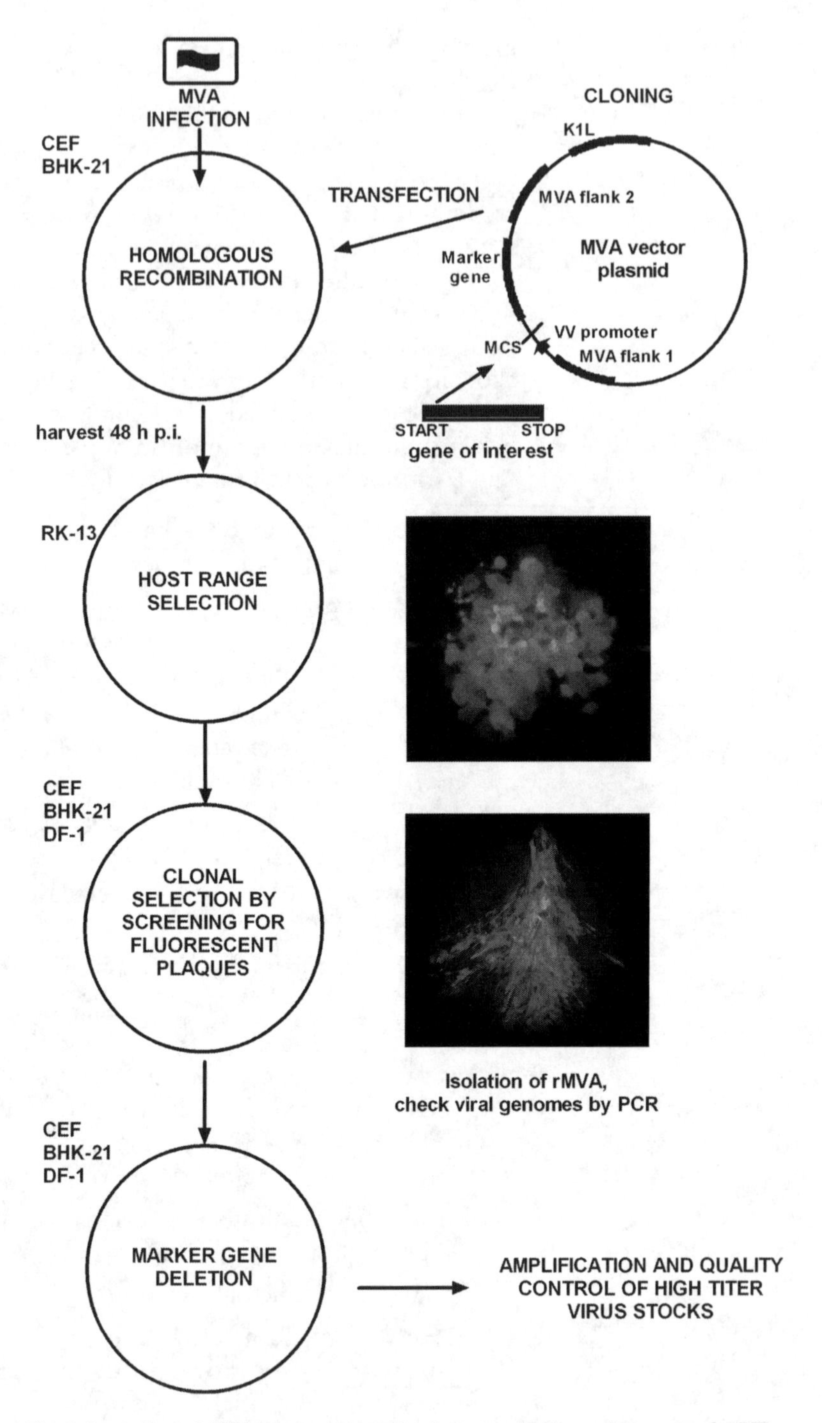

Fig. 5. Generation of rMVA using combined host-range selection and screening for RFP.

12. 2 h after infection of cell monolayers, mix equal amounts of 2% LMP-agarose and 2× medium.
13. Remove inoculum from cells and overlay cell monolayers with 1 ml of LMP-agarose/2× medium mixture. Allow agar to solidify at room temperature, and incubate for 48 h at 37°C.
14. Select green or red fluorescent viral foci under a fluorescent microscope. Mark foci with a permanent marker on the bottom of the culture well (see Note 21).
15. Add 0.5 ml virus growth medium to sterile microcentrifuge tubes.
16. Pick marked foci of cells by inserting the tip of a sterile cotton-plugged Pasteur pipette through the agarose into the marked fluorescent viral foci. Scrape and aspirate cells together with agarose plug into the Pasteur pipette tip, and transfer contents to the tube containing 0.5 ml medium by squeezing the rubber bulb on the Pasteur pipette. Pick 5–15 foci, using separate sterile pipettes and placing contents into separate tubes.
17. Freeze–thaw, sonicate, and replate virus material obtained from plaque picks as described in steps 9–16 or store at –80°C.
18. Repeat steps as described in steps 9–17 until clonally pure rMVA/GFP or rMVA/RFP is obtained. Use PCR analysis of viral DNA to monitor for absence of wild-type MVA (Subheading 3.4.7).
19. Continue plaque purification now picking *nonfluorescent* viral foci. Repeat steps as described in 9–17 until all viral isolates fail to produce any fluorescent foci.
20. Amplify isolated virus (Subheading 3.4.9) and analyze (Subheading 3.5) the cloned rMVA (see Note 20).

3.4.7. Characterization of Recombinant MVA Genomes by PCR

MVA-DNA is analyzed by PCR using oligonucleotide primers that are designed to amplify DNA fragments at the specific insertion site used within the MVA genome (see Table 1). Thus, genomes of rMVA and wild-type MVA can be easily identified and distinguished in DNA preparations from infected cell cultures. Elimination of wild-type MVA during plaque purification of rMVA can be monitored (see Note 22), and correct insertion of foreign DNA within the MVA genome can be ascertained. An example of a PCR analysis is described for using deletion III of MVA as an insertion site and transient K1L selection or RFP screening to generate recombinant viruses. Primers MVA-III-5′ and MVA-III-3′ (Table 1) anneal to template MVA-DNA sequences adjacent to insertion site III and PCR will produce DNA fragments that are specific for wild-type MVA, rMVA/K1L expressing the K1L marker gene or rMVA/RFP expressing the RFP marker gene, or for the final rMVA. Expected fragments are shown in Table 3. The amplification product for wild-type MVA has a defined size of 0.8 kb indicating

Table 3
PCR fragment sizes from rMVA using primers MVA-III-5′ and MVA-III-3′

	Expected size PCR fragment
Wild-type MVA	0.8 kb, corresponding to empty deletion III
rMVA/K1L	2.2 kb (Del III + K1L) + x kb (x = inserted gene sequence)
rMVA/RFP	1.9 kb (Del III + RFP) + x kb (x = inserted gene sequence)
rMVA	0.8 kb (rMVA-marker gene) + x kb (x = inserted gene sequence)

that no foreign DNA is inserted into insertion site III. The expected molecular weight of the PCR product for rMVA/K1L or rMVA/RFP can be calculated by adding the size of the recombinant insert to 2.2 kb (empty MVA and K1L marker cassette) or 1.9 kb (empty MVA and RFP marker cassette). DNA extracted from cells infected with wild-type MVA and plasmid DNA from transfer vectors are used as control templates. The size of the PCR fragment specific for the final recombinant virus rMVA results from the molecular weight of rMVA/K1L or rMVA/RFP being reduced by 1.36-kb or 1.1-kb DNA corresponding to the desired loss of K1L or RFP reporter gene sequences. Further testing for the loss of K1L can also be performed (see Note 23).

3.4.8. Extraction of Viral Genomic DNA

1. Infect cell monolayer of one well in six-well or 12-well plate with 1 ml of 10^{-1} dilution of the virus suspension obtained from the last round of plaque purification, and incubate for 3 days at 37°C (see Notes 24 and 25).
2. Discard medium, harvest cell monolayer in 400 µl Distilled water and transfer into 1.5-ml microcentrifuge tube.
3. Add 50 µl 10× TEN, and freeze–thaw 3×.
4. Mix by vortexing, then centrifuge at 1,800 × *g* for 5 min at room temperature to remove cellular debris.
5. Transfer supernatant into fresh 1.5-ml microcentrifuge tube. Add 50 µl Proteinase K and 23 µl 20% SDS.
6. Vortex and incubate for 2 h at 56°C.
7. Extract DNA twice with phenol–chloroform by adding equal volumes of phenol–chloroform 1:1, mix and microcentrifuge at top speed for 5 min at room temperature. Pipette supernatant into new 1.5-ml microcentrifuge tube.
8. Add 1/10 volume of 3 M NaAc followed by two volumes of ice-cold absolute EtOH. Mix gently and cool for 15 min at –80°C or for 30 min at –20°C. Centrifuge at top speed for 15 min at 4°C.

9. Aspirate off supernatant, wash DNA pellet with 1 ml 70% EtOH and centrifuge at top speed for 15 min at 4°C.
10. Remove EtOH and air-dry DNA pellet for at least 10 min. Resuspend dried pellet in 50 μl distilled water (see Note 26).
11. Prepare PCR reaction mix on ice by adding 39 μl distilled water, 5 μl primer 1, 5 μl primer 2, 1 μl template DNA, and 50 μl PCR master mix to obtain total volume of 100 μl.
12. Perform PCR using the conditions in Table 4 (see Note 27).
13. Use 20-μl aliquot of each PCR reaction to perform agarose gel electrophoresis to visualize amplified DNA fragments and determine molecular weights in comparison to double-stranded DNA standards (e.g., 1-kb DNA Ladder) (see Notes 28–30).

3.4.9. Growth of Virus Stocks from Single Plaque Isolates (See Note 31)

1. Infect confluent CEF, DF-1 or BHK-21 cell monolayer grown in a ~35-mm tissue culture dish, adding to the medium 250 μl virus suspension of isolated rMVA obtained from the last plaque purification. Incubate at 37°C for 2 days or until CPE is obvious.
2. Discard medium, harvest cell monolayer in 1 ml virus growth medium, transfer into 1.5-ml microcentrifuge tube, freeze–thaw, and sonicate. Either continue or store at –20 to –80°C as first passage of rMVA.
3. Infect cell monolayer grown in ~60-mm tissue culture dish, by adding to 1 ml medium 0.5 ml virus suspension obtained from first passage of rMVA. Allow virus to adsorb for 1 h at 37°C, add 2 ml virus growth medium and incubate at 37°C for 2 days or until CPE is obvious.
4. Scrape cells, transfer to 15-ml conical centrifuge tube, centrifuge 5 min at 500 × *g*, discard medium and resuspend cells in

Table 4
PCR conditions for characterization of clonal rMVA isolates

Step	Cycle		Temp (°C)	Time
1		Denaturation	94	2 min
2	1–30	Denaturation Annealing Elongation	94 55 72	30 s 40 s 3 min
3		Final Storage	72 4	7 min

2 ml virus growth medium, freeze–thaw and sonicate. Either continue or store at −20 to −80°C as second passage of rMVA.

5. Infect cell monolayer of one 75-cm^2 tissue culture flask by adding 0.5 ml of virus material from second passage of rMVA and 1.5 ml virus growth medium. Allow virus adsorption for 1 h at 37°C, rocking flask at 20 min intervals (see Note 11). Overlay with 10 ml virus growth medium and incubate at 37°C for 2 days or until CPE is obvious.
6. Scrape cells, transfer to 15-ml conical centrifuge tube, centrifuge 5 min at 500 × *g*, discard medium and resuspend cells in 5 ml virus growth medium, freeze–thaw and sonicate. Either continue or store at −20 to −80 °C as third passage of rMVA.
7. Infect cell monolayer of one 175-cm^2 tissue culture flask with 2 ml of virus material from the third passage of rMVA. Allow virus adsorption for 1 h at 37°C, rocking flask at 20-min intervals, add 30 ml virus growth medium, and incubate at 37°C for 2 days or until CPE is obvious.
8. Scrape cells in medium, transfer to 50-ml conical centrifuge tube, centrifuge 5 min at 500 × *g*, discard medium and resuspend cells in 15 ml virus growth medium, freeze–thaw and sonicate and store at −20 to −80 °C as fourth passage of rMVA. The titer of this material should be approximately 10^9 IU, but as a general rule material of one well-infected T175 culture flask can be used to infect ten T175 flasks.
9. Generate a high-titer seed virus stock following procedure outlined in Subheading 3.2 (see Note 32).

3.5. Quality Control of rMVA

3.5.1. PCR Assay for MVA Identity

The genetic identity of MVA can be checked by PCR monitoring for the characteristic six major deletions in the MVA DNA (12). MVA-DNA is analyzed by PCR using oligonucleotide primers (Table 1) that are designed to amplify specific DNA fragments that extend over the six major deletions sites within the MVA genome. Thus, genomes of rMVA and wild-type MVA can be easily identified and also distinguished in DNA preparations from infected cell cultures. PCR of wild-type MVA results in unique PCR products of 291 bp, 354 bp, 447 bp, 502 bp, 603 bp, and 702 bp, respectively (Fig. 6a). Correct insertion of the recombinant gene within the MVA genome can be also verified. As an example PCR with DNA from MVA-GFP (15) was used with a *gfp* gene inserted into deletion III of the MVA genome (Fig. 6b). This analysis produces DNA fragments specific for Del I, II, IV, V and VI of the same size as for wild-type MVA. The expected molecular weight of the PCR product for rMVA can be calculated by adding the size of the recombinant insert to the product size of the respective insertion site as in our example is Del III with a specific PCR product of 447 bp. Thus, PCR of MVA-GFP DNA will result in a Del III specific product of 1,488 bp.

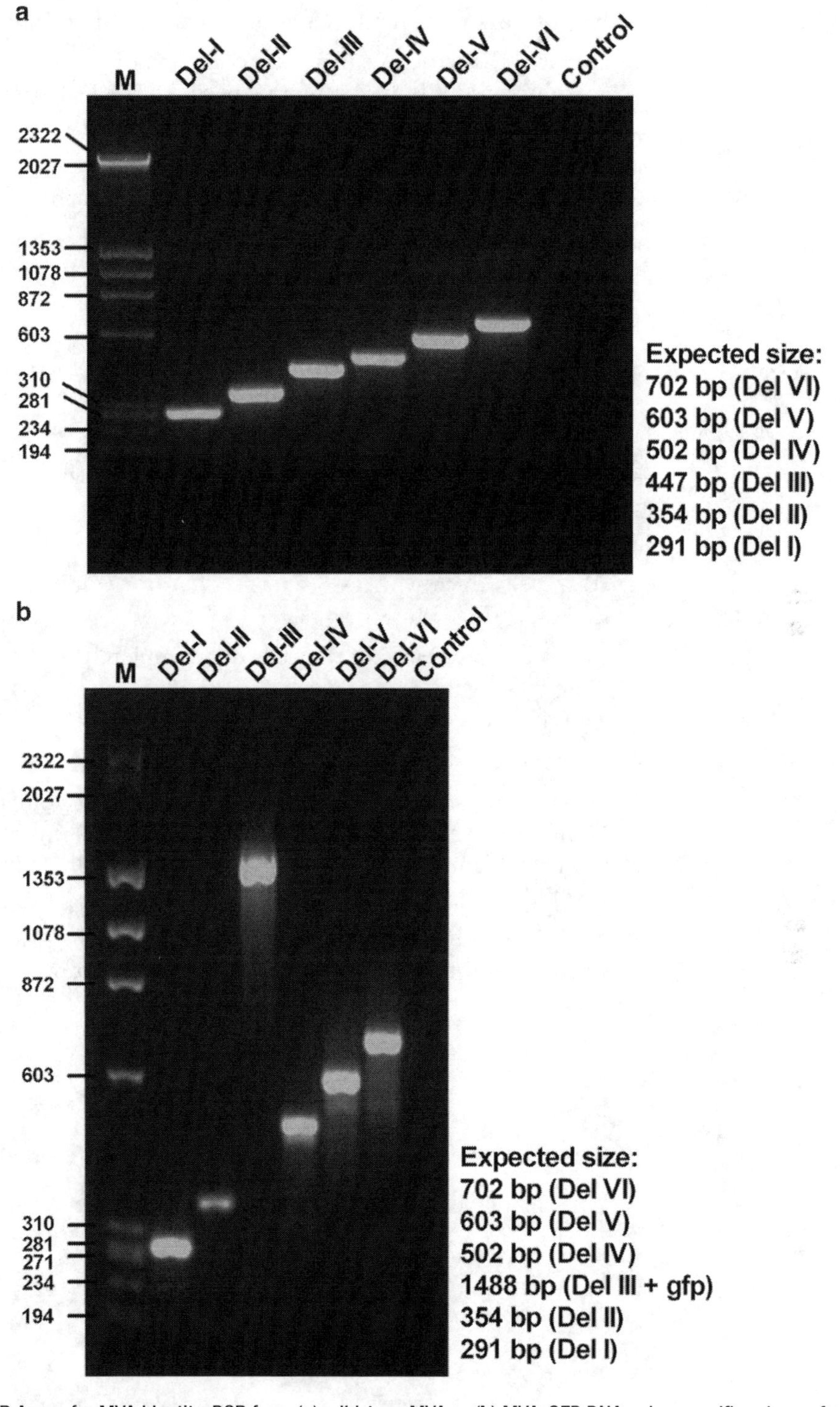

Fig. 6. PCR Assay for MVA Identity. PCR from (**a**) wild-type MVA or (**b**) MVA-GFP DNA using specific primers for Deletion I to VI. Control is without DNA.

1. Extract viral genomic DNA as described in Subheading 3.4.8.
2. Prepare PCR reaction mix on ice in a total volume of 25 μL containing PCR buffer, 200 nM specific MVA Deletion primer pairs Del I—Del VI (see Table 1), 50 ng viral DNA and 0.2 U DyNAzyme™ II.
3. Perform PCR using conditions outlined in Table 5.
4. Separate PCR products on a 1.7% agarose gel. Phage X174 DNA-*Hae*III digest and λDNA-*Hin*dIII digest serves as a molecular weight marker.

3.5.2. Assay for Expression of Recombinant Gene by Immunocytochemistry

1. Infect subconfluent (80–90%) CEF cell layer of a six-well plate with the recombinant MVA at an MOI of 0.01 in 1 ml virus growth medium (see Note 33).
2. Incubate for 2 h at 37°C, remove inoculum and add 2 ml fresh virus growth medium.
3. Incubate for 2–3 days until small plaques are visible.
4. Wash cells once with PBS and fix them for 3 min with 2 ml ice-cold fixing solution.
5. Remove fixative and let the cells dry completely at the air (see Note 34).
6. Add 2 ml blocking buffer per well and incubate 1 h at room temperature or overnight at 4°C (see Note 35).
7. Dilute your primary antibody (directed against your protein of interest (e.g., Rabbit anti-HA antibody)) in blocking buffer (If antibody is rabbit derived, generally dilution is 1:5,000–1:10,000).
8. Replace the blocking buffer in each well with 1 ml of the antibody solution and incubate for 1 h at room temperature (rocking gently).

Table 5
PCR conditions for characterization of clonal rMVA isolates

Step	Cycle		Temp (°C)	Time
1		Denaturation	95	3 min
2	1–35	Denaturation Annealing Elongation	95 57 72	30 s 45 s 45 s
3		Final Storage	72 4	5 min

9. Remove primary antibody solution and wash cells three times with 2 ml blocking buffer.
10. Dilute your secondary antibody (directed against your first antibody (e.g., anti-Rabbit HRP)) in blocking buffer and add 1 ml of antibody solution to the wells and incubate 45 min at room temperature (rocking gently).
11. Remove secondary antibody solution and wash cells three times with 2 ml blocking buffer.
12. Add 0.5 ml TrueBlue Substrate per well and monitor development of your staining while rocking the plate gently.
13. Remove substrate and add 2 ml distilled water to the well (or glycerine solution for longer preservation).
14. Qualify antigen-specific staining.

3.5.3. Assay rMVA Homogeneity by Double-Immunostaining

1. Perform immunocytochemistry as described in Subheading 3.5.2 until step 12. However at step 7 use antibodies against vaccinia virus instead of the protein of interest. Then proceed as follows to stain for the antigen that is expressed by the recombinant virus.
2. Remove substrate and wash cells two times with 2 ml blocking buffer.
3. Add 2 ml blocking buffer per well and incubate 1 h at room temperature or overnight at 4°C (see Note 35).
4. Replace the blocking buffer in each well with 1 ml of the diluted primary antibody in blocking buffer and incubate for 1 h at room temperature (rocking gently) (see Note 36).
5. Remove antibody solution and wash cells three times with 2 ml blocking buffer.
6. Dilute the secondary antibody (directed against your primary antibody (e.g., anti-Rabbit HRP)) in blocking buffer and add 1 ml of antibody solution to the wells. Incubate 45 min at room temperature (rocking gently) (see Note 36).
7. Remove antibody solution and wash cells three times with 2 ml blocking buffer.
8. Prepare fresh substrate solution for peroxidase staining (see Note 3).
9. Add 1 ml of substrate solution per well and monitor color development of stained viral foci.
10. Qualify antigen-specific staining, aim for >95% plaques positive both for the protein of interest and anti-vaccinia staining.

3.5.4. Assay for Expression of Recombinant Gene by Western Blot Analysis

1. Infect subconfluent (80–90%) CEF cell layer or BHK-21 cells of a six-well plate with the recombinant MVA at an MOI of 5 in 1 ml medium (see Note 33).

2. Incubate for 2 h at 37°C, remove inoculum, and add 2 ml fresh virus growth medium.
3. Incubate for 2–3 days or until 90–100% CPE is visible.
4. Wash cells once with PBS.
5. Add 1 ml PBS and scrape cells in this volume.
6. Pellet cells by centrifugation for 5 min at 500 × *g* and 4°C.
7. Resuspend pellet in ice-cold 200 μl RIPA buffer containing protease inhibitors.
8. Incubate for 20 min on ice and centrifuge for 10 min at top speed at 4°C.
9. Collect supernatant to a new microfuge tube.
10. Proceed with western blot analysis of the lysates for the recombinant protein as in references (18, 19). Generally 1/10 of your supernatant should provide enough protein.

3.5.5. Assay for Stability of Recombinant Gene expression

Occasionally, expression of a recombinant gene product can hamper the replication of the rMVA. Strong suppressive effects on virus growth will result in the occurrence of non-expressing virus mutants that may outgrow the rMVA upon amplification. To test for the stability of recombinant gene expression, primary stocks of rMVA should be evaluated for maintained target protein synthesis after serial cell culture passage at low MOI.

1. Infect subconfluent (80–90%) CEF cell layer of a well of six-well plate with the recombinant MVA at an MOI of 0.05 in 1 ml virus growth medium.
2. Incubate for 48 h and harvest cells and supernatant together.
3. Freeze–thaw three times (vortex in between).
4. Sonicate three times (vortex in between).
5. Dilute virus suspension 1:1,000 by tenfold serial dilutions at a final volume of 1 ml.
6. Infect subconfluent (80–90%) CEF cell layer of a well of six-well plate with 1 ml of the 1:1,000-diluted virus material.
7. Incubate for 48 h and repeat steps 3–6 three additional times.
8. Check genetic stability by comparing DNA from the starting material and DNA from the 5th passage in PCR assay for MVA identity (see Subheading 3.5.1).
9. Monitor for recombinant gene product by immunocytochemistry (see Subheading 3.5.2) or double-immunostaining (see Subheading 3.5.3) of infected monolayers.

3.5.6. Replication Deficiency Assay

The inability of MVA to productively replicate in cell lines of human origin is a characteristic feature of MVA and considered a prerequisite for handling the virus under laboratory conditions of minimal potential biohazard to laboratory personnel and the environment. Thus, rMVA containing new foreign gene sequences are routinely tested for growth properties in human HeLa cells. Monolayers are infected with low MOI of MVA and recombinant MVA for 72 h. A replication-competent vaccinia virus or MVA grown on permissive cells may serve as control. Cells and supernatants are harvested 0 h and 72 h post infection and infectivity is determined on CEF, DF-1, or BHK-21 cells.

1. Infect confluent HeLa cells grown on six-well plates with 0.01 PFU per cell of the respective virus in 1 ml of virus growth medium. Infect two wells per time point and virus.
2. Incubate at 37°C, 5% CO_2 for 1 h.
3. Remove inoculum and wash twice with fresh medium.
4. Harvest 0 h time points by scraping monolayers into medium and transferring cells and supernatants into microcentrifuge tubes. Store at –80°C.
5. Incubate remaining wells for 72 h at 37°C, 5% CO_2. Harvest wells (cells and supernatants) as described in step 4.
6. Titrate samples on confluent CEF, DF-1, or BHK-21 six-well plates as described in Subheading 3.3.
7. Calculate the replication efficiency of each virus as a ratio t_{72}/t_0 using the mean titer for each time point (four values: two samples for each time point, titrated in duplicates).

4. Notes

1. DF-1 cells are grown in DMEM.
2. Note that not all low-melting-point agaroses work well. Some are toxic to cell lines. We have found the product from Life Technologies to work with the cell lines we routinely use.
3. Owing to the toxicity of dianisidine and the need for such a tiny amount, we typically do not weigh it out, but just use an amount that fills the tip of a small spatula. To remove small clumps of o-dianisidine, filter the PBS/dianisidine mix through a 0.2-μm filter into a new tube before adding the H_2O_2 >30%.
4. Recipe is enough for two 6-well plates.
5. For all steps, remember, when using frozen virus stocks always sonicate 30 s in cup sonicator after thawing to disrupt clumps of virus.
6. When infecting cell monolayers grown in larger tissue culture flasks (e.g., 175-cm^2 flasks), avoid drying of the cell monolayer by rocking flask by hand at 20 min intervals.

7. The crude material has typical titers of approximately 3×10^9 IU per infected tissue culture flask.
8. In order to homogenize virus material most efficiently after amplification, we recommend the use of a sonication needle instead of the cup sonicator. Place tube containing virus material in small beaker with ice water and plunge sterilized sonication needle into virus suspension. Sonicate 4× for 15 s. at maximal power. Take care to avoid heating of the sample.
9. Usual titers for purified material are approximately 1×10^9 IU per infected tissue culture flask (i.e. titers of ~10^{10} IU/ml after concentration of virus material).
10. In order to obtain highly (i.e., band) purified viruses, material obtained from Subheading 3.2.2, step 6 can be centrifuged through a 25 to 40% sucrose gradient for 50 min at $28{,}000 \times g$ and 4°C. Harvest virus band that appears at the lower middle of the tube. To concentrate the virus in the band and to remove remaining sucrose, fill an ultracentrifuge tube with > 3× volume of 1 mM Tris–HCl, pH 9.0 and pellet virus material at $38{,}000 \times g$ for 1 h at 4°C. Resuspend the pellet in this Tris buffer and store at –80°C.
11. If available, the proper removal of cell debris can be easily monitored by electron microscopy. Negative stain is a very rapid method to control the quality of virus purification. Place a small amount of virus (1–3 µl) onto a copper grid (Square 300 mesh, Plano GmbH) and 30 s later stain with 20 µl of Tungstophosphoric acid (2% aqueous solution, Merck) for 1 min. Then, absorb the stain with Whatman filter paper and allow the grid to dry before examination by electron microscopy.
12. We describe two different methods to determine titers of virus stocks. An advantage of the first method (amount of PFU/ml, Subheading 3.3) is the relative ease of using six-well rather than 96-well plates. Also, the handling for culture, plating of virus dilutions and immunostaining is simpler. On the other hand, determining the tissue culture infectious dose 50 ($TCID_{50}$) (Subheading 3.2.2) of virus stocks may result in more reproducible titers since a choice between infected or not-infected is independent of the number of plaques per well.
13. Before titration, virus material *must* be homogenized by sonication. Sonicate aliquots of maximal 1.5 ml virus suspension as described in Subheading 3.2.1, step 9.
14. Example for calculating $TCID_{50}$: If all eight wells are counted positive in dilution 10^{-7}, $x=7$. If, five infected wells are found in dilution 10^{-8}, and the number of infected wells in dilution 10^{-9} (the highest dilution in which positive wells can be

found) is 2. Then, the log 50% end-point dilution would be: $7 - 1/2 + (8/8 + 5/8 + 2/8) = 7 - 0.5 + (1.875) = 7 + 1.375 = 8.375$. Therefore, the end-point dilution that will infect 50% of the wells inoculated is $10^{-8.375}$. The reciprocal of this number yields the titer in terms of infectious dose per unit volume. Since the inoculum added to an individual well was 0.1 ml, the titer of the virus suspension would therefore be: $10^{8.375}$ $TCID_{50}/0.1$ ml = $10^{9.375}$ $TCID_{50}/ml$.

15. If recombinant gene of interest contains putative viral transcription termination signals (TTTTTNT) or runs of more than four G's or C's, sequences might be altered by silent mutations to enhance foreign gene expression. Custom-made target gene sequences can be prepared by gene synthesis available from different companies.
16. For best transfection efficiencies prepare clean, super coiled DNA either by centrifugation through cesium chloride gradients or using plasmid purification kits (e.g., QIAGEN).
17. In general, transfection of BHK-21 cells yields the best efficiencies. For the generation of rMVA for clinical use, CEF cells prepared under serum-free conditions should be used for transfection of MVA-infected cells.
18. In order to distinguish rMVA-specific RK-13 aggregates, it may be helpful to use mock and wild-type MVA infected control wells for comparison.
19. To allow for most efficient plaque cloning, it may be helpful to perform plaque passages under agar (for overlay follow Subheading 3.4.5).
20. Having obtained a rMVA stock virus, the following procedures are recommended:

    - It is recommended to prepare a first seed virus stock as *primary* stock (material of about ten infected T175 culture flasks), which is used to amplify *working* stocks of rMVA. This is done to minimize multiple passages of the virus.
    - Titer the virus stock on CEF, DF-1, or BHK-21 cell monolayers (see Subheading 3.3).
    - Check clonal purity and genomic identity of rMVA by PCR (see Subheading 3.5.1) of viral DNA.
    - Characterize synthesis of recombinant antigen (see Subheading 3.5.2).
    - Check for stability of recombinant gene expression (see Subheading 3.5.5).
    - Evaluate growth properties of rMVA on human cells (see Subheading 3.5.6).

21. Since the expression cassette containing the *gfp or rfp* marker gene is designed to be efficiently deleted from the rMVA genome, non-staining MVA foci may be observed during plaque purification even after all wild-type MVA has been successfully eliminated. To avoid needless plaque passages, it is important that the absence of parental MVA is confirmed by PCR analysis.
22. To monitor the presence of wild-type MVA during plaque purification, viral DNA sufficient for PCR analysis is isolated from cell monolayers infected with the 10^{-1}-dilution of virus suspensions plated out for plaque passage.
23. One can also check for deletion of K1L marker cassette from virus genomes by K1L-specific PCR using primers K1L-int-1 and K1L-int-2 (Table 1) and 55°C annealing temperature. If K1L is still present in viral genomes, this PCR yields a product of 290 bp. If K1L has successfully been recombined out, there will not be any PCR product. Always use controls (e.g., transfer plasmid and wild-type MVA) and perform PCR for specific insertion site in parallel to make sure that template DNA was present.
24. Virus material harvested from cell monolayers grown in 6-well/12-well plates and infected for 24 h with an MOI of 10 IU/cell will yield a good amount of viral DNA for PCR/Southern blot analysis. If infectivity is to low, a second round of amplification may be necessary. Harvest amplification, freeze–thaw, sonicate and replate on cell monolayers.
25. After infecting cells, avoid sonication of infected tissue culture material that will be used for DNA extraction because unpackaged viral DNA will be destroyed and lost due to sonication.
26. Carefully air-dry the pelleted DNA material to remove all ethanol.
27. Always use DNA of wild-type MVA and respective plasmid as control templates for PCR analysis.
28. As DNA preparations might contain variable quantities of viral DNA, the amount of template DNA used for PCR may be optimized.
29. PCR conditions (temperatures and number of cycles) may need to be optimized according to the size of the expected fragment to be amplified. Conditions as stated in the protocol have been used for amplification of up to 4 kb DNA inserted into the MVA genome.
30. If template DNA is derived from mixed virus populations containing both, rMVA/K1L or rMVA/RFP as well as wild-type MVA, PCR may amplify preferentially the fragment for wild-type MVA because of its smaller size and rMVA/K1L or

rMVA/RFP may not be detectable. A signal for rMVA (having already lost the K1L or RFP gene) may also be detected, as the K1L/RFP marker cassette is designed to be efficiently deleted from recombinant genomes, and this process, can also occur in RK-13 cells.

31. The key to the amplification of a stock of virus from a single plaque is to follow a step-wise growth of virus on small then larger amounts of cells. If one initially puts a low-titer growth on a high cell number, there is a risk of losing the virus.
32. The titer of this material should be approx. 10^9 IU. As a general rule, in growing up a seed stock material from one well-infected T175 culture flask can be used to infect 10 T175 flasks.
33. For proper controls remember to also infect wells with wild-type MVA as well as include uninfected wells.
34. If staining is not performed immediately after fixation, cells should be stored at 4°C with 2 ml PBS per well (staining is possible at least up to 2 days after fixation).
35. Staining of recombinant protein might result in high background staining. To minimize unspecific staining, blocking might be performed for 1 h at 37°C instead of room temperature. Make also sure to use subconfluent cell layer as more confluent layers might enhance background staining (especially under serum-free conditions).
36. Recommended dilutions range is from 1:5,000 to 1:10,000 in blocking buffer.

Acknowledgments

This work was supported by the European Community (FP7 2010; VECTORIE grant No. 261466).

References

1. Moss B (1996) Genetically engineered poxviruses for recombinant gene expression, vaccination, and safety. Proc Natl Acad Sci USA 93:11341–11348
2. Sutter G, Moss B (1992) Nonreplicating vaccinia vector efficiently expresses recombinant genes. Proc Natl Acad Sci USA 89:10847–10851
3. Tartaglia J et al (1992) NYVAC: a highly attenuated strain of vaccinia virus. Virology 188:217–232
4. Acres B, Bonnefoy J-Y (2008) Clinical development of MVA-based therapeutic cancer vaccines. Expert Rev Vaccines 7:889–893
5. Gomez CE, Najera JL, Krupa M, Esteban M (2008) The poxvirus vectors MVA and NYVAC as gene delivery systems for vaccination against infectious diseases and cancer. Curr Gene Ther 8:97–120
6. Rimmelzwaan GF, Sutter G (2009) Candidate influenza vaccines based on recombinant modified vaccinia virus Ankara. Expert Rev Vaccines 8:447–454
7. Kennedy JS, Greenberg RN (2008) IMVAMUNE®: modified vaccinia Ankara strain as an attenuated smallpox vaccine. Expert Rev Vaccines 8:13–24

8. Mackett M, Smith GL, Moss B (1984) General method for production and selection of infectious vaccinia virus recombinants expressing foreign genes. J Virol 49:857–864
9. Cottingham MG, Gilbert SC (2010) Rapid generation of markerless recombinant MVA vaccines by en passant recombineering of a self-excising bacterial artificial chromosome. J Virol Methods 168:233–236
10. Domi A, Moss B (2005) Engineering of a vaccinia virus bacterial artificial chromosome in Escherichia coli by bacteriophage [lambda]-based recombination. Nat Meth 2:95–97
11. Wyatt LS et al (2009) Elucidating and minimizing the loss by recombinant vaccinia virus of human immunodeficiency virus gene expression resulting from spontaneous mutations and positive selection. J Virol 83:7176–7184
12. Meyer H, Sutter G, Mayr A (1991) Mapping of deletions in the genome of the highly attenuated vaccinia virus MVA and their influence on virulence. J Gen Virol 72:1031–1038
13. Staib C et al (2000) Transient host range selection for genetic engineering of modified vaccinia virus Ankara. Biotechniques 28:1137–1142
14. Antonis AFG et al (2007) Vaccination with recombinant modified vaccinia virus Ankara expressing bovine respiratory syncytial virus (bRSV) proteins protects calves against RSV challenge. Vaccine 25:4818–4827
15. Staib C, Lowel M, Erfle V, Sutter G (2003) Improved host range selection for recombinant modified vaccinia virus Ankara. Biotechniques 34:694–696, 698, 700
16. Staib C, Drexler I, Sutter G (2004) Construction and isolation of recombinant MVA. Methods Mol Biol 269:77–100
17. Wong YC, Lin LC, Melo-Silva CR, Smith SA, Tscharke DC (2011) Engineering recombinant poxviruses using a compact GFP-blasticidin resistance fusion gene for selection. J Virol Methods 171:295–8
18. Renart J, Reiser J, Stark GR (1979) Transfer of proteins from gels to diazobenzyloxymethyl-paper and detection with antisera: a method for studying antibody specificity and antigen structure. Proc Natl Acad Sci USA 76:3116–3120
19. Towbin H, Staehelin T, Gordon J (1979) Electrophoretic transfer of proteins from polyacrylamide gels to nitrocellulose sheets: procedure and some applications. Proc Natl Acad Sci USA 76:4350–4354

Chapter 5

Isolation of Recombinant MVA Using F13L Selection

Juana M. Sánchez-Puig, María M. Lorenzo, and Rafael Blasco

Abstract

Modified vaccinia virus Ankara (MVA) has become a widely used vector for vaccine and laboratory purposes. Despite significant advances in recombinant MVA technology, the isolation of recombinant viruses remains a tedious and difficult process. This chapter describes the use of an efficient and easy-to-use selection system adapted for MVA. The system is based on the requirement of the viral gene F13L for efficient virus spread in cell culture, which results in a severe block in virus transmission when F13L gene is deleted (Blasco R, Moss B. J Virol 65:5910–5920, 1991; Blasco R, Moss B. J Virol 66:4170–4179, 1992). The insertion of foreign genes in the MVA genome is accomplished by recombination of a transfected plasmid carrying the foreign genes and the F13L with the genome of an F13L knockout virus. Subsequently, selection of virus recombinants is carried out by serial passage and/or plaque purification of viruses that have recovered the F13L gene.

Key words: Virus recombinants, Vaccinia virus MVA, Genetic selection, F13L gene

1. Introduction

Modified Vaccinia Ankara (MVA) has proven to be a useful vaccine vector with excellent immunogenicity and safety characteristics. MVA can be handled under biosafety 1 conditions (instead of the biosafety 2 conditions required for most vaccinia virus strains) and therefore its use may be advantageous not only for vaccine use, but also in many other situations such as laboratory expression of proteins for functional studies.

A major factor in MVA biology is its restricted host range, a consequence of the long passage history of the virus in chick embryo fibroblasts (CEFs). In the laboratory, MVA replicates in some avian cells, such as primary CEFs, but also in some mammalian cell lines, like the baby hamster kidney cell line BHK-21 (1, 2).

Stuart N. Isaacs (ed.), *Vaccinia Virus and Poxvirology: Methods and Protocols*, Methods in Molecular Biology, vol. 890, DOI 10.1007/978-1-61779-876-4_5,

The limited host range and growth properties of MVA increase the safety of this vector but at the same time constitute a disadvantage for the isolation of recombinant viruses and, as a consequence, the isolation of MVA recombinants has remained relatively slow and inefficient. Major factors in the difficulties associated with manipulation of MVA are the limited number of permissive cell lines and the fact that virus plaques are slow to develop and difficult to recognize.

Because of the interest in generating MVA-based vaccines, during the last years, a number of selection procedures have been devised to facilitate MVA selection. First, methods for recombinant MVA isolation relied on selective staining of virus foci expressing β-galactosidase or β-glucuronidase (3–5). Alternative selection methods included insertion into the thymidine kinase or hemagglutinin (A56) genes (6, 7), or the insertion of selective markers such as *E. coli gpt* gene encoding xanthine-guanine-phosphoribosyltransferase, zeocin, or blasticidin resistance gene (4, 8–10). Among the genetic markers for MVA, host range selection has been used extensively and has been the subject of a previous chapter in this series (11). This method is based on the fact that insertion of the K1L gene, which is partially deleted in MVA, allows the virus to grow in the rabbit cell line RK13 (12–14). Therefore, after integration of a transfected plasmid into the MVA genome, the K1L gene can be used to select for virus recombinants on the basis of virus growth in RK13 cells (14). To preserve the K1L-deficient genotype in the final recombinant, direct repeats flanking K1L have been included to subsequently eliminate the gene (15, 16).

In addition or in conjunction with host range selection, further technical improvements have been described in recent years. For instance, bacterial artificial chromosomes have been used to generate recombinant MVAs (17, 18). Also, autofluorescent proteins have been used to facilitate the process by microscopy or cell sorting (9, 19, 20).

Despite the above technical improvements, the isolation of MVA recombinants remains a problematic, time-consuming, and inefficient process compared to other vaccinia strains. In this chapter, an alternative selective marker, the F13L gene, is used as a means to facilitate the isolation process of recombinant MVA.

2. Materials

1. Tissue culture plates, 6-, 24-, and 96-well tissue culture plates.
2. Tissue culture flasks, 25-, 75-, and 150-cm^2 tissue culture flasks.
3. BHK-21 cells (ATCC-CCL10).

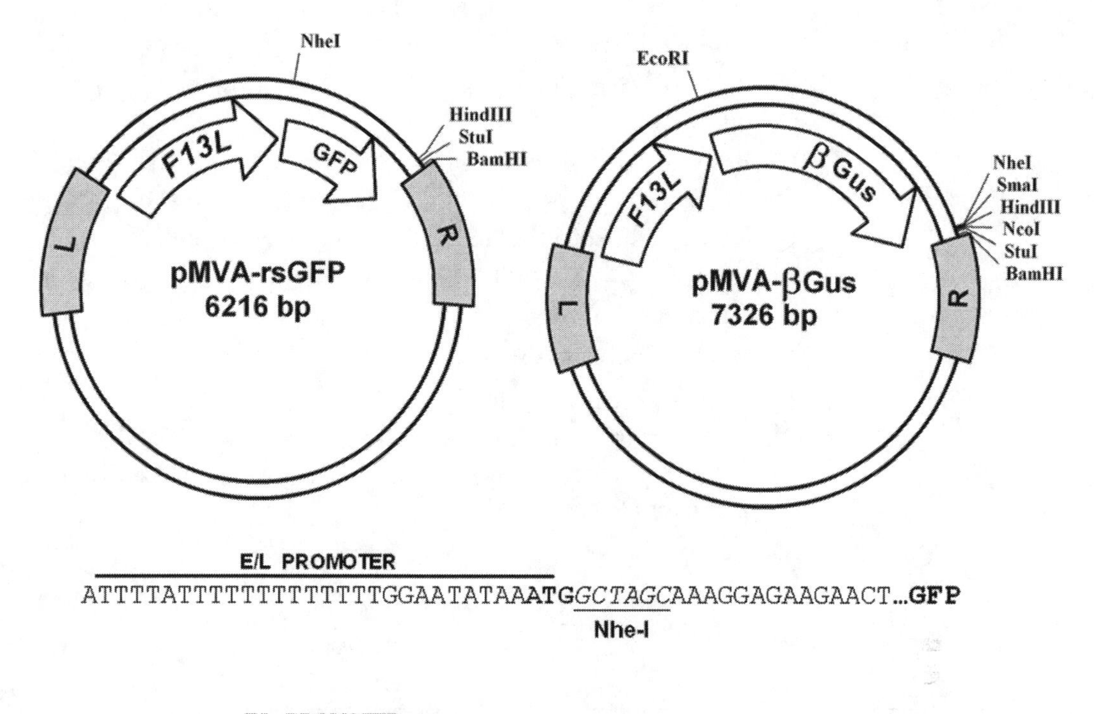

Fig. 1. Plasmids for F13L selection. Schematic diagram showing relevant features of plasmids pMVA-rsGFP and pMVA-βGus. Relative locations of F13L coding sequence and unique restriction sites are shown. *Left* (L) and *right* (R) MVA recombination flanks are shown as *gray boxes*. In the lower part, the sequences of promoters and the location of restriction sites useful for insertion at the 5′ end of foreign genes are shown. A foreign gene can be (1) subcloned without the ATG initiation codon into pMVA-rsGFP by replacing the GFP gene, (2) subcloned *in frame* in the NheI site of pMVA-rsGFP to fuse the foreign protein with GFP, or (3) subcloned with inclusion of a starting ATG codon into pMVA-βGus by replacing the βGus gene. Figure adapted from reference *25* © 2011 BioTechniques. Used with permission.

4. Complete BHK medium: BHK medium supplemented with 3 g/ml tryptose phosphate broth, 10 mM HEPES, 2 mM glutamine, 0.1 μg/ml penicillin, and 0.1 μg/ml streptomycin.
5. Fetal bovine serum (FBS).
6. BHK/5% FBS: Complete BHK medium containing 5% FBS.
7. BHK/2% FBS: Complete BHK medium containing 2% FBS.
8. CO_2 incubator: Humidified incubator at 37 °C and 5% CO_2.
9. MVA-ΔF13L deletion mutant.
10. Plasmid pMVA-rsGFP (see Note 1 and Fig. 1).
11. Plasmid pMVA-βGus (see Note 2 and Fig. 1).
12. TE buffer: 10 mM Tris–HCl, pH 8, 1 mM EDTA.
13. Fugene 6 transfection reagent (Roche).
14. Water bath sonicator.

15. Disposable sterile rubber scraper.
16. Poly propylene tubes: 1.5-, 5-, 15-, and 50-ml polypropylene tubes.
17. Inverted fluorescence microscope.
18. 2× EMEM: 2× EMEM supplemented with 4 mM glutamine, 0.2 μg/ml penicillin, 0.2 μg/ml streptomycin, and 4% FBS.
19. 2% LMP agarose: 2% (w/v) low-melting-point agarose in water, autoclaved.
20. Fluorescence microscope.
21. Pasteur pipettes, sterile.
22. Tetramethyl rhodamine isothiocyanate (TRITC) filter set (excitation 515–560 nm; dichroic mirror 575 nm; and emission over 590 nm).
23. Centrifuge.
24. Crystal violet solution: 0.5%(w/v) crystal violet, 20% ethanol in distilled water.

3. Methods

General procedures to work and modify vaccinia MVA have been previously detailed in a previous volume of this series (11). The system described here takes advantage of the requirement of the viral F13L gene for cell-to-cell transmission (21, 22), and has been adapted from a cloning system previously designed for vaccinia virus strain WR (23, 24). Selection of recombinants relies on the reintroduction of the F13L gene into an MVA-ΔF13L knockout mutant, which is severely impaired in plaque formation (25). As a consequence of the insertion, recovery of the normal plaque phenotype of the virus is accompanied by the insertion of the foreign gene downstream of the F13L gene.

The advantages of this system are that the whole process can be done in the permissive BHK-21 cell line and does not require the use of drugs or antibiotics. Also, since selection can be carried out by batch passaging of the population (25, 26), it can potentially be used for complex populations like gene libraries. In addition, since the genetic background of the final MVA recombinant is identical to the normal MVA, there is no need to eliminate a selection gene.

MVA-ΔF13L was obtained by replacing most of the F13L coding sequence by an expression cassette containing the *dsRed* gene, which greatly facilitated the isolation process (25). Because of *dsRed* expression, the tiny MVA-ΔF13L plaques can be easily recognized by their red fluorescence. In addition, the presence or absence of the *dsRed* gene can be taken as an indication of the genetic arrangement of the virus recombinant along the isolation process (see Fig. 2).

To allow for insertion of the foreign gene, we developed plasmids that contain, together with the MVA F13L gene, recombination flanks and an early/late synthetic promoter to drive the foreign gene protein expression (see Fig. 1 and Notes 1 and 2). These plasmids are similar to plasmid pRB21 and derivatives (see Note 3) and can be used to insert a foreign gene under the control of a vaccinia promoter. These plasmids have been designed for isolation of MVA recombinants after they are transfected into cells previously infected with the MVA-ΔF13L deletion mutant.

3.1. Growth of MVA-ΔF13L Virus Stocks

MVA-ΔF13L virus is severely restricted in the acquisition of the outer virus envelope and in virus egress (see Fig. 2). Since MVA-ΔF13L is not readily exported to the medium or transmitted between cells, efficient amplification of the virus requires mechanical release of the virus replicated within cells and consecutive rounds of amplification. Therefore, amplification should be carried out using sufficient amount of virus to infect most of the cells in the culture (see Note 4). The following procedure is designed to show the successive amplification of a small amount of virus by successive passage through increasing number of cells. If larger viral stocks are already available, go directly to step 6, 8, 10, or 12.

1. Seed BHK-21 cells in BHK/5% FBS medium in a 24-well tissue culture plate and incubate in a CO_2 incubator until the cells are about 50% confluent.
2. Thaw a small crude stock or a resuspended virus plaque of MVA-ΔF13L and sonicate in an ice/water bath sonicator for at least three cycles of 15 s or until the material in the suspension is dispersed (see Note 5).
3. Remove culture medium and inoculate the monolayer in a well of the 24-well plate with 200 μl of the crude stock. Place the plate in a CO_2 incubator for 2 h.
4. After the 2-h adsorption period, remove virus inoculum and add 0.5 ml of BHK/2% FBS medium. Place in a CO_2 incubator until cytopathic effect (CPE) is complete (usually 48–72 h).
5. Detach the cells from the plastic by pipetting repeatedly on top of the monolayer. Transfer the cell suspension to a 1.5-ml polypropylene tube. Freeze–thaw three times and sonicate to make a homogeneous cell lysate.
6. Dilute the cell lysate to 1 ml with BHK/2% FBS medium to infect BHK-21 cells in a well of a 6-well plate. After 2 h of adsorption in a CO_2 incubator, remove virus inoculum, add 2 ml BHK/2% FBS medium, and leave in the incubator until CPE is complete (typically 48–72 h).

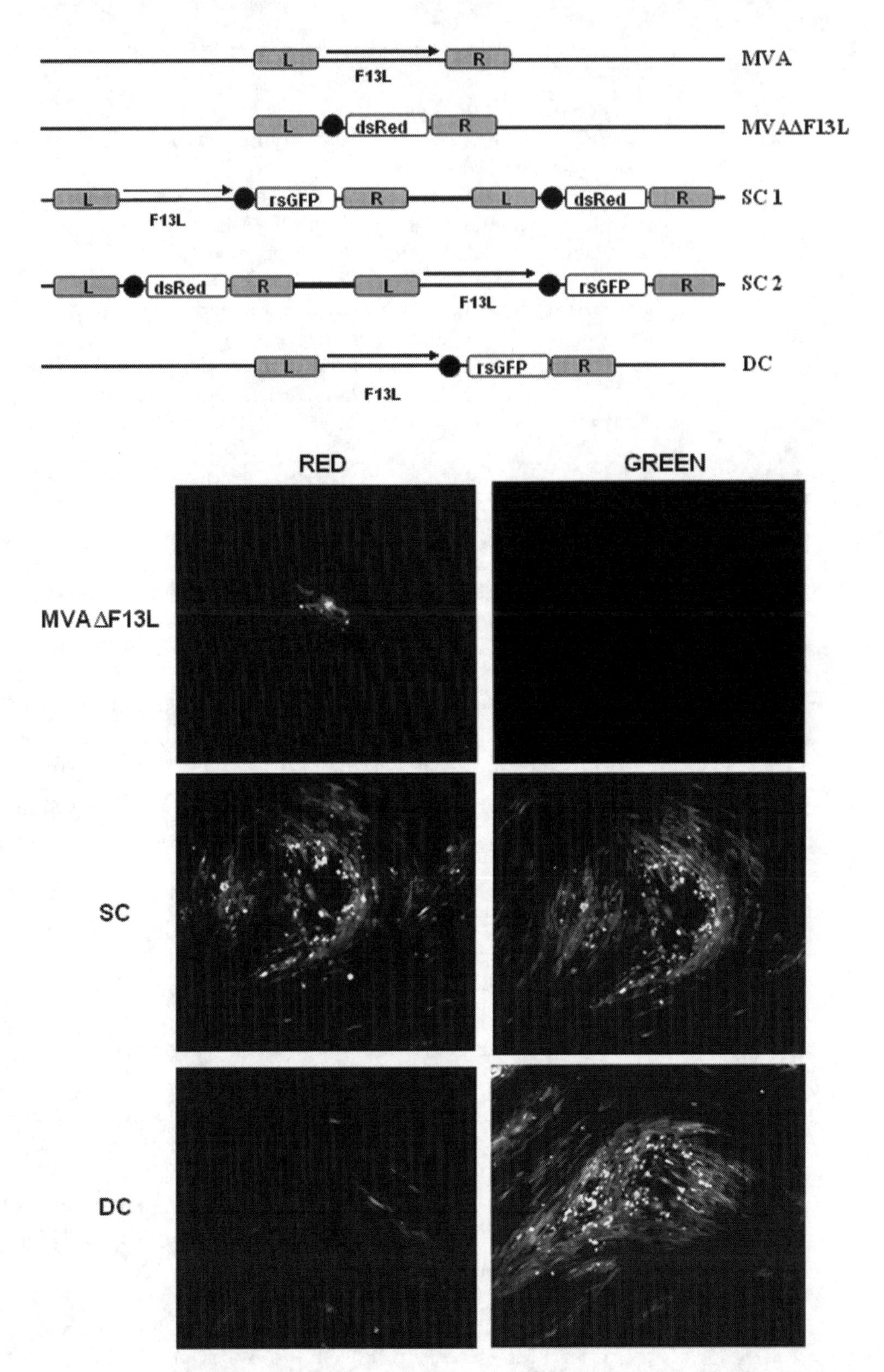

Fig. 2. Genotypes and phenotypes of different MVA viruses. *Upper panel*: Representation of different recombination products. The relative position of different elements in the MVA virus, the F13L MVA deletion mutant (MVAΔF13L), and the different products obtained by recombination between MVAΔF13L and plasmid pMVA-rsGFP are shown. SC1 and SC2 are

7. Detach the cells from the plastic by pipetting repeatedly on top of the monolayer. Transfer the cell suspension to a 15-ml polypropylene tube. Centrifuge for 5 min at 1800 x g, discard supernatant, and resuspend the cells in 1 ml BHK/2% FBS medium. Freeze–thaw three times and sonicate.
8. Use half of the cell lysate, sonicated and diluted to 2 ml in BHK/2% FBS medium to infect confluent BHK-21 monolayer culture in a 25-cm^2 flask. After 2 h of adsorption in a CO_2 incubator, remove virus inoculum, add 5 ml BHK/2% FBS medium, and place in the CO_2 incubator until CPE is complete (typically 48–72 h).
9. Detach infected cells from the flask with a disposable scraper and transfer to a 15-ml centrifuge tube. Centrifuge for 5 min at 1,800 × *g*, discard supernatant, and resuspend the cells in 2 ml of BHK/2% FBS medium. Freeze–thaw three times and sonicate.
10. Use half of the cell lysate, sonicated and diluted to 4 ml in BHK/2% FBS medium to infect confluent BHK-21 monolayer culture in a 75-cm^2 flask. After 2 h of adsorption in a CO_2 incubator, remove virus inoculum, add 12 ml BHK/2% FBS medium, and place in the CO_2 incubator until CPE is complete (typically 48–72 h).
11. Detach infected cells from the flask with a disposable scraper and transfer to a 15-ml centrifuge tube. Centrifuge for 5 min at 1,800 × *g*, discard supernatant, and resuspend the cells in 4 ml of BHK/2% FBS medium. Freeze–thaw three times and sonicate.
12. Use half of the cell lysate, sonicated and diluted to 9 ml in BHK/2% FBS medium to infect confluent BHK-21 monolayer culture in a 150-cm^2 flask. After 2 h of adsorption in a CO_2 incubator, remove virus inoculum, add 20 ml BHK/2% FBS medium, and place in the CO_2 incubator until CPE is complete (typically 48–72 h).
13. Detach infected cells from the flask with a disposable scraper and transfer to a 50-ml centrifuge tube. Centrifuge for 5 min at 1,800 × *g*, discard supernatant, and resuspend the cells in 4 ml of BHK/2% FBS medium. Freeze–thaw three times and

Fig. 2. (continued) single-crossover (SC) products, arising from recombination in one of the two F13L flanking sequences. Double crossover (DC) is a stable double crossover, arising from recombination in the two F13L flanking sequences. L and R are the MVA flanks for the F13L gene used for homologous recombination. *Lower panel*: Identification of recombinant viruses by plaque phenotype and fluorescence. Representative plaques formed by parental virus MVAΔF13L, SC recombinants and DC recombinants, are shown. If the figure was in color, plaques would appear with red or green fluorescence. Figure adapted from reference *25* © 2011 BioTechniques. Used with permission.

sonicate. This virus stock can be aliquoted and stored for long-term use in a –80°C freezer.

14. Determine the titer of the virus stock (see Note 4).

3.2. Generation of MVA Recombinants by Infection/Transfection

Gene insertion into the MVA genome is carried out by recombination between the plasmid and the replicating MVA-ΔF13L virus DNA. For this, cells infected with virus MVA-ΔF13L are transfected with a plasmid carrying the F13L gene and the foreign gene placed downstream of a poxvirus promoter. Suitable plasmids for constitutive expression of a foreign gene throughout the replication cycle can be obtained by replacing the GFP or β-Glucuronidase gene in the plasmids depicted in Fig. 1.

1. About 24 h before infection, split an overly confluent flask of BHK-21 cells 1:5 and seed each well of a 6-well tissue culture plate with cells suspended in 2 ml BHK/5% FBS medium per well. Incubate the cells in a CO_2 incubator for 20–24 h, when the cells should be 60–80% confluent (see Note 6).
2. Prepare virus inoculum by diluting an MVA-ΔF13L crude virus stock in BHK/2% FBS medium. The amount of virus should be adjusted to give a multiplicity of infection of 0.05 PFU/cell.
3. Remove the BHK/5% FBS medium from the cultures and immediately add virus inoculum (1 ml per well). Place in a CO_2 incubator for 1–2 h.
4. About 45 min before the end of the viral adsorption period, prepare transfection mix in 1.5-ml polypropylene tubes. For each transfection, dilute 10 μl of FUGENE 6 in 100 μl serum-free medium and incubate at room temperature for 5 min. Transfer the medium–FUGENE mix to a tube containing 2 μg of plasmid DNA in TE buffer, and mix gently by pipetting up and down 2–3 times. Leave at room temperature for 30 min.
5. Aspirate virus inoculum from the BHK-21 cell cultures and wash the cells once with 1 ml of serum-free complete BHK medium. Finally, add 2 ml of BHK/2% FBS medium.
6. Immediately add the DNA–FUGENE solution drop by drop on top of the culture medium. Mix by gently moving the plate on a flat surface. Incubate for 72 h in a CO_2 incubator.
7. Detach the cells from the plastic with a disposable rubber scraper or by pipetting repeatedly on top of the monolayer. Collect the cell suspension in a sterile 15-ml polypropylene tube and pellet cells by centrifugation.
8. Resuspend cells in 1 ml BHK/2% FBS medium, lyse the cells by freezing in a dry ice/ethanol bath, thaw in a 37°C water

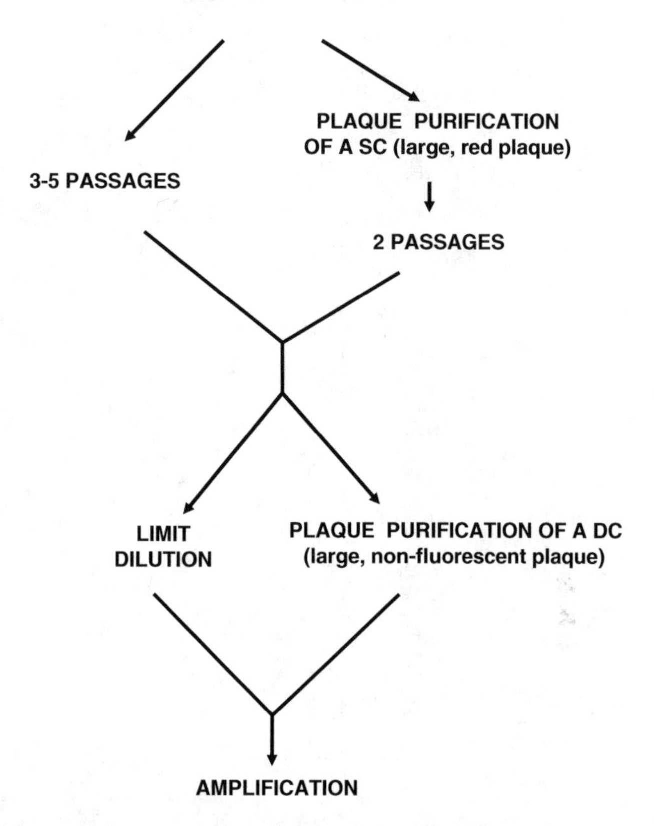

Fig. 3. Strategies for the isolation of MVA recombinants. MVA recombinants are generated by recombination between transfected plasmids and MVAΔF13L. The isolation of F13L$^+$ viruses can be done by enrichment and/or plaque isolation. Two strategies leading to the successful isolation of recombinant MVA are shown.

bath, and vortex. Repeat the freeze–thaw cycling three times. The cell lysate can be stored at –80°C until next use.

3.3. Isolation of MVA Recombinants

After infection/transfection, the fraction of recombinant virus is expected to be about 10^{-4} to 10^{-3} of the total virus population. The crude cell lysate prepared from the culture contains different virus genotypes, including an overwhelming majority (>99%) of the parental MVA-ΔF13L virus along with the relatively unstable single recombination events (F13L$^+$, dsRed$^+$) and the stable double recombinants (F13L$^+$, dsRed$^-$). Figure 2 depicts the genotypes and plaque phenotypes of the different viruses in the mixture. Different procedures can be applied to this mixture to enrich and/or isolate virus recombinants. The F13L$^+$ virus recombinants form large virus plaques in susceptible cells, and also have a competitive advantage with respect to MVA-ΔF13L in low m.o.i. infections. Of the possible isolation procedures that take advantage on the better transmissibility of the F13L$^+$ virus, we have successfully used the strategies depicted in Fig. 3.

Passage	Parental	SC	DC	% DC
0	$1.0\ 10^6$	$2.0\ 10^2$	$1.0\ 10^3$	0,1
1	$4.2\ 10^5$	$1.1\ 10^3$	$6.0\ 10^4$	12,5
2	$4.0\ 10^4$	$1.1\ 10^4$	$6.0\ 10^6$	99,2
3	$1,8\ 10^4$	$1.0\ 10^4$	$3.6\ 10^7$	99,9
4	$< 10^3$	$1.0\ 10^4$	$9.0\ 10^6$	99,9
5	$< 10^3$	$5.0\ 10^3$	$1.1\ 10^7$	100,0

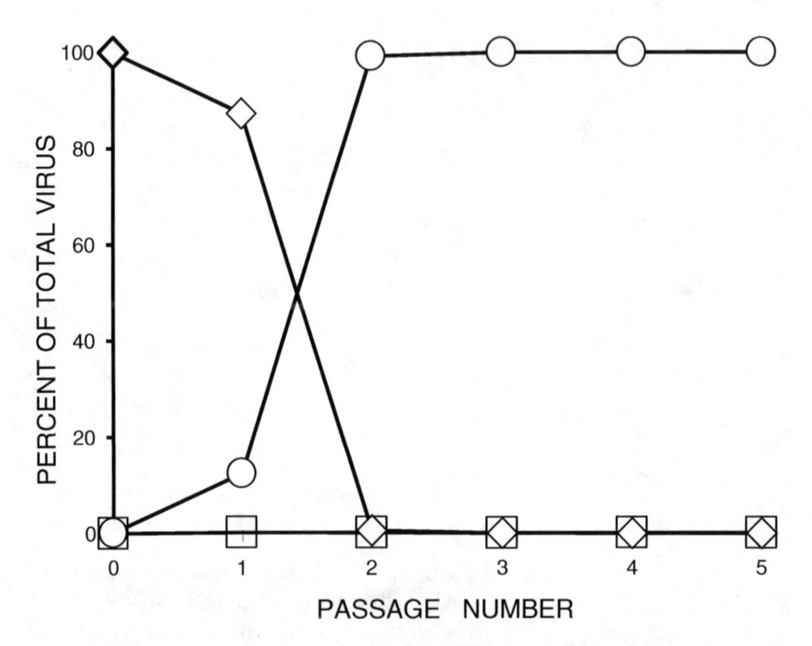

Fig. 4. Enrichment of virus recombinants by serial passages. Virus obtained from the infection/transfection procedure using plasmid pMVA-βGus was passaged five times as specified in Subheading 3.3.1. Progeny virus from each passage was analyzed by plaquing and βGus staining. Passage "0" indicates the progeny virus from the infection/transfection. Different viruses were identified by their phenotypes: parental virus produces small infection foci showing red fluorescence, single-crossover viruses (SC) form large plaques showing red fluorescence and βGus-positive staining, and double-crossover virus (DC) form large plaques showing βGus-positive staining, but not red fluorescence. *Lower panel*: The fraction of parental MVAΔF13L virus (*diamonds*), SC (*squares*), and DC (*circles*) virus after each passage is represented as the percentage of the total virus population.

F13L⁺ viruses can be enriched from the infection/transfection progeny by simply passing repeatedly the virus mixture. This procedure (Subheading 3.3.1) results, under normal conditions, in virus stocks in which most of the virus should be recombinants after 3–5 consecutive passages (see example in Fig. 4). After enrichment, the virus stocks can be used directly when only crude expression of the foreign protein is required. However, in most instances, clonal purification is carried out from this enriched population to ensure homogeneity of the virus population.

As an alternative to enrichment by passaging, plaque purification can be done directly from the infection/transfection mixture (Subheading 3.3.3). In this case, one initially isolates a single crossover product (identified as a large, red fluorescent plaque (see Fig. 2)), followed by passaging and virus cloning by plaque isolation or limit dilution to obtain a pure growth of a stable double-crossover recombinant virus.

3.3.1. Enrichment by Serial Passage in BHK-21 Cells

1. Seed BHK cells in 6-well tissue culture plates and incubate until the cells are 80% confluent.
2. Prepare virus inoculum by diluting 100 μl of the virus stock from Subheading 3.2, step 8, to a final volume of 1 ml with BHK/2% FBS.
3. Remove medium from one well and immediately add virus inoculum. Place in a CO_2 incubator to allow virus adsorption.
4. After a 1-h adsorption period, remove virus inoculum and add 2 ml of fresh BHK/2% FBS medium. Incubate in CO_2 incubator for 48 or 72 h, when CPE should be evident.
5. Detach the cells from the plastic by repeatedly pipetting the medium on the monolayer. Transfer the cells to a 15-ml polypropylene tube. Make a crude virus stock by lysing the cells by three cycles of freeze–thawing.
6. Repeat steps 1–4 using 20 μl of the passaged virus to prepare the inoculum of next passage.

3.3.2. Isolation of MVA Recombinants by Plaque Isolation

The different genetic arrangements and phenotypes of the viruses generated in the system are shown in Fig. 2. Recombination of the plasmid with the vaccinia MVA-ΔF13L genome renders virus recombinants that produce large plaques displaying red fluorescence ($F13L^+$, $dsRed^+$) or not ($F13L^+$, $dsRed^-$). One of these viruses, presumed to be the result of single recombination events ($F13L^+$, $dsRed^+$), can be easily identified using an inverted fluorescence microscope. Direct inspection of the plaques allows marking the position of fluorescence-positive plaques that can be subsequently isolated. For dsRed visualization, we have routinely used a standard Rhodamine filter set (Excitation: 515–560 nm; Dichroic mirror: 575 nm Emission: BA590).

3.3.3. Isolation of Single Crossover by Screening of Plaques with Fluorescence Microscopy

1. Seed BHK cells in 6-well tissue culture plates and incubate until the cells are 80% confluent.
2. Sonicate the thawed cell lysate from the infection/transfection (step 8 in Subheading 3.2) in an ice/water bath sonicator for at least three 15-s cycles or until the material in the suspension is dispersed (see Note 5).
3. Make tenfold serial dilutions of the infection/transfection cell lysate in BHK/2% FBS medium (see Note 7). Aspirate medium

from the BHK cell monolayers and add 1 ml of the corresponding dilution per well (use at least dilutions 10^{-2}, 10^{-3}, and 10^{-4}). Incubate for 1–2 h in the CO_2 incubator.

4. Melt the 2% LMP agarose by heating in a microwave oven and cool in a 37°C water bath. Warm complete 2× EMEM medium at 37°C.
5. Immediately before use, prepare EMEM-agarose overlay by mixing equal volumes of complete 2× EMEM and 2% LMP agarose solution.
6. Remove virus inoculum from each well, add 3 ml of the warm EMEM-agarose overlay, and allow to solidify at room temperature (see Note 8).
7. After the overlay solidifies, place the plates in the CO_2 incubator for 72 h to allow for vaccinia MVA plaques to develop.
8. Identify large dsRed-positive virus plaques by inspection in an inverted fluorescence microscope and mark their position with a permanent marker.
9. Pick 4–6 well-separated plaques by plunging a sterile Pasteur pipette through the agarose/medium all the way to the plastic, rock the pipette tip slightly to break the agarose, and detach infected cells from the plastic. Aspirate the small agarose piece in the tip of the pipette and transfer to a microcentrifuge tube containing 0.5 ml of BHK/5% FBS medium. Freeze–thaw the tube three times and sonicate.
10. Carry out a second plaque purification using the material from two to three plaques by repeating steps 1–9.

3.3.4. Amplification of the Single Crossover Population

To allow for intramolecular recombination to occur and generate double-crossover recombinant viruses, the unstable single recombinant virus (SC1 or SC2 in Fig. 2) is passaged on BHK-21 cells. During passage, this virus can produce, by intramolecular recombination, the parental MVA-ΔF13L virus and the stable double recombinant (DC in Fig. 2). Since MVA-ΔF13L is severely restricted in cell-to-cell transmission, successive passage leads to the accumulation of the double recombinant.

1. Seed BHK-21 cells in 6-well tissue culture plates and incubate until the cells are about 80% confluent.
2. Thaw and sonicate the cell lysate from Subheading 3.3.3, step 9, in an ice/water bath sonicator for at least three 15-s cycles or until the material in the suspension is dispersed.
3. Prepare virus inoculum by diluting 200 μl of the cell lysate in a final volume of 1 ml of BHK/2% FBS medium.
4. Aspirate medium from the BHK-21 cell monolayers, add the inoculum, and incubate for 1–2 h in CO_2 incubator.

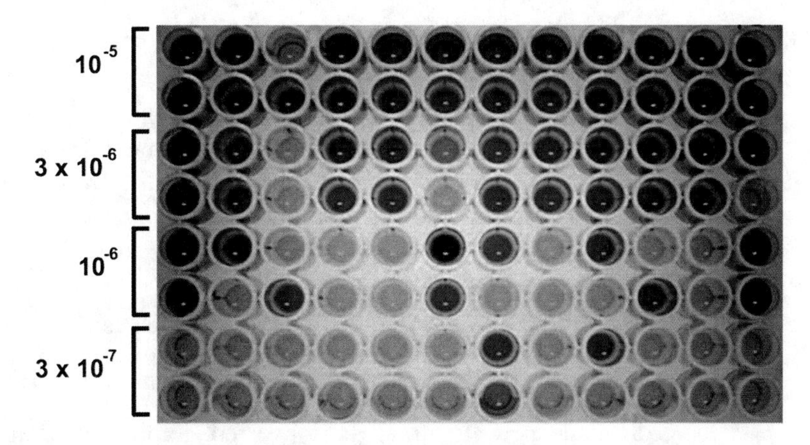

Fig. 5. Example of a limiting dilution cloning. A virus stock derived from the infection/transfection procedure using plasmid pMVA-βGus was amplified by repeated passaging and then subjected to cloning by limiting dilution (Subheading 3.3.5). For this, BHK-21 cells in 96-well plates were infected with 100 μl per well of different dilutions of the virus sock. The dilutions used here are shown on the left side, and 24 wells (two rows of wells) were infected per dilution. After 72 h of infection, βGus-positive wells were visualized after adding X-Gluc, a chromogenic βGus substrate. In this example, after βGus staining, the three positive wells in the bottom two rows (the samples that are most diluted) would be selected for amplification. In the absence of specific staining, i.e., when other genes are inserted, wells that display cytopathic effect and lack any red fluorescent cells would be selected.

5. Remove virus inoculum from each well and add 2 ml of BHK/2% FBS medium. Place the plates in the CO_2 incubator for 48–72 h until complete CPE is evident.
6. Detach the cells from the plastic by repeated pipetting. Transfer to a 15-ml polypropylene tube.
7. Lyse the cells by three cycles of freeze–thawing.
8. Repeat steps 1–7 using 200 μl of the cell lysate for the next round of infection. After two or three consecutive rounds of amplification, double-crossover virus recombinants should constitute the majority of the virus in the population.

3.3.5. Limiting Dilution of Amplified Virus Stock

MVA plaques are often difficult to recognize. As an alternative to plaque isolation, we have also used limiting dilution method. This procedure allows the clonal isolation of virus recombinants after enrichment of the virus population for $F13L^+$ viruses with or without previous isolation of single-crossover plaques (see Note 9). In this procedure, dilutions of the virus stock are used to infect a number of wells per dilution (see example in Fig. 5). To identify suitable recombinant clones ($F13L^+$, $dsRed^-$), wells that show CPE and are devoid of any red fluorescent cells are selected. Note that the presence of red fluorescent cells would be indicative of the presence of parental or single-crossover virus, and therefore that well should be discarded.

1. Seed BHK-21 cells in 96-well tissue culture plates and incubate until the cells are about 50% confluent.
2. Thaw and sonicate the cell lysate from the final step in Subheadings 3.3.1 or 3.3.4 in an ice/water bath sonicator for

at least three cycles of 15 s until the material in the suspension is dispersed.

3. Make two- or threefold serial dilutions of the virus stock, starting at a 10^{-3} dilution in BHK/2% FBS medium (see Note 10). Aspirate medium from the BHK-21 cell monolayers and add 100 μl of the corresponding dilution per well. Infect 12–24 wells per dilution. Incubate for 1–2 h in CO_2 incubator to allow virus adsorption.
4. Remove virus inoculum from each well and add 0.2 ml of BHK/2% FBS medium. Place the plates in the CO_2 incubator for 72 h.
5. Inspect wells under an inverted fluorescence microscope. Identify wells that display extensive CPE, indicative of the presence of an $F13L^+$ virus. Inspect wells for fluorescence and discard wells that contain any red fluorescent cells. Select virus clones from the last dilution that contains infected cells and no red fluorescent cells.
6. Recover infected cells from selected wells and transfer them to a 1.5-ml polypropylene tube(s). Prepare a small crude stock by three cycles of freeze–thawing.

3.3.6. Isolation of a Double-Crossover Virus by Plaque Purification

1. Seed BHK cells in 6-well tissue culture plates and incubate until the cells are 80% confluent.
2. Thaw the cell lysate obtained from enrichment (Subheading 3.3.1) or from an amplified single recombinant plaque (Subheading 3.3.4) and sonicate for at least three 15-s cycles or until the material in the suspension is dispersed.
3. Make tenfold serial dilutions of the lysate in BHK/2% FBS medium. Aspirate medium from the BHK cell monolayers and add 1 ml of the corresponding dilution per well (use at least dilutions 10^{-4} to 10^{-7}). Incubate for 1–2 h in a CO_2 incubator.
4. Melt the 2% LMP agarose by heating in a microwave oven and cool in a 37°C water bath. Warm complete 2× EMEM medium at 37°C.
5. Immediately before use, prepare EMEM-agarose overlay by mixing equal volumes of complete 2× EMEM and 2% LMP agarose solution.
6. Remove virus inoculum from each well and immediately add 3 ml of warm EMEM-agarose overlay medium. Allow to solidify at room temperature (see Note 8). Place the plates in the CO_2 incubator for 72 h to allow for vaccinia MVA plaques to develop.
7. Identify large virus plaques by inspection in an inverted fluorescence microscope by using phase contrast and mark

their position with a permanent marker (and to check that there is no contaminating parental virus by fluorescence) (see Note 11).

8. Pick well-separated plaques by plunging a sterile Pasteur pipette through the agarose medium all the way to the plastic, rock the pipette tip slightly to break the agarose, and detach infected cells from the plastic. Aspirate the small agarose piece in the tip of the pipette and transfer to a 1.5-ml microcentrifuge tube containing 0.5 ml of BHK/5% FBS medium. Freeze–thaw three times and sonicate.
9. Carry out a second plaque purification using the material from four to six plaques by repeating steps 1–8.

3.4. Virus Amplification

1. Seed BHK-21 cells in a 6-well tissue culture plate and incubate in a CO_2 incubator until the cells are about 50% confluent.
2. Thaw the cell lysate from the final plaque purification (final step, Subheading 3.3.6) or final step of limiting dilution (Subheading 3.3.5) and sonicate for three 15-s pulses or until the material in the suspension is dispersed (see Note 5).
3. Prepare virus inoculum by diluting 100 μl from the limiting dilution protocol or 250 μl of a resuspended plaque to a final volume of 1 ml with BHK/2% FBS medium.
4. Remove the medium from the 6-well dish and add the virus inoculum. Place the plates in a CO_2 incubator for 2 h to allow the virus to adsorb to cells.
5. After the 2-h adsorption period, remove virus inoculum and add 2 ml of BHK/2% FBS medium. Place 48–72 h in a CO_2 incubator until CPE is evident.
6. Detach the cells from the plastic by pipetting repeatedly on top of the monolayer. Transfer the cell suspension to a 1.5-ml polypropylene tube. Freeze–thaw three times and sonicate.
7. Use half of the cell lysate, sonicated and diluted to 2 ml in BHK medium 2% FBS to infect confluent BHK-21 monolayer culture in a 25-cm^2 flask. After 2 h of adsorption in a CO_2 incubator, remove virus inoculum, add 5 ml BHK/2% FBS medium, and place in the CO_2 incubator until CPE is complete (typically 48–72 h).
8. Detach infected cells from the flask with a disposable scraper and transfer to a 15-ml centrifuge tube. Centrifuge for 5 min at $1{,}800 \times g$, discard supernatant, and resuspend the cells in 2 ml of BHK/2% FBS medium. Freeze–thaw three times and sonicate.
9. Use half of the cell lysate, sonicated and diluted to 9 ml in BHK medium 2% FBS to infect a confluent BHK-21 monolayer culture in a 150-cm^2 flask. After 2 h of adsorption in a

CO_2 incubator, remove virus inoculum, add 20 ml BHK/2% FBS medium, and place in the CO_2 incubator until CPE is complete (typically 48–72 h).

10. Detach infected cells from the flask with a disposable scraper and transfer to a 50-ml centrifuge tube. Centrifuge for 5 min at 1,800×*g*, discard supernatant, and resuspend the cells in 4 ml of BHK/2% FBS medium. Freeze–thaw three times and sonicate. Titer the virus stock (11) (see the next subheading and Note 4).

3.5. Estimation of the Virus "Titer" by Staining with Crystal Violet (See Note 12)

In many instances, MVA plaques are difficult to be recognized by simple inspection of the cultures. Thus, accurate virus titer should be determined by immunostaining (11). However, we have found that it is possible to estimate the titer/infectivity of a stock by infection of monolayers of BHK-21 followed by staining with crystal violet (see Note 12).

1. About 24 h before infection, split a culture of BHK-21 cells and seed 6-well tissue culture plates with cells suspended in 2 ml BHK/5% FBS medium per well. Incubate the cells in a CO_2 incubator until the cells are 60–80% confluent.
2. Thaw the virus stock and sonicate in a water-bath sonicator until dispersion of the material is complete (see Note 5).
3. Prepare virus inoculum by making tenfold dilutions of the virus stock in BHK/2% FBS medium (see Note 10).
4. Remove the BHK/5% FBS medium from the wells and immediately add consecutive dilutions of virus inoculum (1 ml per well), starting from the most diluted inoculum. Place in a CO_2 incubator for 1–2 h.
5. Aspirate virus inoculum from the wells and add 2 ml of BHK/2% FBS medium. Incubate for 72 h in a CO_2 incubator.
6. Stain the cells by adding 0.5 ml per well of crystal violet staining solution. Incubate at room temperature for 5–10 min.
7. Aspirate medium from the wells and let dry before counting the plaques.

4. Notes

1. pMVA-rsGFP was designed to mediate the insertion of F13L gene and GFP (rsGFP; Quantum Biotechnologies) cassette into the F13L locus of vaccinia MVA. The GFP gene, which is placed downstream of the vaccinia promoter, can be replaced by using restriction sites NheI (placed immediately after the

initiator ATG codon), and HindIII or StuI (see Fig. 1). Alternatively, in-frame cloning into the NheI site results in the fusion of GFP to the C-terminus of the protein.

2. pMVA-βGus was designed to mediate the insertion of F13L gene and the β-Glucuronidase gene. In this plasmid, the β-Glucuronidase gene, which is placed downstream of the vaccinia promoter, can be replaced by a foreign gene (remembering to include the ATG of the gene) using the EcoRI and restriction sites at the 3′ end.
3. The plasmids described here are derived entirely from MVA sequences, i.e., the recombination flanks and F13L gene were amplified from the MVA genome. We have tested similar plasmids used with vaccinia virus strain WR, like plasmid pRB21 and derivatives (23, 24, 27), and found that they can also be used successfully to generate recombinants in the MVA background. However, if WR-based plasmids are used, the MVA recombinants will contain the WR version of the F13L and of possibly part of the recombination flanks.
4. Despite the small size of the virus plaques, the titer of MVA-ΔF13L can be easily determined by taking advantage of the dsRed expression by the virus. To titrate, dilutions of the virus stock are made and used to infect BHK-21 monolayers. After 2 days, red fluorescent cells or small groups of cells are counted under the fluorescence microscope, and the virus titer calculated.
5. We have found that sonication is best carried out in a small volume in rigid-wall tubes, for example 0.5 ml in a 5-ml polypropylene tube.
6. An overly confluent flask is one where we allow the cells to grow for 1–2 days after they become confluent. The cells become even more packed and may have twice the number of cells as a confluent culture. So splitting an overly confluent flask 1:5 will provide the proper amount of cells to seed wells in a plate.
7. When isolating single recombinants directly from the infection/transfection, the first plaquing step is a crucial step. Usually, large plaques are clearly visible in the monolayers, but one should bear in mind that many more cells, infected by the parental virus, are present and replicate in the cells. If the amount of parental virus is high, a diffuse CPE appears, which impacts the development of the wild type-sized plaques by the recombinant virus. Because of that, making two- or threefold dilutions is highly recommended for the first plaquing cycle.
8. One can also put the plate at 4°C for 10–15 min to ensure proper gelification of the agarose.

9. Successful cloning by limiting dilution critically depends on the percentage of the recombinant virus in the starting virus population. Therefore, limiting dilution should be carried out after enrichment of F13L+ virus or in conditions where the virus recombinant constitutes a significant fraction of the virus population.
10. The titer of starting virus is crucial when deciding what dilutions should be used for the infection. Expect ~10^2 to 10^4 infectious units from a plaque isolation, and 10^5 to 10^7 infectious units from amplified cultures. It is advisable to make dilutions in the range of 10^{-3} to 10^{-5} when working with a stock from picked plaques and dilutions in the range of 10^{-6} to 10^{-8} for virus from amplified stocks.
11. Depending on the condition of the cell monolayers, MVA plaques are sometimes difficult to discern. However, in most instances, the expert eye is able to locate virus plaques using only CPE as seen in an inverted microscope equipped with phase contrast. Alternatively, immune staining of virus plaques can be used prior to plaque isolation (11).
12. While the standard in the field for obtaining a titer of MVA is immunostaining in CEF cells, we use BHK-21 cells to get a rough titer/infectivity estimate. In BHK-21 cells, MVA makes long comets, which also make accurate plaque counting difficult.

Acknowledgments

This work was supported by grant BIO2008-03713 from Plan Nacional de Investigación Científica y Técnica, Spain, grant S2009/TIC-1476 from Comunidad de Madrid, and contract CT2006-037536 from the European Commission. We thank Stuart Isaacs for critical reading of the manuscript and Maite Ostalé for excellent technical support.

References

1. Carroll MW, Moss B (1997) Host range and cytopathogenicity of the highly attenuated MVA strain of vaccinia virus: propagation and generation of recombinant viruses in a nonhuman mammalian cell line. Virology 238: 198–211
2. Drexler I, Heller K, Wahren B, Erfle V, Sutter G (1998) Highly attenuated modified vaccinia virus Ankara replicates in baby hamster kidney cells, a potential host for virus propagation, but not in various human transformed and primary cells. J Gen Virol 79:347–352
3. Carroll MW, Moss B (1995) E. coli beta-glucuronidase (GUS) as a marker for recombinant vaccinia viruses. Biotechniques 19 (352–354):356
4. Sutter G, Moss B (1992) Nonreplicating vaccinia vector efficiently expresses recombinant

genes. Proc Natl Acad Sci USA 89: 10847–10851

5. Drexler I, Antunes E, Schmitz M, Wolfel T, Huber C, Erfle V, Rieber P, Theobald M, Sutter G (1999) Modified vaccinia virus Ankara for delivery of human tyrosinase as melanoma-associated antigen: induction of tyrosinase- and melanoma-specific human leukocyte antigen A*0201-restricted cytotoxic T cells in vitro and in vivo. Cancer Res 59:4955–4963
6. Scheiflinger F, Falkner FG, Dorner F (1996) Evaluation of the thymidine kinase (tk) locus as an insertion site in the highly attenuated vaccinia MVA strain. Arch Virol 141:663–669
7. Antoine G, Scheiflinger F, Holzer G, Langmann T, Falkner FG, Dorner F (1996) Characterization of the vaccinia MVA hemagglutinin gene locus and its evaluation as an insertion site for foreign genes. Gene 177: 43–46
8. Scheiflinger F, Dorner F, Falkner FG (1998) Transient marker stabilisation: a general procedure to construct marker-free recombinant vaccinia virus. Arch Virol 143:467–474
9. Garber DA, O'Mara LA, Zhao J, Gangadhara S, An I, Feinberg MB (2009) Expanding the repertoire of Modified Vaccinia Ankara-based vaccine vectors via genetic complementation strategies. PLoS One 4:e5445
10. Wong YC, Lin LC, Melo-Silva CR, Smith SA, Tscharke DC (2010) Engineering recombinant poxviruses using a compact GFP-blasticidin resistance fusion gene for selection. J Virol Methods 171:295–298
11. Staib C, Drexler I, Sutter G (2004) Construction and isolation of recombinant MVA. Methods Mol Biol 269:77–100
12. Perkus ME, Limbach K, Paoletti E (1989) Cloning and expression of foreign genes in vaccinia virus, using a host range selection system. J Virol 63:3829–3836
13. Sutter G, Ramsey Ewing A, Rosales R, Moss B (1994) Stable expression of the vaccinia virus K1L gene in rabbit cells complements the host range defect of a vaccinia virus mutant. J Virol 68:4109–4116
14. Smith KA, Stallard V, Roos JM, Hart C, Cormier N, Cohen LK, Roberts BE, Payne LG (1993) Host range selection of vaccinia recombinants containing insertions of foreign genes into non-coding sequences. Vaccine 11:43–53
15. Staib C, Drexler I, Ohlmann M, Wintersperger S, Erfle V, Sutter G (2000) Transient host range selection for genetic engineering of modified vaccinia virus Ankara. Biotechniques 28:1137–1142, 1144–1146, 1148
16. Zhu LX, Xie YH, Li GD, Wang Y (2001) High frequency of homologous recombination in the genome of modified vaccinia virus ankara strain (MVA). Sheng Wu Hua Xue Yu Sheng Wu Wu Li Xue Bao (Shanghai) 33:497–503
17. Cottingham MG, Andersen RF, Spencer AJ, Saurya S, Furze J, Hill AV, Gilbert SC (2008) Recombination-mediated genetic engineering of a bacterial artificial chromosome clone of modified vaccinia virus Ankara (MVA). PLoS One 3:e1638
18. Cottingham MG, Gilbert SC (2010) Rapid generation of markerless recombinant MVA vaccines by en passant recombineering of a self-excising bacterial artificial chromosome. J Virol Methods 168:233–236
19. Di Lullo G, Soprana E, Panigada M, Palini A, Agresti A, Comunian C, Milani A, Capua I, Erfle V, Siccardi AG (2010) The combination of marker gene swapping and fluorescence-activated cell sorting improves the efficiency of recombinant modified vaccinia virus Ankara vaccine production for human use. J Virol Methods 163:195–204
20. Di Lullo G, Soprana E, Panigada M, Palini A, Erfle V, Staib C, Sutter G, Siccardi AG (2009) Marker gene swapping facilitates recombinant Modified Vaccinia Virus Ankara production by host-range selection. J Virol Methods 156: 37–43
21. Blasco R, Moss B (1991) Extracellular vaccinia virus formation and cell-to-cell virus transmission are prevented by deletion of the gene encoding the 37,000-Dalton outer envelope protein. J Virol 65:5910–5920
22. Blasco R, Moss B (1992) Role of cell-associated enveloped vaccinia virus in cell-to-cell virus spread. J Virol 66:4170–4179
23. Blasco R, Moss B (1995) Selection of recombinant vaccinia viruses on the basis of plaque formation. Gene 158:157–162
24. Lorenzo MM, Galindo I, Blasco R (2004) Construction and isolation of recombinant vaccinia virus using genetic markers. Methods Mol Biol 269:15–30
25. Sanchez-Puig JM, Blasco R (2005) Isolation of vaccinia MVA recombinants using the viral F13L gene as the selective marker. Biotechniques 39:665–666, 668, 670 passim
26. Sanchez-Puig JM, Blasco R (2000) Puromycin resistance (pac) gene as a selectable marker in vaccinia virus. Gene 257:57–65
27. Galindo I, Lorenzo MM, Blasco R (2001) Set of vectors for the expression of histidine-tagged proteins in vaccinia virus recombinants. Biotechniques 30(524–526):528–529

Chapter 6

Screening for Vaccinia Virus Egress Inhibitors: Separation of IMV, IEV, and EEV

Chelsea M. Byrd and Dennis E. Hruby

Abstract

Concerns about the possible use of variola virus as a biological weapon as well as the need for therapeutics for the treatment or prevention of naturally acquired poxvirus infections or vaccination complications have led to the search for small molecule inhibitors of poxvirus replication. One unique and attractive target for antiviral development is viral egress. Part of understanding the mechanism of action of viral egress inhibitors involves determining which virion form is being made. This can be accomplished through buoyant density centrifugation.

Key words: Vaccinia virus, Orthopoxvirus, Egress inhibitors, Antivirals, Extracellular virus, Intracellular virus, ST-246, Tecovirimat, Buoyant density centrifugation

1. Introduction

Smallpox (variola) virus has garnered considerable attention in the last few years as a potential biological threat agent. To counter this threat, as well as to provide countermeasures for naturally occurring poxvirus infections, or complications from vaccination, extensive research has been done to identify and develop small molecule inhibitors of orthopoxvirus replication.

Vaccinia virus (VACV) is a large cytoplasmically replicating DNA virus and a member of the *Orthopoxvirus* genus that can be used safely in the laboratory to screen for orthopoxvirus inhibitors. Other orthopoxviruses include cowpox virus, monkeypox virus, camelpox virus, ectromelia virus, taterapox virus, and raccoonpox virus, which are all morphologically similar. VACV produces four infectious virion forms, which include intracellular mature virus

Stuart N. Isaacs (ed.), *Vaccinia Virus and Poxvirology: Methods and Protocols*, Methods in Molecular Biology, vol. 890,
DOI 10.1007/978-1-61779-876-4_6, © Springer Science+Business Media, LLC 2012

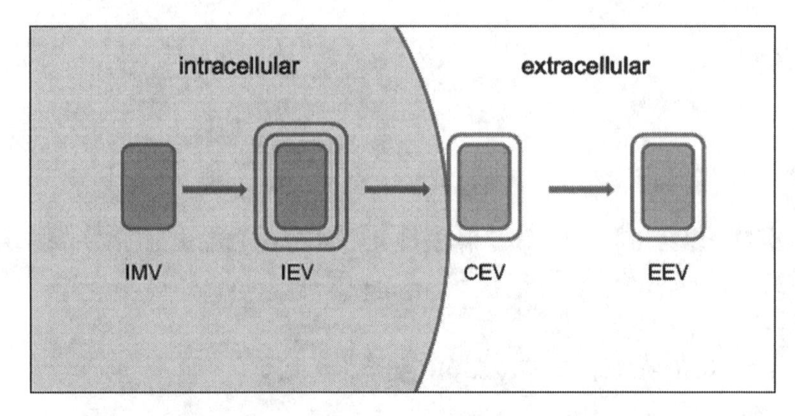

Fig. 1. Schematic of the four infectious virion forms of VACV produced during the virus life cycle. *IMV* intracellular mature virus, *IEV* intracellular enveloped virus, *CEV* cell-associated enveloped virus, *EEV* extracellular enveloped virus.

(IMV), intracellular enveloped virus (IEV), cell-associated enveloped virus (CEV), and extracellular enveloped virus (EEV) (see Fig. 1) (1–5). These virions have an identical core structure but different envelopes, location, and roles in the virus life cycle and can be separated by their buoyant densities.

To facilitate the development of small molecule inhibitors of viral replication, the mechanism of action of the compound needs to be understood. One attractive approach has been to screen compounds for their ability to inhibit the egress of virus from the cell. ST-246 is a novel small molecule orthopoxvirus egress inhibitor being developed by SIGA Technologies for the treatment and prevention of orthopoxvirus infection of humans, and is used in this chapter as a prototype compound (6, 7).

Indications that antiviral compounds inhibit the release of extracellular virus are when they inhibit viral plaque formation in vitro and prevent systemic viral spread in vivo, both of which are EEV dependent. In order to determine whether the block in EEV release is due to a lack of production of the earlier forms of virus such as a reduction in IEV, CEV, or IMV maturation, virus propagated in the presence and absence of inhibitor can be radiolabeled with tritiated thymidine and fractionated by equilibrium centrifugation. In the absence of the inhibitor, radiolabeled cell-associated virus from cell lysates can be separated into three distinct peaks of radioactivity corresponding to IMV, CEV, and IEV, based on the presence of one, two, or three membranes, respectively. Radiolabeled extracellular virus from the culture medium will form one distinct peak corresponding to EEV. In the presence of an inhibitor, the radioactive peak formation can be analyzed to determine where the block in viral production occurs.

2. Materials

2.1. Cells and Virus

1. Virus: VACV-IHDJ (vaccinia virus strain IHD-J, see Note 1).
2. Cells: RK13 (rabbit kidney epithelial cell line; ATCC #CCL-37) (see Note 2).
3. Cell growth medium: Minimum Essential Medium (MEM) supplemented with 10% fetal bovine serum (FBS), 2 mM l-glutamine, and 10 μg/mL gentamicin.
4. Cell infection medium: MEM supplemented with 5% FBS, 2 mM l-glutamine, and 10 μg/mL gentamicin.
5. Viral inhibitor compound: In this example, ST-246 at a concentration of 10 μM.

2.2. Metabolic Radiolabeling Virions

1. [methyl-^{3}H] thymidine.
2. Thymidine-deficient MEM.

2.3. Buoyant Density Centrifugation

1. 10 mM Tris–HCl, pH 8.0.
2. Hypotonic buffer: 50 mM Tris–HCl, pH 8.0, 10 mM KCl.
3. Phosphate-buffered saline (PBS) (see Note 3).
4. Dounce homogenizer.
5. 36% Sucrose solution: 36 g of sucrose and bring the volume up to 100 mL with 10 mM Tris–HCl, pH 8.0.
6. Beckman ultracentrifuge tubes (e.g., Cat #344060).
7. Ultracentrifuge.
8. Cesium-chloride (CsCl) solutions with densities of 1.20, 1.25, and 1.30 g/mL (see Note 4).
9. Whatman glass microfiber filters 21 mm.
10. Scintillation fluid (e.g., Microscint 20).
11. Scintillation counter.

2.4. Immunoblot Analysis of Proteins

1. Pre-poured 4–12% bis–tris polyacrylamide gel.
2. Running buffer.
3. Pre-stained molecular weight markers.
4. Nitrocellulose membrane.
5. Transfer buffer.
6. Tris-buffered saline (TBS): 20 mM Tris–HCl, pH 7.5, 500 mM NaCl.
7. Tris-buffered saline with Tween (TTBS): 20 mM Tris–HCl, pH 7.5, 500 mM NaCl, 0.05% Tween 20.
8. Blocking buffer: 3% gelatin in TBS.

9. Antibody buffer: 1% gelatin in TTBS.
10. Primary antisera, for example anti-L4 (core) and anti-B5 antibodies.
11. Secondary antibodies: Anti-mouse-HRP or anti-rabbit-HRP.

3. Methods

The methods outlined here describe how to use buoyant density centrifugation of radiolabeled virions to separate the different forms of VACV (IMV, IEV, and EEV) in order to determine whether an inhibitor compound is blocking production of some forms of the virus.

3.1. Growth to Metabolically Radiolabel Virus

1. Seed two 150-cm^2-diameter tissue culture dishes with RK13 cells at 1×10^7 cells per dish in cell growth media. Incubate overnight at 37°C in a 5% CO_2 atmosphere.
2. Infect cells with VACV-IHDJ at a multiplicity of infection (MOI) of 10 pfu/cell in 10 mL infection media in the absence or presence of inhibitor compound.
3. Incubate at 37°C.
4. At 3 h post infection (hpi), aspirate the culture media and replace with 10 mL of thymidine-deficient MEM containing 12 μCi/mL of [methyl-^{3}H]-thymidine, either in the presence or absence of inhibitor compound.
5. Incubate at 37°C for 24 h.

3.2. Separation of Intracellular and Extracellular Viral Particles

1. Following step 5 above, remove the culture supernatants from the cells (this sample contains mainly the extracellular virus) and centrifuge at low speed (4,000 × *g* at 25°C for 5 min) to remove the cell debris.
2. Layer the supernatant onto a 7-mL cushion of 36% sucrose in PBS and centrifuge at 40,000 × *g* at 4°C for 80 min.
3. Remove supernatant for proper disposal (see Note 5).
4. Resuspend the remaining pellet in the tube (containing the extracellular virus) in 1 mL PBS and store on ice.
5. Begin to process the infected cells by first gently washing the cell monolayer with PBS (see Note 3).
6. Harvest the infected cells by scraping and pellet cells by low-speed centrifugation as in step 1 (but this time, discarding the supernatant and keeping the cell pellet).
7. Suspend the cell pellet in 1 mL of hypotonic buffer (this is the sample that contains the cell-associated virus forms).

8. Allow the cells to swell on ice for 10 min.
9. Freeze–thaw the cells two times by putting the tube in dry ice and then once frozen thawing the tube at 37°C (see Note 6).
10. Homogenize by 20 strokes in a Dounce homogenizer using a type-A pestle (see Note 6).
11. After douncing, remove the cellular debris by centrifugation at 700 × *g* for 10 min at 4°C.
12. Apply the supernatant to a 7-mL cushion of 36% sucrose in PBS and centrifuge at 40,000 × *g* at 4°C for 80 min.
13. Remove supernatant for proper disposal (see Note 5).
14. Resuspend the remaining pellet in the tube (containing the cell-associated virus) in 1 mL PBS and store on ice.
15. The night prior to banding of virus by equilibrium centrifugation, prepare the CsCl step gradient. The step gradient is made by sequentially pipetting the following solutions into an ultracentrifuge tube: 3.5 mL of CsCl solution with a density of 1.30 g/mL, followed by 4.0 mL CsCl solution with a density of 1.25 g/mL, and finally, 3.5 mL CsCl solution with a density of 1.20 g/mL (see Notes 4 and 7).
16. Layer both the extracellular virus sample (from step 4) and the cell-associated virus sample (from step 14) over individual CsCl step gradients by carefully pipetting the virus-containing samples to the top of the ultracentrifuge tube containing CsCl from step 15.
17. Make sure that the centrifuge tubes are balanced.
18. Centrifuge at 100,000 × *g* for 3 h at 15°C.
19. Gently remove the tubes from the centrifuge and look for white bands which should correspond to the different forms of virus. IMV bands at 1.27 g/mL and EEV bands at 1.23 g/mL (8) (see Fig. 2 and Note 8).
20. Collect 0.5-mL fractions from the bottom of the tube drop wise. The density of the fraction can be determined by weighing each fraction (see Note 9).
21. Add 50 μL of each fraction to Whatman paper and let dry overnight.
22. Quantify CPM using liquid scintillation counting.
23. An example of the results of such a procedure is shown in Fig. 3.

3.3. Immunoblot Analysis of IEV and IMV Proteins

In order to confirm the identity of the type of viral particle in the peak fractions, immunoblot analysis of fractions from the equilibrium centrifugation can be performed with antisera to proteins specific to particle type. There are several proteins that are associated with the IEV, CEV, and EEV particles that are not associated

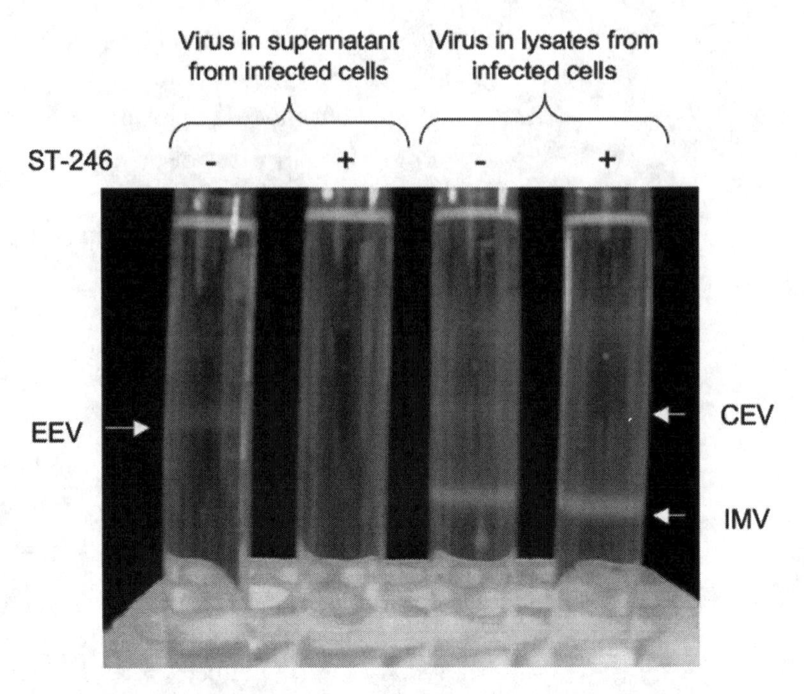

Fig. 2. CsCl gradient separation of the various forms of VACV found in the media of infected cells or found in infected cell lysates (in the presence or absence of ST-246).

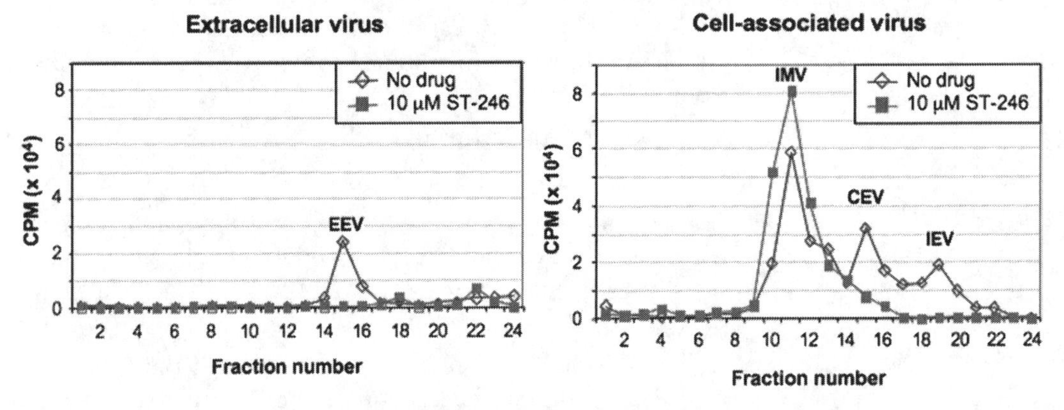

Fig. 3. Equilibrium centrifugation of radiolabeled virus measured by liquid scintillation counting in the absence of drug (*open diamonds*) and in the presence of drug (*closed rectangles*). Adapted from ref. 6 with permission from BioMed Central.

with IMV. IEV proteins include A33R, A34R, A36R, A56R, B5R, F13L, and F12L (5). In the example shown here, we used an anti-L4 antiserum to detect viral cores (present in all forms of virus particles) and anti-B5 antiserum (which detects the B5 protein found only on the wrapped viral particles).

1. Add sample buffer to 20 μL of virus-containing fractions obtained at the end of Subheading 3.2.
2. Boil fractions in a 100°C heat block for 3 min.

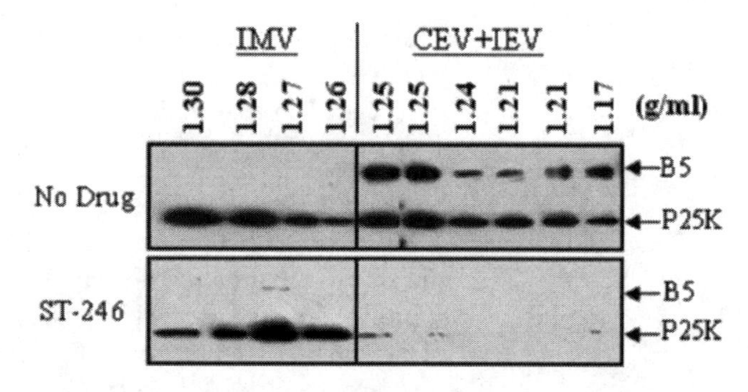

Fig. 4. Equilibrium centrifugation of radiolabeled virus. The viral proteins in different fractions (based on the CsCl density (in g/mL) of the fraction) were detected by immunoblot analysis using antisera against the L4 (P25K) and B5 proteins. Reproduced from ref. 6 with permission from BioMed Central.

3. Load SDS-PAGE gel and run at 125 V for ~90 min (see Note 10).
4. Prepare to transfer gel to nitrocellulose by first presoaking transfer membrane in methanol for 20 min.
5. Transfer proteins to nitrocellulose membranes in a western blotting apparatus for 1 h at 400 mA.

   Block membrane with blocking buffer for 2–4 h.
7. Wash membrane 2× with TTBS.
8. Apply antibody buffer with primary antibody (in this example, either anti-L4 or anti-B5 antibodies) for at least 2 h.
9. Wash membrane 2× with TTBS.
10. Apply antibody buffer with secondary antibody for 1 h.
11. Wash membrane 2× with TTBS.
12. Wash membrane 1× with TBS.
13. Visualize bands using standard techniques.
14. An example of a western blot using this procedure is shown in Fig. 4.

4. Notes

1. The amount of EEV released from an infected cell varies significantly depending on the strain of vaccinia virus that is used. The IHD-J strain makes almost 40 times more EEV than the WR strain of virus (9).
2. Other cell lines that VACV grows well on may be used instead of RK-13, such as Vero or BSC40 cells. RK-13 cells are thought to be less "sticky" and thus release more EEV into the media.

This amplifies the high EEV producing IHD-J phenotype, which allows enough EEV to be released into the media to detect it and separate it from the other forms of virus.

3. Make sure that PBS is without calcium and magnesium.

4. Note that the g/mL listed are densities of a CsCl solution. Thus, to make 100 mL of each of the CsCl solutions needed for the step gradient, do the following. For the 1.30 g/mL CsCl solution, weigh out 31.15 g of CsCl and dissolve in 68.85 mL of 10 mM Tris–HCl, pH 8.0. For the 1.25 g/mL CsCl solution, weigh out 26.99 g of CsCl and dissolve in 73.01 mL of 10 mM Tris–HCl, pH 8.0. For the 1.20 g/mL CsCl solution, weigh out 22.49 g of CsCl and dissolve in 77.5 mL of 10 mM Tris–HCl, pH 8.0. Filter to sterilize solutions. The densities of each solution can be confirmed by measuring the refractive index with a refractometer (see Note 9).

5. Discuss proper disposal of radioactive infectious waste with your institution's environmental health and radiation safety office. Since VACV is a membraned virus, detergent can be added to the radioactive waste to make the material noninfectious. Obviously, radioactive infectious waste should NOT be autoclaved!

6. Freeze–thawing breaks open cells and dounce homogenizing helps to release the virus from the cells.

7. Pour the gradient the night before and keep at 4°C before use to allow some equilibration between the layers.

8. Placing a black card behind the tubes will help to see the bands.

9. Alternatively, density of fractions can be measured using a refractometer. Using this instrument, one can measure the refractive index of a CsCl solution and then use available tables to convert the refractive index to the density of the CsCl solution.

10. There are various gel and buffer systems that can be used for SDS-PAGE.

References

1. Blasco R, Moss B (1992) Role of cell-associated enveloped vaccinia virus in cell-to-cell spread. J Virol 66:4170–4179

2. Moss B (1996) Poxviridae: the viruses and their replication. In: Fields BN, Knipe DM, Howley PM (eds) Fields virology, 3rd edn. Lippincott-Raven, Philadelphia, PA, pp 2637–2671

3. Moss B (2001) Poxviridae and their replication. In: Knipe B, Howley P (eds) Fields virology, 4th edn. Raven, New York, NY, pp 2849–2884

4. Moss B (2006) Poxvirus entry and membrane fusion. Virology 344:48–54

5. Smith GL, Vanderplasschen A, Law M (2002) The formation and function of extracellular enveloped vaccinia virus. J Gen Virol 83:2915–2931

6. Chen Y, Honeychurch KM, Yang G et al (2009) Vaccinia virus p37 interacts with host proteins associated with LE-derived transport vesicle biogenesis. Virol J 6:44
7. Yang G, Pevear DC, Davies MH et al (2005) An orally bioavailable antipoxvirus compound (ST-246) inhibits extracellular virus formation and protects mice from lethal orthopoxvirus challenge. J Virol 79:13139–13149
8. Payne LG, Norrby E (1976) Presence of haemagglutinin in the envelope of extracellular vaccinia virus particles. J Gen Virol 32:63–72
9. Blasco R, Sisler JR, Moss B (1993) Dissociation of progeny vaccinia virus from the cell membrane is regulated by a viral envelope glycoprotein, effect of a point mutation in the lectin homology domain of the A34R gene. J Virol 67:3319–3325

Chapter 7

Imaging of Vaccinia Virus Entry into HeLa Cells

Cheng-Yen Huang and Wen Chang

Abstract

The recently developed technique of live cell imaging has found numerous applications, including the detection of virus movements in living cells. To monitor virus motility, viruses or cellular proteins are fused with fluorescence markers and then detected by time-lapse fluorescence microscopy. These techniques allow kinetic analyses of individual virus particles in motion during the virus entry process as well as monitoring of dynamic interactions between viruses and cellular structures in real time. The methods presented here describe how to construct a fluorescent recombinant vaccinia virus expressing the core protein A4L fused to mCherry, and how to detect the virus movement on actin-EYFP-expressed HeLa cells.

Key words: Vaccinia virus, mCherry, Actin-EYFP, Time lapse, Fluorescence microscopy

1. Introduction

Virus entry involves complex processes that require multiple and dynamic interactions between virions and cellular structures. Therefore, it is important to establish experimental approaches that are sensitive enough to monitor the entry process of individual virus particles and to detect dynamic interactions between viruses and cellular proteins. Imaging analyses of fixed infected cell samples using immunofluorescence microscopy and confocal microscopy have been used extensively in virological studies to understand how viruses bind to cellular surface receptors and how viruses traffic inside cells using various routes ((1–13) and review articles (14–18)). Recent advancements in imaging instrumentation and analysis as well as in the development of specific fluorescent probes have opened up new possibilities to investigate the dynamics of infection processes (see refs. 19, 20 for reviews). However, it is an important issue that the appropriate method is selected among the

Stuart N. Isaacs (ed.), *Vaccinia Virus and Poxvirology: Methods and Protocols*, Methods in Molecular Biology, vol. 890,
DOI 10.1007/978-1-61779-876-4_7,

various developed methods to address and pursue particular research questions.

Vaccinia virus has a wide range of infectivity and vaccinia mature virions (MV) enter cells through different routes (21), including plasma membrane fusion (5, 22–24) and macropinocytosis/fluid-phase endocytosis (3, 7, 25, 26). After binding to glycosaminoglycans (27, 28) and the extracellular matrix laminin (29), vaccinia MVs were internalized through fluid-phase endocytosis/marcopinocytosis into HeLa cells (3, 7). Here, we describe our experimental protocols that we employed to generate a fluorescent recombinant vaccinia virus, mCherry-VV, which expresses a viral A4 core protein fused to the red fluorescent protein mCherry (30). We also describe how we applied mCherry-VV in real-time imaging analyses to visualize vaccinia virus MV particles and actin dynamics (3), and to determine vaccinia virus entry process.

2. Materials

2.1. Cell Culture

1. Complete Dulbecco's Modified Eagle's Medium (DMEM): DMEM supplemented with 10% fetal bovine serum (FBS) or 10% calf serum (CS).
2. Phenol red-free DMEM medium: Supplemented with 100 mM HEPES and then add FBS to 2% (see Note 1).
3. Microscope round coverslips, 42 mm × 0.17 mm (e.g., Carl Zeiss).
4. A cell culture chamber system for *p*erfusion and "*o*pen" and "*c*losed" cultivation (POC-R) (e.g., Carl Zeiss).

2.2. Virus and Cells

1. Vaccinia virus strain WR, mCherry-VV.
2. HeLa cells.
3. 293T cells.
4. BSC40 cells.

2.3. Generation of Recombinant Virus

1. Vaccinia virus strain WR genomic DNA.
2. Plasmid pmCherry (30).
3. Oligonucleotides designed for overlap PCR (see Table 1 and Fig. 1).
4. GeneAmp PCR System 9700.
5. QIAquick gel extraction kit.
6. TOPO pCDNA3.1 plasmid.

Table 1
Primers to produce the PCR product that is used to generate a recombinant vaccinia virus expressing mCherry-A4 fusion protein

Primer name	Sequence	Description
Oligonucleotide 1	5′-CTC CGT TGA ATT CGA TGA CTA TAG GAC AAG AAC CCT CCT C-3′	Used to generate a 463-bp PCR product (fragment a) located upstream of the A4L gene open reading frame (ORF). Underlined bases represent an EcoRI restriction site
Oligonucleotide 2	5′-ATC CTC CTC GCC CTT GCT CAC CAT TTA AGG CTT TAA AAT TGA ATT GCG-3′	
Oligonucleotide 3	5′-GGC ATG GAC GAG CTG TAC AAG GAC TTC TTT AAC AAG TTC TCA CAG GGG-3′	Used to generate a 1,215-bp PCR fragment (fragment b) containing the A4L gene ORF and 346 bp downstream of the A4L gene ORF. Underlined bases represent an HindIII restriction site
Oligonucleotide 4	5′-CGT ACT CCA AGC TTG TGT AGA TGC TAC TTC GTC GAT GG-3′	
Oligonucleotide 6	5′-ATG GTG AGC AAG GGC GAG GAG GAT-3′	Used to generate a 711-bp PCR fragment (fragment c) of the mCherry gene
Oligonucleotide 7	5′-CTT GTA CAG CTC GTC CAT GCC-3′	

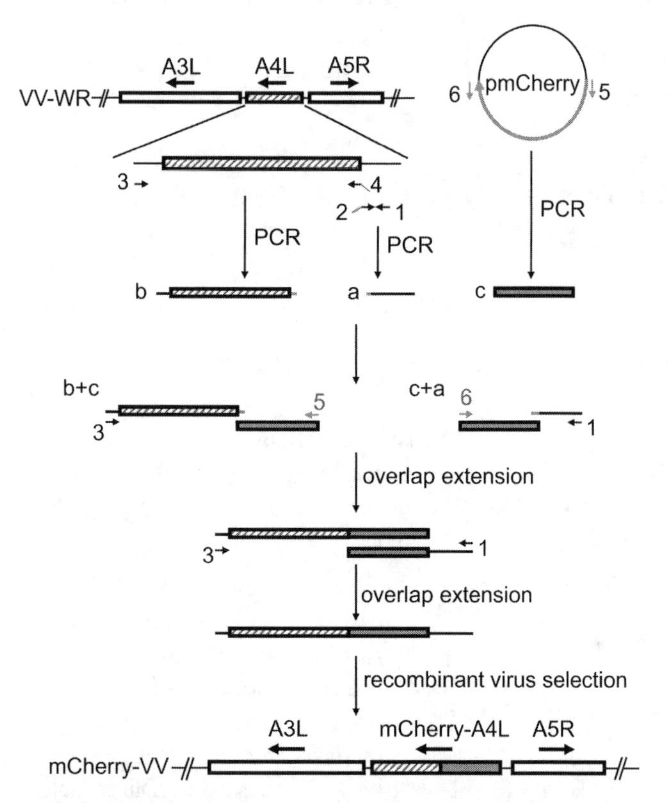

Fig. 1. Construction of mCherry-VV. The coding sequence of mCherry was fused at the N-terminus to vaccinia A4L ORF by overlapping extension methods. *Black arrows* indicate transcription direction of each ORF, and *small arrows* indicate the primers for PCR.

2.4. High-Titer Virus Purification

1. Phosphate-buffered saline (PBS).
2. TM buffer: 10 mM Tris, pH 7.4, 5 mM $MgCl_2$.
3. Dounce homogenizer, 50-mL capacity.
4. 36% Sucrose solution: 36 g of sucrose weighed out and water is added until the total volume of 100 mL.
5. Optima L-90K Ultracentrifuge and SW28 rotor.
6. SONIFIER 450 sonicator.

2.5. Transient Transfection

1. CMV promoter-driven pEYFP-actin plasmid (e.g., Clontech).
2. Lipofectamine transfection reagent.

2.6. Live Cell Imaging

1. Inverted wide-field fluorescent microscope: Zeiss Axiovert 200 M microscope with oil immersion plan-NEOFluar 63×, 1.25 NA objective.
2. Cell observer: Incubator XL S1, Heating Unit XL S1, TempModule S1, CO_2 Module S1, CO_2-Cover PMS1, Heating Device Humidity S1.
3. Filter sets: YFP shift-free Ex 500/20, Em 535/30; mCherry shift-free Ex 531/40, Em 593/40.
4. Zeiss AxioVision Rel. 4.6 software.

3. Methods

3.1. Cell Culture

1. HeLa cell and 293T cell monolayers are maintained in DMEM supplemented with 10% FBS, cultured in 100-mm tissue culture dishes, and passaged upon reaching 80% confluence.
2. BSC-40 cell monolayers are maintained in DMEM supplemented with 10% CS, cultured in 100-mm tissue culture dishes, and passaged upon reaching 80% confluence.

3.2. Construction of a Fluorescent Recombinant Vaccinia Virus Expressing the Core Protein A4L Fused to mCherry (mCherry-VV)

Fusion of mCherry to the N-terminus of A4L is achieved by PCR splicing by overlap extension (31) using vaccinia virus strain WR genomic DNA and pmCherry plasmid as the template (see Fig. 1 and Table 1).

1. Oligonucleotides 1 and 2 generate a 463-bp PCR product (fragment a) located upstream of the A4L gene open reading frame (ORF) and containing the 5′ region (24 bp) of the mCherry gene. Oligonucleotides 3 and 4 generate a 1,215-bp PCR fragment (fragment b) containing the 3′ region (21 bp) of the mCherry gene, the A4L gene ORF, and 346 bp downstream of the A4L gene ORF. Oligonucleotides 5 and 6 generate a 711-bp PCR fragment (fragment c) of the mCherry gene. Each PCR

amplification is performed for 25 cycles at 94°C for 1 min, 50°C for 1 min, and 72°C for 3 min. PCR fragments are then gel purified by QIAquick gel extraction kit.

2. The three PCR fragments are subsequently assembled and extended pair-wise (b + c and a + c), PCR amplified (see Note 2), and gel purified. The resulting two fragments are extended again to obtain the final 2,389-bp PCR fragment (see Note 2). Overlap extension is performed by five cycles at 92°C for 1 min, 50°C for 2 min, and 72°C for 7 min.
3. The final product is gel purified and then cloned into TOPO pCDNA3.1 plasmid to form pA4L-mCherry-N (see Note 3).
4. Recombinant virus is generated by first infecting 293T cells with vaccinia virus (strain WR) at MOI of 5 pfu per cell followed by transfection using Lipofectamine with the DNA fragment produced by EcoRI and HindIII digestion of pA4L-mCherry-N (see Note 4).
5. The cell lysates are collected at 24 h post infection and recombinant viruses expressing mCherry are identified by fluorescence.
6. Recombinant virus then underwent three rounds of plaque purification on BSC-40 cells.

3.3. Virus Purification (See Note 5)

Purification of recombinant mCherry-VV is done as described previously (28, 32).

1. One hundred 100 × 100-mm dishes of BSC40 cells are infected with mCherry-VV at an MOI of 0.05 pfu per cell.
2. The infected cells are harvested when the cytotoxic pathologic effect is complete (see Note 6).
3. The harvested cells are centrifuged at 850 × *g* for 15 min to pellet the cells.
4. The medium is removed and the cell pellet is washed three times with PBS.
5. The pelleted cells are resuspended in 20 mL TM buffer, and the suspension is dounced 20 times in a Dounce homogenizer to break open the cells.
6. Nuclei are removed by low-speed centrifugation at 850 × *g* for 10 min, and the supernatant is collected.
7. Steps 5 and 6 are repeated for a total of three times, each time saving and combining the supernatants.
8. The combined supernatant containing virions (typically ~20 mL) is placed on top of 16 mL 36% sucrose solution and virions are pelleted at 45,000 × *g* for 80 min in a Beckman SW28 rotor.

9. The virus pellet is resuspended in 3 mL of TM buffer, sonicated, and further purified by centrifugation through a 33 mL continuous 25–40% sucrose gradient at 27,500 × *g* for 40 min in a Beckman SW28 rotor.
10. Fractions are collected slowly from the top of the gradient, and fractions containing virions are diluted fivefold in TM buffer (see Note 7).
11. The fractions containing virions (MV) are re-pelleted by centrifugation in TM buffer to wash out the sucrose.
12. The virion infectivity is analyzed by a plaque formation assay, and the MV particle number is determined by electron microscopy (see Note 8).

3.4. Transient Transfection

To visualize actin in infected cells, we transfect cells with a plasmid expressing a YFP-actin fusion protein.

1. Confluent HeLa cells are trypsinized and seeded at a density of 6×10^5 cells per 100-mm culture dishes to obtain 60–70% confluence within 24 h.
2. 3 μg of pEYFP-actin plasmid DNA is diluted into 900 μL DMEM, mixed with 900 μL DMEM containing 30 μL lipofectamine, and incubated for 20 min at RT.
3. While preparing the DNA–lipofectamine mixture, replace 7.2 mL fresh DMEM (without FBS) onto HeLa cells for 20 min.
4. The DNA and lipofectamine mixture is added drop-wise to HeLa cells and incubation is continued for 4 h at 37°C.
5. The medium is aspirated off and replaced with 10 mL complete DMEM. Cells are incubated overnight or grown until sufficient cell numbers are reached.

3.5. Time-Lapse Microscopy of Cells During Vaccinia Virus Infection

1. One day before microscopy analysis, pEYFP-actin-expressing HeLa cells (which were transfected one day earlier, see Subheading 3.4) are trypsinized and seeded at a density of 2×10^5 cells in 60-mm culture dishes containing sterile coverslips, and allowed to adhere overnight and reach 70% confluence (see Notes 9 and 10).
2. Coverslips with adherent cells are picked up by forceps and placed on POC-R (see Note 11) containing 1 mL phenol red-free DMEM medium (see Note 12).
3. All sets of cells are placed on the microscope stage inside the cell observer system and equilibrated for 1 h at 37°C (see Note 13).
4. Purified VV-mCherry-expressing MVs are sonicated and then loaded onto HeLa cells in POC-R at 37°C aiming for 50–100 MV particles/cell (see Note 14).

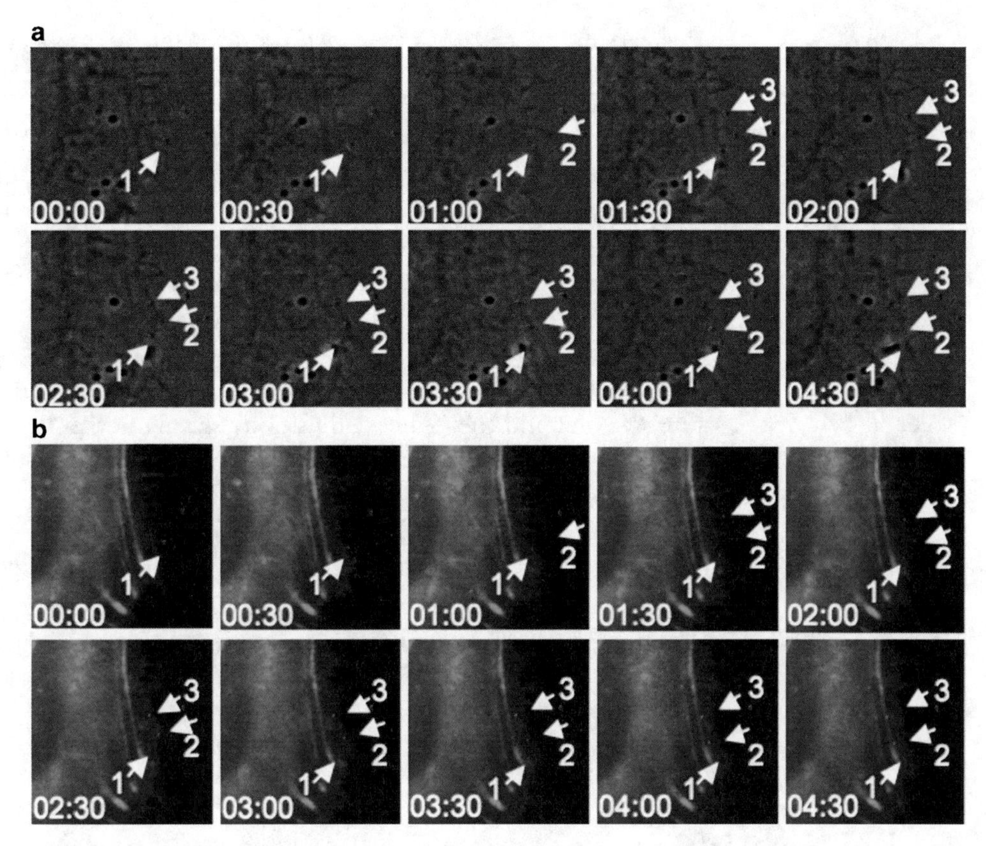

Fig. 2. Vaccinia MV entry into HeLa cells. Live imaging of mCherry-VV movement obtained by time-lapse immunofluorescence microscopy. HeLa cells expressing EYFP-actin were infected with purified mCherry-VV MV, and time-series images were immediately collected at 30-s intervals. *Panel A* shows merged images of phase contrast and mCherry fluorescent channels, while *panel B* shows merged images of EYFP and mCherry fluorescent channels. It can be seen that mCherry-VV particles (*white dots*) move along the cell surface of actin protrusions before internalization. *Numbered arrows* track the progression of individual virions toward the cell surface. In color version of this figure (available in the electronic version of this chapter), EYFP-actin is pseudocolored in green and mCherry-VV is pseudocolored in red. It can be seen that mCherry-VV particles (*red dots*) move along the cell surface of actin protrusions (*green*) before internalization.

5. Cells are observed on a Zeiss Axiovert 200 M microscope at 37°C with an oil immersion plan-NEOFluar 63×, 1.25 NA objective.
6. Fluorescence images are acquired and processed using the Zeiss AxioVision Rel. 4.6 software. The software can be used to overlay phase contrast and fluorescent virus, or to overlay the fluorescence actin-EYFP and mCherry-VV (see Fig. 2).

3.6. Image Acquisition

1. The condition for image acquisition must be set up before virus is dropped on cells. After setup, the POC-R set containing HeLa cells is placed on the stage of the microscope, and the Zeiss AxioVision Rel. 4.6 software is opened.
2. The focus can be adjusted by using epifluorescence: find a fluorescent cell or fluorescent virus, center it in the field of view, and turn the focusing knob.

3. Purified mCherry-VV, dropped on a coverslip, is used as a control to adjust the exposure time for mCherry fluorescence acquisition.
4. Subsequently, select the "Multidimensional acquisition" function, and then select "Channel definition" and the ph3, YFP, and mCherry buttons. The "Fixed mode" is chosen and the exposure time is measured by checking on the "Measure" button. Afterwards, adjust the exposure time and then chick "OK" to define the condition for the three channels.
5. Make sure that the "Hardware settings" are correct. When the ph3 channel is turned on, the "During acquisition" setting must be ph3 and the "After acquisition" setting is TL shutter off. When the YFP or mCherry channel is turned on, the "During acquisition" setting must be YFP or mCherry and the "After acquisition" setting should be RL off.
6. The "Autofocus" function can be chosen to avoid loss of focus during image acquisition.
7. The time interval and total acquisition time can be set by selecting "Time lapse," setting the "Time interval" to 30 s and "Duration" to 30 min (see Note 15).
8. After finishing all settings, add virus to cells and then click the "Start" button to acquire images.

3.7. Data Processing

1. The Zeiss AxioVision Rel. 4.6 software is useful for data processing. Serial images of selected regions can be cut out by choosing the "Edit" function and "Define ROI." After selecting a region of interest, click "Copy ROI" and paste to get the selected region.
2. When serial time points must be extracted from the whole time points, choose the "Gallery view" function, select "Time region," and then click the "Extract selection" button.
3. Text and arrows can be added to images by selecting first "Annotations" and then "Draw annotations."
4. To add serial time points to images, click "Annotations," select "Frequent annotations," and then choose "Relative time."
5. The images can be exported as jpg, tif, or movie file for data presentation.

4. Notes

1. Protect from light and store up to several months at 4°C.
2. If necessary, the product of overlap extension can be amplified by PCR again, but may introduce more mutations during the PCR amplification.

3. The insert in the resulting plasmid should be sequenced to confirm that no unintended errors are introduced. We usually picked ten colonies from each transformation and can typically find at least one with no errors.
4. Transfecting linear DNA instead of the full plasmid prevents isolating "single" crossover recombinants that insert the full plasmid into the virus. Such a single crossover event with a plasmid results in a virus that has a tandem repeat that will ultimately recombine out of the virus and result in either wild-type virus or the desired virus.
5. After harvest of virus-infected cells, all steps during virus purification are performed at 4°C.
6. For this recombinant virus, this takes about 3 days.
7. The fraction containing virions is visualized as an opaque white band near the center of the gradient.
8. Mix purified virions with 2% uranyl acetate at equal volumes for 30 s at room temperature. Perform several tenfold dilutions with TM buffer. Sonicate and spot on metal grids (400 mesh N-FC) for particle counting under electron microscopy. The same virus preparation was also used to determine plaque formation units. For this recombinant virus (mCherry-VV), we typically get a particle-to-PFU ratio of 11.
9. Reseeding HeLa cells after transfection reduces the nonspecific fluorescent background caused by transfection liposomes during imaging acquisition.
10. We have been able to use cells that were transiently transfected with YFP-actin for up to 3 days after transfection.
11. The POC-R system must be tightly closed to avoid loss of medium, which reduces cell viability and instability of focal plane during image acquisition.
12. Phenol red-free DMEM medium reduces nonspecific fluorescence backgrounds, but may not be necessary, if the target fluorescent proteins are expressed at high levels in cells.
13. The cell observer system should be turned and adjusted to 37°C at least 1 day before the experiments for better equilibrium.
14. Application of virus to cells requires steady hands and swift action. Alternatively, one could use the injection tube to inject viruses into the POC-R; however, this would waste a large amount of virus.
15. To avoid any toxic effects of the light source and fluorescence photobleaching, the overall exposure time should be minimized by choosing appropriate time intervals.

References

1. Forzan M, Marsh M, Roy P (2007) Bluetongue virus entry into cells. J Virol 281:4819–4827
2. Helenius A, Kartenbeck J, Simons K, Fries E (1980) On the entry of Semliki forest virus into BHK-21 cells. J Cell Biol 84:404–420
3. Huang CY, Lu TY, Bair CH, Chang YS, Jwo JK, Chang W (2008) A novel cellular protein, VPEF, facilitates vaccinia virus penetration into HeLa cells through fluid phase endocytosis. J Virol 82:7988–7999
4. Iyengar S, Hildreth JE, Schwartz DH (1998) Actin-dependent receptor colocalization required for human immunodeficiency virus entry into host cells. J Virol 72:5251–5255
5. Locker JK, Kuehn A, Schleich S et al (2000) Entry of the two infectious forms of vaccinia virus at the plasma membane is signaling-dependent for the IMV but not the EEV. Mol Biol Cell 11:2497–2511
6. Marechal V, Prevost MC, Petit C, Perret E, Heard JM, Schwartz O (2000) Human immunodeficiency virus type 1 entry into macrophages mediated by macropinocytosis. J Virol 75:11166–11177
7. Mercer J, Helenius A (2008) Vaccinia virus uses macropinocytosis and apoptotic mimicry to enter host cells. Science 320:531–535
8. Patterson S, Russell WC (1983) Ultrastructural and immunofluorescence studies of early events in adenovirus-HeLa cell interactions. J Gen Virol 64:1091–1099
9. Pernet O, Pohl C, Ainouze M, Kweder H, Buckland R (2009) Nipah virus entry can occur by macropinocytosis. Virology 395: 298–311
10. Superti F, Derer M, Tsiang H (1984) Mechanism of rabies virus entry into CER cells. J Gen Virol 65:781–789
11. Vanderplasschen A, Smith GL (1997) A novel virus binding assay using confocal microscopy: demonstration that the intracellular and extracellular vaccinia virions bind to different cellular receptors. J Virol 71:4032–4041
12. Vanderplasschen A, Hollinshead M, Smith GL (1998) Intracellular and extracellular vaccinia virions enter cells by different mechanisms. J Gen Virol 79:877–887
13. Wang QY, Patel SJ, Vangrevelinghe E et al (2009) A small-molecule dengue virus entry inhibitor. Antimicrob Agents Chemother 53:1823–1831
14. Chazal N, Gerlier D (2003) Virus entry, assembly, budding, and membrane rafts. Microbiol Mol Biol Rev 67:226–237 (table of contents)
15. Helle F, Dubuisson J (2008) Hepatitis C virus entry into host cells. Cell Mol Life Sci 65:100–112
16. Lakadamyali M, Rust MJ, Zhuang X (2004) Endocytosis of influenza viruses. Microbes Infect 6:929–936
17. Marsh M, Helenius A (2006) Virus entry: open sesame. Cell 124:729–740
18. Sieczkarski SB, Whittaker GR (2002) Dissecting virus entry via endocytosis. J Gen Virol 83:1535–1545
19. Brandenburg B, Zhuang X (2007) Virus trafficking—learning from single-virus tracking. Nat Rev Microbiol 5:197–208
20. Greber UF, Way M (2006) A superhighway to virus infection. Cell 124:741–754
21. Bengali Z, Townsley AC, Moss B (2009) Vaccinia virus strain differences in cell attachment and entry. Virology 389:132–140
22. Armstrong JA, Metz DH, Young MR (1973) The mode of entry of vaccinia virus into L cells. J Gen Virol 21:533–537
23. Chang A, Metz DH (1976) Further investigations on the mode of entry of vaccinia virus into cells. J Gen Virol 32:275–282
24. Doms RW, Blumenthal R, Moss B (1990) Fusion of intra- and extracellular forms of vaccinia virus with the cell membrane. J Virol 64:4884–4892
25. Dales S, Kajioka R (1964) The cycle of multiplication of vaccinia virus in Earle's Strain L cells. I. Uptake and penetration. Virology 24:278–294
26. Townsley AC, Weisberg AS, Wagenaar TR, Moss B (2006) Vaccinia virus entry into cells via a low-pH-dependent endosomal pathway. J Virol 80:8899–8908
27. Chung CS, Hsiao JC, Chang YS, Chang W (1998) A27L protein mediates vaccinia virus interaction with cell surface heparan sulfate. J Virol 72:1577–1585
28. Hsiao JC, Chung CS, Chang W (1999) Vaccinia virus envelope D8L protein binds to cell surface chondroitin sulfate and mediates the adsorption of intracellular mature virions to cells. J Virol 73:8750–8761
29. Chiu WL, Lin CL, Yang MH, Tzou DL, Chang W (2007) Vaccinia virus 4c (A26L) protein on intracellular mature virus binds to the extracellular cellular matrix laminin. J Virol 81: 2149–2157
30. Shaner NC, Campbell RE, Steinbach PA, Giepmans BN, Palmer AE, Tsien RY (2004)

Improved monomeric red, orange and yellow fluorescent proteins derived from Discosoma sp. red fluorescent protein. Nat Biotechnol 22:1567–1572

31. Ho SN, Hunt HD, Horton RM, Pullen JK, Pease LR (1989) Site-directed mutagenesis by overlap extension using the polymerase chain reaction. Gene 77:51–59

32. Jensen ON, Houthaeve T, Shevchenko A et al (1996) Identification of the major membrane and core proteins of vaccinia virus by two-dimensional electrophoresis. J Virol 70:7485–7497

Chapter 8

New Method for the Assessment of Molluscum Contagiosum Virus Infectivity

Subuhi Sherwani, Niamh Blythe, Laura Farleigh, and Joachim J. Bugert

Abstract

Molluscum contagiosum virus (MCV), a poxvirus pathogenic for humans, replicates well in human skin in vivo, but not in vitro in standard monolayer cell cultures. In order to determine the nature of the replication deficiency in vitro, the MCV infection process in standard culture has to be studied step by step. The method described in this chapter uses luciferase and GFP reporter constructs to measure poxviral mRNA transcription activity in cells in standard culture infected with known quantities of MCV or vaccinia virus. Briefly, MCV isolated from human tissue specimen is quantitated by PCR and used to infect human HEK293 cells, selected for ease of transfection. The cells are subsequently transfected with a reporter plasmid encoding firefly luciferase gene under the control of a synthetic early/late poxviral promoter and a control plasmid encoding a renilla luciferase reporter under the control of a eukaryotic promoter. After 16 h, cells are harvested and tested for expression of luciferase. MCV genome units are quantitated by PCR targeting a genome area conserved between MCV and vaccinia virus. Using a GFP reporter plasmid, this method can be further used to infect a series of epithelial and fibroblast-type cell lines of human and animal origin to microscopically visualize MCV-infected cells, to assess late promoter activation, and, using these parameters, to optimize MCV infectivity and gene expression in more complex eukaryotic cell culture models.

Key words: Molluscum contagiosum virus, Luciferase reporter construct, Eukaryotic cells, Infection, Transfection, Quantitative PCR

1. Introduction

Molluscum contagiosum virus (MCV) does not produce a quantifiable cytopathic effect and does not produce viral progeny in infected standard cell cultures. But small amounts of viral mRNA and protein expression can be detected indicating that MCV virions are transcriptionally active (1–3). Many investigators have observed

Stuart N. Isaacs (ed.), *Vaccinia Virus and Poxvirology: Methods and Protocols*, Methods in Molecular Biology, vol. 890, DOI 10.1007/978-1-61779-876-4_8, © Springer Science+Business Media, LLC 2012

that poxvirus transcription complexes can drive luciferase reporters under the control of poxviral promoters in plasmids in poxvirus-infected cells. A recent paper uses this as a method to diagnose orthopoxvirus infections (4).

In the assay described in this chapter, the same principle is used. We introduce a luciferase reporter expression in trans as a new surrogate marker of infectivity and gene expression for MCV. To compare the infectivity of MCV with other poxviruses (11), the number of virions must be determined. Quantitation by EM or OD300 can be used (5), but requires relatively large amounts of gradient purified virons. However, currently, MCV can only be isolated from clinical specimens and thus is difficult to obtain amounts sufficient for gradient purification (5). PCR is an alternative method of quantitation of smaller amounts of poxviruses from clinical specimens, which is both reliable and highly specific for individual poxviruses. The method described in this chapter uses a novel PCR target in an area with significant DNA homology (~65%) between MCV and vaccinia virus strain WR (VACV-WR). The MCV gene is mc129R, which is homologous to the VAVWR144 (also called A24R gene encoding RPO132, the large subunit of the DNA-dependent RNA polymerase). The method provides a means to quantitate poxviral genome units in the same virus preparations used to compare transcriptional activity and infectivity of MCV and VACV-WR.

2. Materials

2.1. MCV Luciferase Reporter Assay

2.1.1. Infection–Transfection of Cell Cultures

1. OPTIMEM, stored at 4°C.
2. Plasmids described in Subheading 2.3 (see Note 1).
3. Lipofectamine 2000, stored at 4°C until used.
4. Human HEK 293 cells (ATCC CRL1573) (see Note 2).
5. Dulbecco's modified Eagle medium: DMEM, high glucose with glutamine, stored at 4°C until used.
6. Fetal calf serum: FCS, stored in aliquots at –70°C until used.
7. Cell growth medium: DMEM with 10% FCS.

2.1.2. Luciferase Assay (The Dual-Luciferase® Reporter Assay System from Promega)

1. Dual Luciferase Assay Substrate (lyophilized) stored at –20°C for up to 6 months reconstituted.
2. 10 ml Luciferase Assay Buffer II, stored in a 1-ml aliquots at –20°C for up to 6 months until used.
3. Stop and Glo Substrate (50×) stored at –20°C.

4. 10 ml of Stop & Glo Buffer, stored in a 1-ml aliquots at –20°C for up to 6 months until used.
5. 30 ml of Passive Lysis Buffer (5×) stored at –20°C, then diluted to 1× using sterile water, and kept at 4°C until used.
6. Clear film plate protectors (to prevent evaporation from wells).
7. FLUOStar Luminometer.

2.2. MCV and VACV Quantitative PCR Assay

1. Primers outlined in Table 1 suspended in injection-grade water to a final concentration of 100 pmol/μl and stored at –20°C.
2. MCV isolated from human skin biopsy material as described previously (5) and kept in 100-μl aliquots frozen at –70°C in PBS (see Note 3).
3. VACV-WR, vaccinia virus, strain WR (kind gift of B. Moss) was prepared and purified from infected HeLa cells, titrated in BSC-1 cells, and kept in 100-μl aliquots frozen at –70°C in PBS (see Note 3).
4. DNAse at 1 mg/ml.
5. DNAse/BamHI buffer: 78 μl water, 2 μl DNAse, 20 μl 10× BamHI buffer from New England Biolabs.
6. High Pure viral nucleic acid (HPVNA) kit (e.g., Roche).
7. Nanodrop-Spectrophotometer.

Table 1
MCV–VACV quantitative PCR assay primers

Primer ID	Primer sequence (nhb)	Primer length	Product size (nucleotide position), GenBank Acc. #
Mcv129 1-2F149275	5′-CCGCACTAC TCCTGGATGCAGAA-3′	23	576 bp (149,275–149,850), U60315
Mcv129 1-3R149850	5′-CTGGATGTC GGAGAAGGTCATG-3′	22	
VACV-WR 1-2F132482	5′-CCTCACTAT TCATGGATGCAGAA-3′ (3)	23	573 bp (132,482–122,054), AY243312
VACV-WR 1-3R133054	5′-CTGAATGTC AGAGAATGTCATG-3′ (3)	22	

nhb nonhomologous bases underlined (number of mismatches)

Primers were designed using BLAST2 (NCBI: http://blast.ncbi.nlm.nih.gov/) alignment of MCV (GenBank accession # U60315) and VACV-WR (GenBank accession # AY243312) genome sequences and Vector NTI vs. 4.0, 1994–1996 InforMax Inc.

8. ImageJ (Wayne Rasband (wayne@codon.nih.gov) Research Services Branch, National Institute of Mental Health, Bethesda, Maryland, USA).
9. Injection-grade water.
10. AmpliTaq 360 polymerase (5 U/μl).
11. dNTP (0.2 mM).
12. 10× PCR buffer.
13. 2% Agarose gel.
14. Ethidium bromide solution: 10 mg/ml stock solution in demineralized water and used at 20 μl per 200 ml.

2.3. Plasmids (See Fig. 1)

1. *PCR control plasmid.* The complete MCV-1 genome was cloned (6) and sequenced (7, 8) and the redundant MCV genome fragment library of MCV type 1 was submitted to

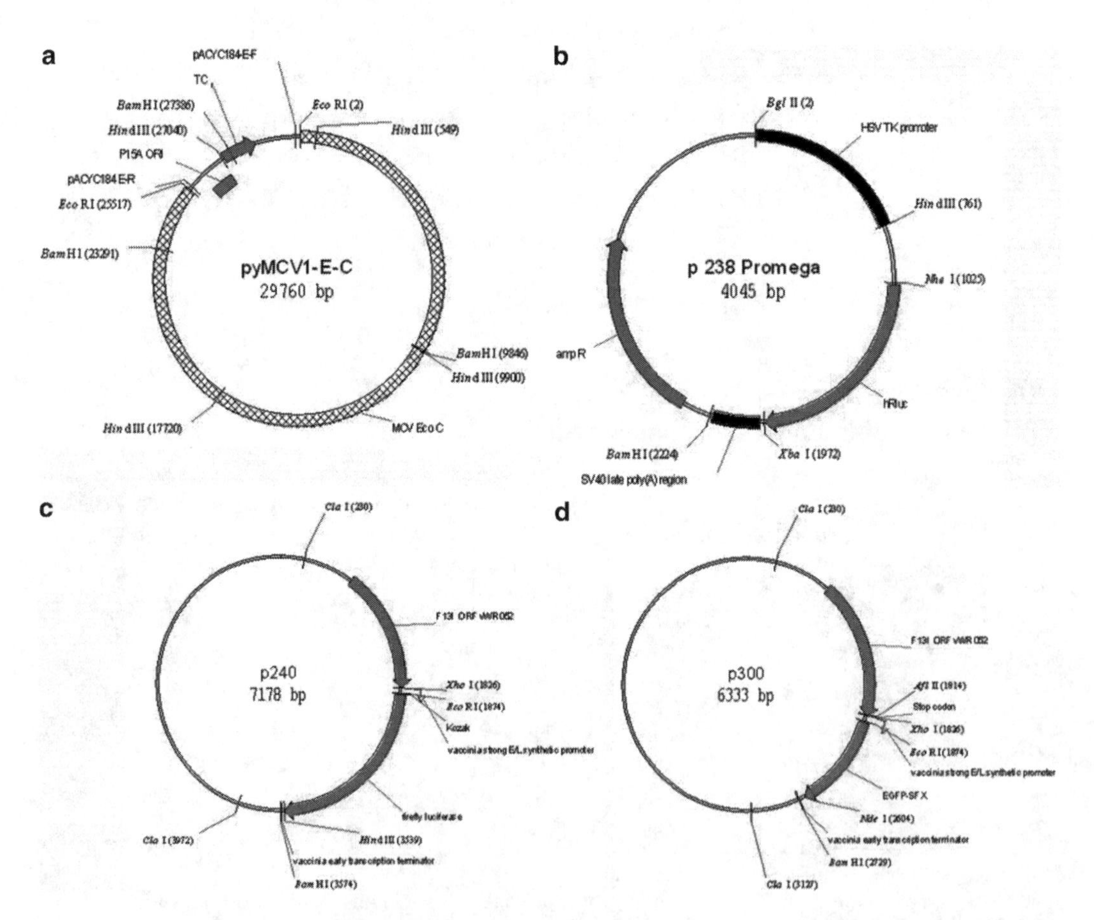

Fig. 1. Plasmid constructs. VectorNTI drawings for the recombinant plasmids (**a**) pyMCV1-EcoRI-fragment C (pyMCV1-E-C; available from ATCC molecular section), (**b**) phRG-TK (Promega; Internal lab reference number p238), (**c**) pRB21-pE/L-FF luciferase (p240), and (**d**) pRB21-pE/L-EGFP-SFX (p300).

ATCC for safekeeping in 2003 and 2008. For the quantitative PCR assay, the genomic MCV-1 *Eco*RI fragment C (25,516 bp) cloned into bacterial plasmid vectorp pACYC184 was used as a MCV target control (pyMCV1-E-C, see Fig. 1a).

2. *Transfection control plasmid*. Plasmid phRG-TK (Promega GenBank accession number AF362545: 4,045 bp), expressing renilla luciferase under the control of the herpes simplex virus TK gene promoter. In this protocol, this plasmid is called p238 and used as plasmid transfection control (p238 Promega; see Fig. 1b).
3. *Poxviral luciferase reporter plasmid*. The coding sequence of firefly luciferase (*Photinus pyralis* GenBank accession number M15077) was amplified with a modified Kozak sequence by PCR and ligated into the pRB21 donor plasmid (kind gift of B. Moss (9, 10)) using the EcoRI and HindIII restriction sites in the donor plasmid multiple cloning site, resulting in the pRB21-E-Koz-Fireflyluciferase-H(alsocalledpRB21-pE/L-FF luciferase) construct of 7,178 bp with the internal lab designation p240 (p240, see Fig. 1c).
4. *Poxviral EGFP reporter plasmid*. The coding sequence of EGFP was amplified from a commercially available plasmid with a modified Kozak sequence by PCR and ligated into the pRB21 donor plasmid using the EcoRI and NheI restriction sites in the donor plasmid multiple cloning site, resulting in the pRB21-E-Koz-EGFP-X-flag-strepII-N construct (also called pRB21-pE/L-EGFP-SFX) of 6,333 bp with the internal lab designation p300 (p300, see Fig. 1d).

3. Methods

3.1. Infection–Transfection: Luciferase Assay

3.1.1. Infection/Transfection

1. Prepare enough 12-well plates containing HEK 293 cells in growth media to allow for infection/transfection in triplicate for each experimental condition (including a mock that will be transfected but not infected, as well as wells that will be harvested at 16 h and wells that will be continued to be incubated for days) (see Note 4).
2. Thaw virus aliquots, sonicate, and keep on ice.
3. Thaw plasmid DNA and keep on ice.
4. Bring OptiMEM and Lipofectamine 2000 to room temperature (RT).
5. Prepare transfection mixes by adding a dilution of 2 μl of Lipofectamine 2000 in 50 μl of OptiMEM to a dilution of 0.3 μg of each plasmid DNA (p240 FF reporter and p238

transfection control plasmid, p300 EGFP reporter) in 50 μl of OptiMEM. Mix gently for 15 min at RT in the dark to allow formation of transfection complexes.

6. Remove growth media from HEK293 cells and put 100 μl of transfection mix in each well.
7. Combine 100 μl each of ice-cold virus in PBS and 100 μl of transfection mix at RT and transfer the mixture into appropriate wells of HEK293 cells (see Note 5).
8. Incubate for 16 h at 37°C in 5% CO_2 atmosphere (see Note 6).

3.1.2. Microscopy and Collection of Cells for Luciferase Assay

1. At 16 h post infection (p.i.), inspect cells transfected with the GFP reporter plasmid using live cell microscopy. Document GFP-positive cells noting that MCV does not show GFP-positive cells after 16 h, whereas WR shows multiple GFP-positive cells.
2. Upon further incubation for another 4 days (5 days p.i.), some individual cells in the MCV-infected wells will show medium to strong GFP signals (see Note 7). At the same time point, the WR-infected wells will show extensive plaques and cell degradation (see Fig. 2a–d).
3. For luciferase assay, at 16 h p.i., wash adherent cells in wells once with PBS and add 100 μl of 1× passive lysis buffer to each well (see Note 8).

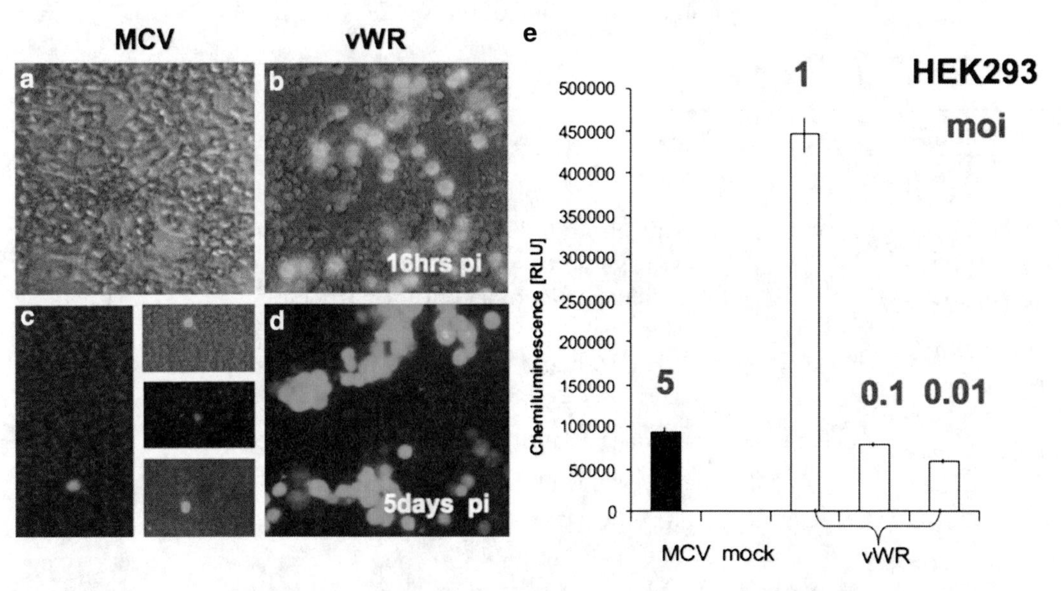

Fig. 2. Images of luciferase and GFP in infected/transfected cells and quantification of luciferase output. *Panels* (**a**–**d**) show HEK 293 cells infected with MCV (**a** and **c**), and vWR (**b** and **d**). Inserts in **c** show individual GFP-positive cells. *Panel* **e** shows a histogram of luciferase data giving chemiluminescence in RLU. HEK 293 cells were infected with MCV or vWR at the indicated moi, and collected at 16 h p.i.

4. Cover the 12-well plate with clear film plate protectors to stop evaporation and incubate with agitation on a belly-dancer at RT for 15 min. Plates are then frozen at –20°C for at least 15 min or stored overnight or for up to 2 weeks before assayed.
5. Cell lysates are tested for luciferase activity by adding 100 μl Dual Luciferase Assay Substrate to each well (see Note 9).
6. Luciferase activity is then measured in a FLUOStar Luminometer.
7. Data is compiled in a Microsoft EXCEL file and evaluated using standard statistical protocols (average, standard deviation, Student's *P* test). A typical result is shown in Fig. 2e (see Notes 6, 10, and 11).

3.2. Quantitative PCR Assay

3.2.1. Virus and DNA Preparation

1. Incubate equal volumes of freshly thawed virions (100-μl aliquot) in 100 μl DNAse/BamHI buffer for 30 min at 37°C.
2. Extract viral genomic DNA using a HPVNA kit following the manufacturer's instructions. The control plasmid pyMCV1-E-C (see Fig. 1a) is prepared using the same procedure.
3. Determine the DNA concentration of the control plasmid using a Nanodrop-Spectrophotometer or a similar device.
4. Calculate molecule numbers using the average molecular weight of DNA molecules and Avogadro's number (6.02×10^{23} per mole). The molecular weight of a plasmid (in Daltons) can be estimated as MW of a double-stranded DNA molecule (http://www.epibio.com/techapp.asp) = (# of base pairs) × (650 Da/base pair). The plasmid pyMCV1-E-C has 29,760 bp. Thus, the molecular weight is calculated as 19.344 MDa and thus 19.344 ng of plasmid would be 6.02×10^{8} mol. The actual plasmid concentration was 21 ng/μl (±1.7) and, thus, represented 6.5×10^{8} mol/μl. From this value, the molecule numbers for the pyMCV1-E-C twofold dilution series are calculated (see Fig. 3d). The molecule numbers are then correlated to the pixel numbers of bands on a gel quantitated by ImageJ (see Fig. 3d).

3.2.2. PCR Reaction (see Note 12)

1. Prepare twofold dilutions of viral genomic DNA and plasmid control in injection-grade water and store at –20°C.
2. Prepare PCR assays as outlined in Table 2. PCR reaction conditions are included in the table.
3. Visualize PCR bands by loading a 2% agarose gel with 10 μl from each PCR reaction and run for 1 h at 100 V (constant voltage). Stain with ethidium bromide, photograph with a digital unit, and export into a jpeg file (see Note 13).
4. To quantitate the PCR product, one can use the captured bands on the jpeg photograph with a series of identical gates

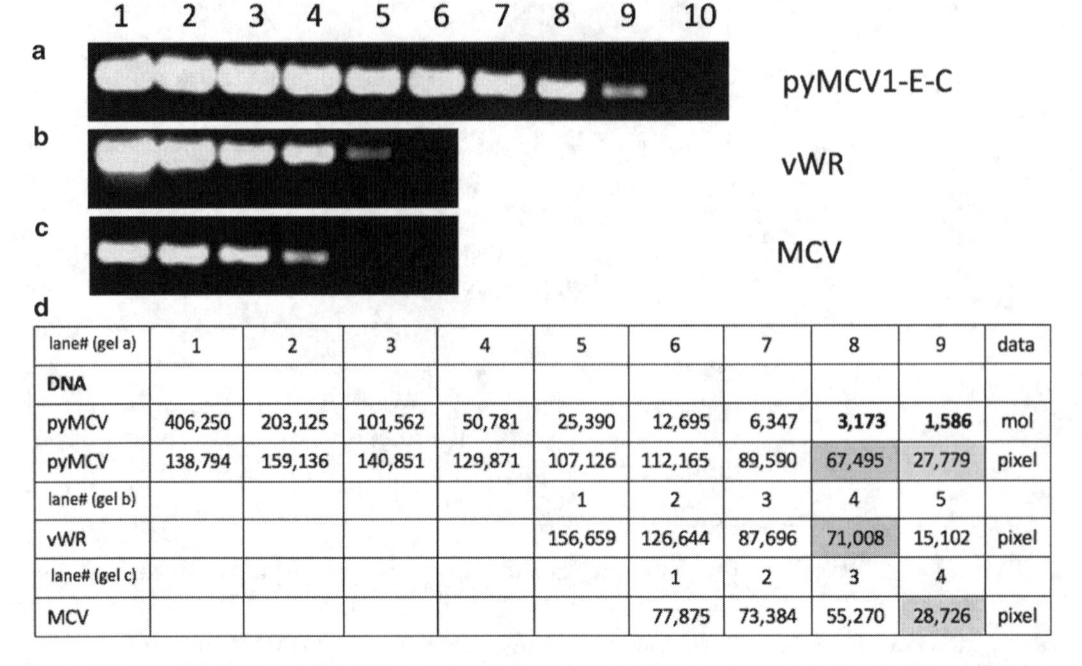

lane# (gel a)	1	2	3	4	5	6	7	8	9	data
DNA										
pyMCV	406,250	203,125	101,562	50,781	25,390	12,695	6,347	**3,173**	**1,586**	mol
pyMCV	138,794	159,136	140,851	129,871	107,126	112,165	89,590	67,495	27,779	pixel
lane# (gel b)					1	2	3	4	5	
vWR					156,659	126,644	87,696	71,008	15,102	pixel
lane# (gel c)						1	2	3	4	
MCV						77,875	73,384	55,270	28,726	pixel

Fig. 3. PCR quantification of purified DNA. *Panels* **a–c** show twofold dilutions of poxviral genomic DNA purified from DNAse-treated virions (**b** and **c**) and plasmid DNA repurified using the HPVNA kit (**a**). Lane numbers at the top of the figure refer to $\log_2$ dilutions from 1 to 10. *Panel* **d** tabulates the calculated molecular number for the reference plasmid pyMCV1-E-C (pyMC) (1,000× by nanodrop in 10 μl of original DNA prep) and the ImageJ pixels for each band in gels **a** to **c**. ImageJ (Wayne Rasband (wayne@codon.nih.gov) Research Services Branch, National Institute of Mental Health, Bethesda, Maryland, USA) was used to quantitate pixel densities in boxes of 212 × 42 pixels.

Table 2
PCR reaction

	Volume (μl)
Primer 1-2F[a] (100 pmol/μl)	0.5
Primer 1-3R[a] (100 pmol/μl)	0.5
Injection-grade water	36.8
10× PCR buffer	5.0
TaKaRa dNTP (0.2 mM)	2.0
Template (series of twofold dilutions)	5.0
AmpliTaq 360 polymerase (1 unit of 5 U/μl)	0.2
Total	50.0

PCR reaction: 2 min of denaturation at 96°C; and then 45 cycles of 1 min at 96°C, 2 min at 55°C, and 3 min at 72°C. Block and then cool to 10°C

[a]For MCV Primer 1-2F and Primer 1-3R, use Mcv129 1-2F149275 and Mcv129 1-3R149850, respectively. For WR Primer 1-2F and Primer 1-3R, use VACV-WR 1-2F132482 and VACV-WR 1-3R133054, respectively

using IMAGEJ software to produce a quantified pixel output that can be imported into a Microsoft EXCEL file.

5. Plot quantitative results of digital imaging (quantified pixel output) of a twofold dilution series of plasmid pyMCV1-E-C against molecule numbers using Microsoft EXCEL software.
6. Take molecule/genome equivalent numbers from the calibration plot and compare to the pixel readings obtained for VACV-WR. Tabulate results and evaluate using standard statistical protocols (average, standard deviation, Student's *P* test).
7. Results from a PCR reaction and the corresponding mole numbers and pixels are shown in Fig. 3d. In that figure, for gel a, lane 9, the signal (27,779 pixels) for the dilution of the MCV control plasmid (pyMCV1-E-C) correlates to 1.586×10^6 plasmid units. The band in gel c (MCV PCR) with comparable pixel density (i.e., within 10%) is in lane 4 with a pixel value of 28,726 pixels. If the 1.586×10^6 mol are multiplied by the dilution factor (16×), the MCV aliquot of 100 μl used for genomic DNA preparation contained 2.5×10^7 mol/genome units. If the pixel values obtained for vWR in gel b, lane 4 (71,008), are used in the same way and related to gel a, lane 8 (67,495 pixels), after multiplying by the dilution factor (16×), a molecule number of 5×10^7 is obtained. This can then be used to relate to the pfu. If WR was at 1.6×10^6 pfu in 100 μl, the pfu-to-genome unit ratio is 1:31 (see Notes 14–16).

4. Notes

1. All plasmid DNA should be purified using 100-μg capacity midiprep-columns (HPVNA) and then stored in elution buffer at –20°C until used.
2. The assay depends to a significant degree on the transfectability of the cell cultures involved. Human keratinocytes and fibroblast cell lines are most interesting as possible natural hosts for MCV, but they are also hard to transfect. We found HEK 293 cells to be the best transfected cell line. However, while this cell line shows robust reporter signals, it is clearly not the type of cell MCV naturally infects.
3. We prepare vaccinia virus and MCV preparations in 1 ml PBS and then immediately make ten 100-μl aliquots and freeze. The vaccinia virus stock was generated by infecting one T150 flask containing adherent HeLa cells. Harvest of the infected cells and resuspending them in 1 ml PBS yielded a titer of 2×10^7 pfu/ml. Thus, each of the ten 100-μl aliquots of vaccinia virus used here contained 2×10^6 pfu. The 100-μl aliquot MCV

contained an unknown number of MCV particles, but is quantified using the described PCR method.

4. Transfection efficiency can vary considerably from cell batch to cell batch. Passage number (best low), cell confluence (best below 60%), and time of culture prior to experiment (best no longer than 24 h) are determinant factors.
5. While in this protocol we transfect adherent cells, we have found that for some harder-to-transfect cells (e.g., human fibroblasts) we can get higher transfection efficiency when cells are in suspension.
6. The incubation time of 16 h allows for a robust signal from the transfection control plasmid p238, so firefly signal readings can be adjusted to renilla transfection efficiency readings between experiments.
7. Transfected plasmids with poxviral transcription signals can be transcribed by the poxviral transcription complex produced by transcriptionally active cores after entry. It is not clear whether the transcription complex is accessed inside partially uncoated virions, with plasmid DNA getting inside cores, or by transcription complex that is released into the cytoplasm. MCV-infected cells produce a robust luciferase signal after 16 h. However, GFP is only visibly expressed in a small number of individual cells, detectable after 5 days of incubation. Potentially, other cells may express GFP at a level undetectable by microscopy. The process where cores are accessible for the reporter plasmids may be delayed in MCV-infected cells.
8. The samples for the luciferase assay are collected at 16 h post infection. This allows the control plasmid ILR#238 to get to the nucleus and be expressed to yield a robust control signal. In 293 cells, the Renilla luciferase signal can be seen in vaccinia-infected cells after 2 h, and is seen in the MCV signal after 8 h.
9. If one does not have instrumentation that can add 100 μl of PROMEGA Dual luciferase firefly substrate one can hand, pipet series of four samples in a row, then load the plate, and read. The reading time, including the initial shake for four samples, is 20 s. Doing this results in a signal loss per reading of $<1\%$, which is less than the sample-to-sample variation in triplicate samples, when compared to machine pipetting sample per sample.
10. The signal from early poxviral promoters can be used as a surrogate parameter of viral infectivity.
11. The signal can be further dissected and used to look at early and late transcription activity using transfected plasmids with a reporter gene behind the respective promoters in isolation.

12. The conventional PCR assay described has the problem of assay-to-assay variation due to agarose gel and staining artifacts. Future advances may be the development of a real-time PCR assay using molecular Taqman probes specific to either VAVC-WR or MCV and binding in the internal section of the rather large PCR product (550 bp).
13. The source template of the PCR products produced can also be determined by XhoI digest, which cleaves the MCV product into 227- and 349-bp subfragments but does not cleave the VAVC-WR PCR product.
14. It is unclear to which extent the different GC content of the two virus genomes would affect the PCR product. This was not further investigated.
15. It is clear that vaccinia plaque-forming units cannot be directly compared to MCV virion units because of the different nature of their biological activity. However, the PCR method described in Subheading 3.2 allows a relative quantification of MCV genome equivalents to VAVC-WR infectious units measured in pfu/ml based on amplifiable genomic DNA units/molecule numbers calculated for a relatively large plasmid containing 25,517 bp of MCV sequence. As described in Subheading 3.2.2, **step 7**, the pfu-to-molecule ratio for vaccinia virus (mature virions) comes out as 1:31, in keeping with previously published ratios (12). The PCR data can be used to calculate an MCV multiplicity of infection equivalents in relation to the control plasmid molecule numbers as well as in the form of pfu equivalents in relation to a titered vaccinia stock for comparison purposes. We have found this approach to be both more reproducible and more specific than electron microscopy or OD quantifications of virions.
16. The biological activity of virions can be assessed using an in vitro transcription reaction (1, 5, 13).

References

1. Bugert JJ, Lohmuller C, Darai G (1999) Characterization of early gene transcripts of molluscum contagiosum virus. Virology 257:119–129
2. Bugert JJ, Melquiot N, Kehm R (2001) Molluscum contagiosum virus expresses late genes in primary human fibroblasts but does not produce infectious progeny. Virus Genes 22:27–33
3. Stefan Mohr S, Grandemange S, Massimi P, Darai G, Banks L, Martinou J-C, Zeier M, Muranyi W (2008) Targeting the retinoblastoma protein by MC007L, gene product of the molluscum contagiosum virus: detection of a novel virus–cell interaction by a member of the poxviruses. J Virol 82:10625–10633
4. Levy O, Oron C, Paran N, Keysary A, Israeli O, Yitzhaki S, Olshevsky U (2010) Establishment of cell-based reporter system for diagnosis of poxvirus infection. J Virol Meth 167: 23–30
5. Melquiot NV, Bugert JJ (2004) Preparation and use of Molluscum Contagiosum Virus (MCV) from human tissue biopsy specimens. In: Isaacs SN (ed) Vaccinia virus and poxvirology: methods and protocols, 1st edn. Humana, Totowa, NJ, pp 371–383

6. Bugert JJ, Darai G (1991) Stability of molluscum contagiosum virus DNA among 184 patient isolates: evidence for variability of sequences in the terminal inverted repeats. J Med Virol 33: 211–217
7. Senkevich TG, Bugert JJ, Sisler JR, Koonin EV, Darai G, Moss B (1996) Genome sequence of a human tumorigenic poxvirus: prediction of specific host response-evasion genes. Science 273:813–816
8. Senkevich TG, Koonin EV, Bugert JJ, Darai G, Moss B (1997) The genome of molluscum contagiosum virus: analysis and comparison with other poxviruses. Virology 233:19–42
9. Blasco R, Moss B (1991) Extracellular vaccinia virus formation and cell-to-cell virus transmission are prevented by deletion of the gene encoding the 37,000-Dalton outer envelope protein. J Virol 65:5910–5920
10. Blasco R, Moss B (1995) Selection of recombinant vaccinia viruses on the basis of plaque formation. Gene 158:157–162
11. Bengali Z, Townsley AC, Moss B (2009) Vaccinia virus strain differences in cell attachment and entry. Virology 389:132–140
12. Payne LG, Kristensson K (1982) Effect of glycosylation inhibitors on the release of enveloped vaccinia virus. J Virol 41:367–375
13. Shand JH, Gibson P, Gregory DW, Cooper RJ, Keir HM, Postlethwaite R (1976) Molluscum contagiosum—a defective poxvirus? J Gen Virol 33:281–295

Chapter 9

An Intradermal Model for Vaccinia Virus Pathogenesis in Mice

Leon C.W. Lin, Stewart A. Smith, and David C. Tscharke

Abstract

Intradermal injection of vaccinia virus in the ear pinnae of mice provides a model of dermal infection and vaccination. The key features of this model are the appearance of a lesion on the surface of the ear that can be measured as a clinical sign of disease and substantial growth of virus in the infected skin in the absence of systemic spread. In addition, infected ears can be easily removed to allow virological, histological, and cellular analyses. Finally, evaluation of the roles of virus (and presumably also host) genes in vaccinia virus pathogenesis in the intradermal model can yield different results than similar experiments using other routes and may reveal otherwise unknown functions.

Key words: Vaccinia virus, Pathogenesis, Mouse models, Intradermal, Skin infection, Immunopathology, Dermal vaccination

1. Introduction

The natural host and therefore site of replication for vaccinia virus (VACV) remain unknown, but laboratory studies have shown that the virus has a wide host range in vitro and in vivo. Recent outbreaks of diseases caused by VACV-like viruses in Brazil and India involving skin infections of livestock and humans suggest that the virus has a natural niche that can include cutaneous infections in multiple species (1, 2). In addition, VACV has been historically administered by skin puncture as a vaccine. These suggest that it is important to understand the consequences of skin infection with this virus. The mouse is the most practical choice of species for modeling virus infections where possible. This is not only due to the availability of facilities and relatively low cost, but also

Stuart N. Isaacs (ed.), *Vaccinia Virus and Poxvirology: Methods and Protocols*, Methods in Molecular Biology, vol. 890,
DOI 10.1007/978-1-61779-876-4_9, © Springer Science+Business Media, LLC 2012

because of the wide range of well-defined strains including transgenics and gene knockouts. In addition, VACV has now been detected in mice during outbreaks from natural sources (3), so mouse models may be more relevant than has been thought in the past. An ideal model of infection should allow many parameters of infection to be monitored, including clinical signs, growth of virus, and histology of infected tissues. Inoculations should also be relatively quick and the inoculation process as well as the ensuing infection should cause as little distress to the mice as possible. For these reasons, a site that does not require shaving is best and when considering sites of inoculation like the foot, tail, or ear pinnae, the latter is preferred.

The ear has been used as a model for immunology for many years. It also has a history of use as a route of infection for viruses such as herpes simplex virus and adenovirus (4–6) and was first published as a site of infection for VACV in the early 1990s (7). An intradermal model for VACV infection using mouse ears was then characterized more carefully some years later (8, 9). As a model of pathogenesis, the method was first established in BALB/c mice and the sign of infection that was monitored was the lesion that forms on the surface of injected ears. Virus titers and infiltrating leukocytes can also be measured. These aspects of infection are sensitive, to varying degrees, to such things as virus strain, dose, mouse strain, and age. The most important point to stress is that the intradermal model reveals aspects of pathogenesis not seen with the intranasal model, which has been used for many decades (10, 11). This is best demonstrated by comparing the range of VACV single-gene deletion mutants that have been found to have inconsistent pathogenesis phenotypes when examined using the different models (9).

Perhaps the biggest barrier to the use of the intradermal model is that while the injections are quick for experienced workers, they can be technically demanding to perfect. In addition, the scoring of lesions and best equipment for grinding the skin to titrate virus can be a concern. Therefore, this chapter focuses on these three aspects of the model, using photographs as well as descriptions.

2. Materials

2.1. Intradermal Injections of Ear Pinnae

1. Hamilton 1710LT 100 μl SYR glass syringe with Teflon plunger (see Note 1 and Fig. 1).
2. A "rubber thumb" or page turner, from a stationery supplier (see Fig. 1).
3. 27-Gauge needles (see Note 2 and Fig. 1).
4. VACV working stock, preferably sucrose cushion purified with a titer greater than 1×10^9 plaque forming units (pfu)/ml.

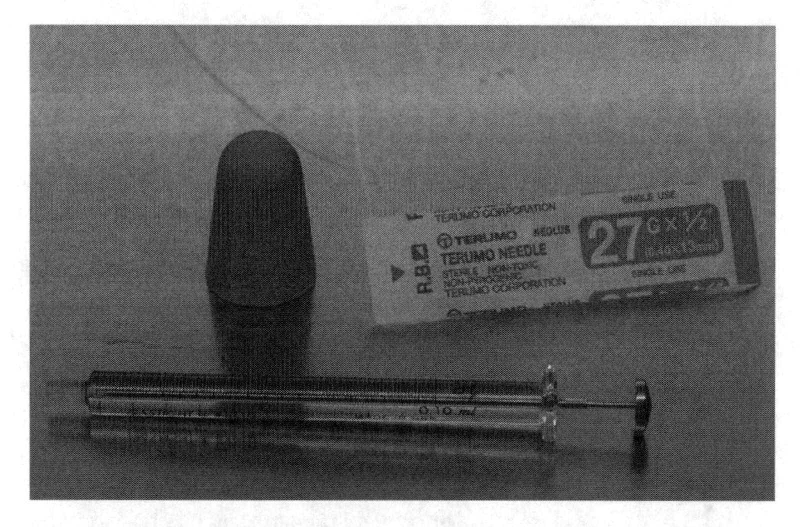

Fig. 1. Equipment for intradermal injections: "rubber thumb" turned inside out, 27-gauge needle, and 100-μl "gastight" Hamilton syringe.

5. Phosphate-buffered saline (PBS) for diluting the virus.
6. Isoflurane and the associated apparatus for induction of anesthesia or alternatively the injected anesthetic of choice.
7. Eight-week-old female BALB/c mice, preferably with a narrow range of ages (see Note 3).
8. 80% Ethanol (v/v).

2.2. Measuring Lesions on Infected Ears

1. Vernier caliper (engineer's micrometer, from a hardware store).

2.3. Titrating Virus from Ears

1. Fine dissection scissors.
2. Tissue grinders, tapered 1 ml with ground glass (not Teflon) pestle (e.g., Wheaton #358103).
3. Cup-horn sonicator.
4. Dulbecco's modified Eagle's medium (DMEM) with 1/100 dilution of penicillin/streptomycin stock, 1× L-glutamine, and either 2% or 10% fetal bovine serum (FBS), referred to as D2 or D10, respectively.
5. BS-C-1 cells (monkey kidney cells, ATCC #CCL-26).
6. Tissue culture grade 6-well plates.
7. Sodium carboxymethylcellulose (CMC): Medium viscosity (e.g., Sigma–Aldrich #C4888-500G).
8. Crystal violet staining solution (0.1% in 15% ethanol). Make this working solution by mixing 20 ml of 2.3% crystal violet solution, 70 ml of ethanol, and 410 ml deionized water (see Note 4).
9. 80% Ethanol (v/v).

3. Methods

3.1. Intradermal Injections of Ear Pinnae

1. Depending on the experience and consistency of the person doing injections and the magnitude of any difference in lesions, group sizes of between 6 and 10 mice should be used. However, it is generally better to use six closely age-matched mice than 8–10 mice where the age might vary by 2 or 3 weeks (see Note 3).
2. Prepare the virus inoculum by diluting the VACV stock with PBS to the desired concentration. Most commonly for pathogenesis experiments, this will be 1×10^6 pfu/ml so that injections deliver 1×10^4 pfu in a 10 μl volume. For determination of virus growth in ears, a lower dose of 1×10^3 pfu gives a larger window to measure virus growth, which reaches a maximum of up to $\sim 1 \times 10^7$ pfu/ear for VACV strain WR irrespective of starting dose.
3. Fill the glass syringe before fitting the needle. To do this, and also remove the air bubble from the void space, draw up a volume of around 30 μl into the barrel and then rapidly expel it by depressing the plunger fully while the end of the syringe remains immersed in the virus suspension. This is repeated several times until all air has been removed. Draw up the required amount of virus (see Note 5) along with an additional 30–40 μl (this is the volume required to fill the void space in the needle) and fit a 27-gauge needle to the syringe. Depress the plunger carefully to fill the void space in the needle with virus suspension. It is important that the virus suspension is just visible at the bevel in the needle to avoid injecting air.
4. Anesthetize the mouse. We use isoflurane at 4% in oxygen with a flow rate of 800 ml/min delivered into a Perspex box. As the mouse goes under, at first, its breaths are quite rapid (and if the mouse is removed from the box at this time, it will wake too quickly). So wait for the mouse to begin a slower breathing pattern. You will then have around a minute of anesthesia, which is enough time to inject an ear (see Note 6).
5. During injections, safety glasses and a mask should also be worn and work be carried out in a class II biosafety cabinet (see Note 7).
6. Assuming the operator is right-handed, turn the rubber page turner inside out and fit onto the left thumb.
7. Once the mouse is anesthetized, to inject the posterior (back) surface of the left ear pinna, place the mouse on its abdomen with the head away from the operator. The operator rotates their wrist so that the thumb (covered by the rubber page turner) is placed in the inside (anterior) of the ear. The forefinger is then

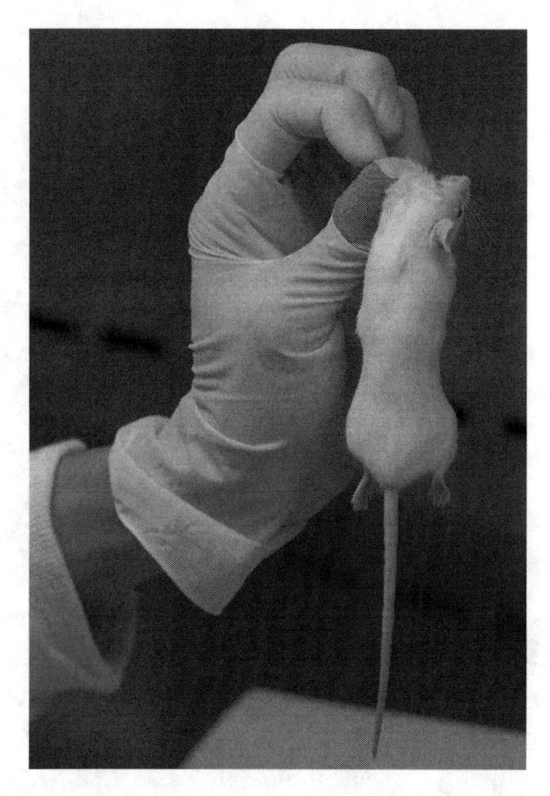

Fig. 2. Holding the mouse to get it into proper position for intradermal injection into the posterior (*back*) of the ear. The thumb (protected by a rubber page turner) is placed inside the ear, the forefinger is then used to pinch the edge of the pinna against the thumb, and the mouse is gently lifted straight up so the weight of the mouse stretches the pinna over the rubber on the thumb.

used to pinch the edge of the pinna against the thumb and the mouse is gently lifted straight up so the weight of the mouse stretches the pinna over the rubber on the thumb (see Fig. 2). This creates a smooth, taut surface to inject and the rubber protects the thumb from needlestick. Doing this correctly is absolutely critical to successful injection (see Notes 8–10).

8. Slide the needle, bevel up, almost parallel to the posterior (back) surface of the ear, starting around one-third of the way from the closest edge (see Fig. 3). Allow the needle to enter the skin (this often happens easily due to the slightly convex shape made by the rubber page turner under the ear) and insert until the whole bevel is under the epidermis by 0.5 mm. Take care not to go too shallow or too deep but follow the surface of the ear (see Note 11).
9. Inject the virus by gently depressing the plunger of the syringe. The epidermis will begin to separate from the underlying cartilaginous tissue and an epidermal bubble is easily seen as this

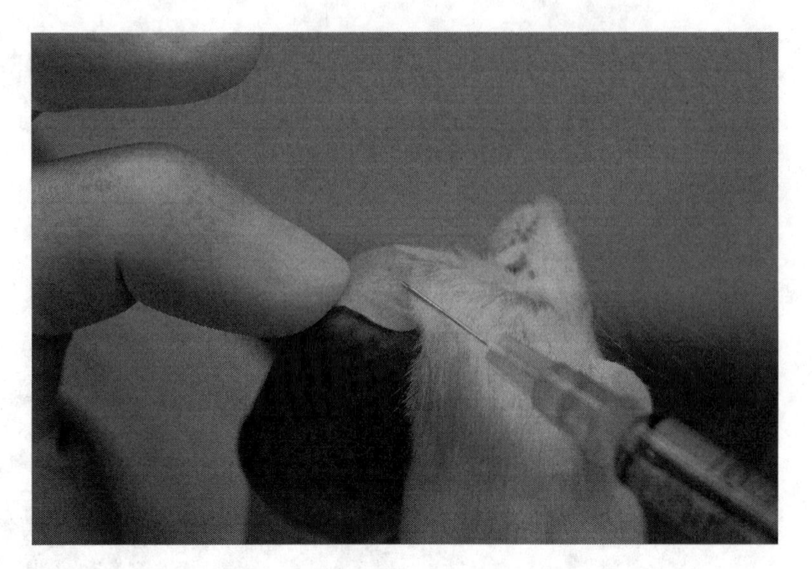

Fig. 3. Holding mouse with needle in position along the posterior (*back*) surface of the ear. The needle could be inserted where shown or anywhere up to 2 mm to the left.

happens. Inject the desired amount of virus, generally 5 or 10 μl (see Notes 12 and 13).

10. Wait a moment after injecting to ensure that the injected virus is not under too much pressure and then withdraw the needle. Usually, a small amount of liquid will come out of the needle hole, but this is to be expected.
11. Put the mouse down, blot away any virus suspension on the surface of the ear, and allow the mouse to recover.
12. The procedure for injecting the right ear pinna is similar, but the mouse is placed so that the head is closest to the operator and its right ear is in a good position to be grasped by the left hand.
13. When finished with the syringe, take extra precautions when removing the needle from the syringe for disposal into a sharps container. Then, rinse and sterilize the syringe with 80% ethanol (v/v).

3.2. Measuring Lesions on Infected Ears

1. Mice should be observed every day after infection and data collected for each individual mouse.
2. The estimation of lesion size can be somewhat subjective, so ideally the person doing the scoring should be blinded to the treatment groups.
3. When a dose of 1×10^4 pfu is used in 8-week-old BALB/c mice, the ear becomes red and thickened from around day 4 after infection and the lesions begin to be visible from day 5 or day 6 (see Note 14). A lesion is shown in Fig. 4 and color

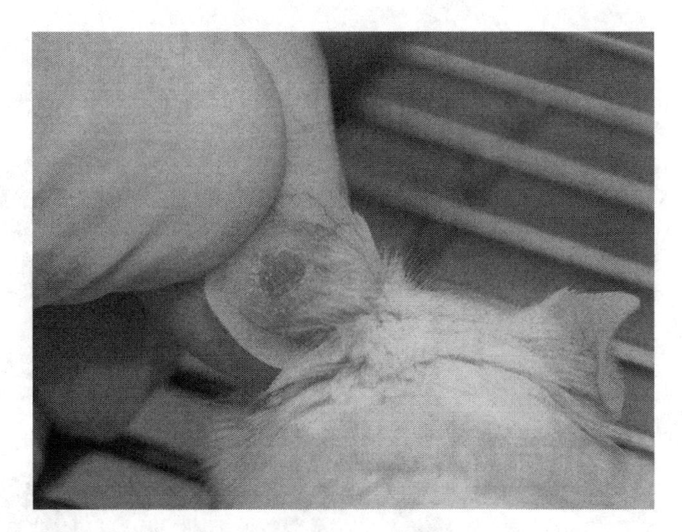

Fig. 4. Appearance of a lesion on the posterior surface of the ear on day 8 after infection of a BALB/c mouse infected with 1×10^4 pfu of VACV strain WR.

photos of lesions and histology at various times after infection have been published (8).

4. First, check for lesions on the posterior (back) and anterior (inside) surfaces of the injected ear. Take care to look for a clear break of the skin and do not score if the pinnae are just red. Often, lesions start with what looks like a pinprick that glistens in the light before a scab begins to form.
5. Assuming a lesion on the posterior (back) of the ear, hold the mouse using the left hand, with its tail between the little and ring fingers and using the thumb and index finger to hold the ear out so that lesions are clear to see (see Fig. 5 and Note 15).
6. With the right hand, use a Vernier caliper as to estimate the diameter of the lesion (see Note 16).
7. If more than one lesion is seen or if lesions are irregular, estimate the diameter of a single round lesion large enough to contain all the affected areas (see Note 17).
8. Lesions on the anterior (i.e., inside) side of pinnae are scored using a similar method as above, but it is easier to see the underside of the ear if the mouse is held with the right hand. If both surfaces were affected, only the larger diameter is recorded (see Note 18).
9. As the lesions begin to resolve, the scabs will fall off. If the skin is completely healed, then a score of zero can be recorded. If a region of broken skin remains (usually seen as a new scab), score the size of this area. Take care to distinguish scabs and scar tissue (see Note 19).

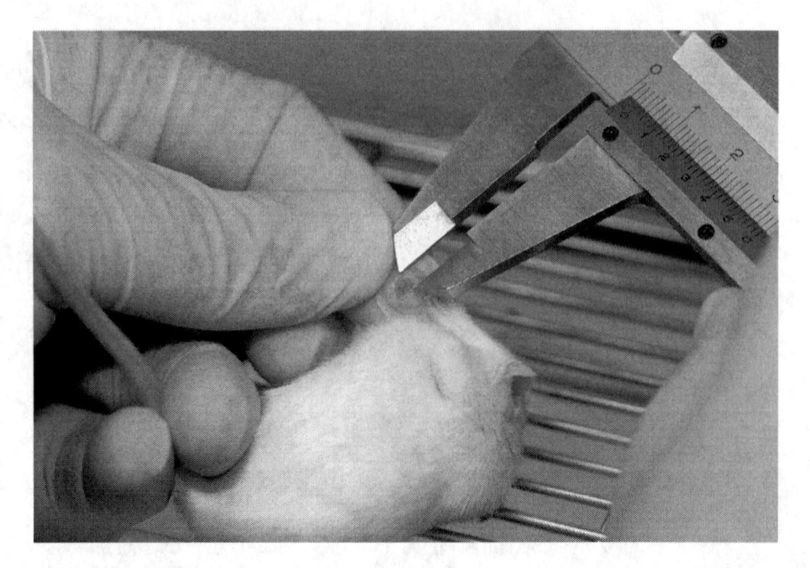

Fig. 5. Estimating lesion size. A way the mouse can be restrained and measurement taken with a caliper.

3.3. Titrating Virus from Ears

1. Euthanize the mouse and remove all of the ear pinnae using dissection scissors that have been sterilized by dipping in 80% ethanol and then dried. Take care to avoid including any fur.
2. Place the ear in 0.5 ml of D2 medium in a 1.5- or 2-ml vial.
3. Either freeze for later processing or keep on ice for immediate grinding.
4. Transfer all the contents of a vial (ear skin and medium) into the mortar of a 1-ml tissue grinder and then use the pestle to grind the skin against the tapered part of the grinder. Continue to grind until the ear is largely all homogenized and further grinding does not reduce the amount of solid left (see Note 20).
5. Once the grinding is done, transfer the homogenate back into the vial and subject it to three rounds of freezing and thawing. Samples may be stored frozen at this point.
6. The 6-well plates of cells required for the titration are prepared one day earlier by seeding BS-C-1 cells at 2×10^5 cells per well in 2 ml D10 and incubating at 37°C in a 5% CO_2 atmosphere.
7. Thaw samples and sonicate for 30 s using a cup horn sonicator (see Note 21). Prior to titering, mix samples well by pipetting up and down (use a filtered tip) and then keep the homogenates on ice.
8. Do a series of tenfold serial dilutions of each sample (typically diluting 100 μl of sample into 900 μl of D2 medium in a 5-ml disposable serum tube), vortexing to mix after each step (see Notes 22 and 23).
9. Aspirate growth medium from the BS-C-1 cells in the 6-well plate, taking care not to damage the monolayer.

10. Add 0.5 ml of sample dilutions to the wells, starting from the most dilute. This allows one to rapidly transfer volume from the dilution tubes to the wells because the same tip can be used for all dilutions of a sample. It is essential that this is done quickly so the cells do not dry, but gently so the monolayer is not damaged by pipetting liquid too vigorously into the well (see Note 24).
11. Incubate plate at 37°C, 5% CO_2, for 90 min with rocking every 10–15 min to ensure that the monolayer does not dry out and that virus falls evenly across the well.
12. Aspirate the inoculum, replace with 2 ml of D2 containing 0.4% CMC (see Note 25), and incubate for 3 days at 37°C, 5% CO_2.
13. Aspirate the D2 with CMC, replace with 1 ml of crystal violet staining solution, and leave at least 10 min.
14. Pipette off stain into a crystal violet waste bottle and leave wells to dry.
15. Count plaques, using a well with between 20 and 100 plaques. Calculate the titer (pfu per ear) based on the dilution factor and remembering that the whole ear was processed in 0.5 ml originally and 0.5 ml of each dilution was plated into each well.

4. Notes

1. The use of a glass Hamilton syringe allows more accurate injection doses, but if one is not available a disposable 1- or 0.5-ml syringe can be used and the dose estimated (see Note 12). A glass syringe is much heavier than the disposable plastic ones, so extra care needs to be taken in handling and injecting. A model with a Teflon tipped plunger (e.g., Hamilton #81001) is superior to syringes with simple stainless steel plungers as the latter tend to wear on the inside of the barrel and jam.
2. A 27-gauge needle with a regular bevel gives the best compromise between being thin enough to help in the delicate nature of the injection while being broad enough to separate the epidermal layer of the ear from the underlying tissue. These features make it possible to begin injecting the virus suspension. While it is tempting to try a finer needle or a shorter bevel (which makes it easier to insert in the right place in the ear), injecting the liquid becomes more difficult. Some manufacturers sell needles with an "intradermal bevel." But these are for human applications and may be difficult to use in the mouse ear pinnae model.
3. BALB/c mice are greatly preferred for pathogenesis experiments, where lesion size is measured. The reasons for this are practical, being that the skin has no pigmentation making lesions

clearer to see and the mice are very docile, which makes measurement of lesions easier. Lesion size on BALB/c mice tends to be a little smaller than on C57Bl/6 mice for the same dose of virus, but this causes no problems (9). The age of mice affects lesion size with older mice having smaller lesions (9). For this reason, it is important to age-match groups very closely and to limit the range of ages of mice used in any single experiment. We have had no experience with the model in male mice.

4. Crystal violet is highly toxic. Thus, buying a 2.3% solution avoids working with the powder.
5. We usually fill the syringe with enough volume to deliver up to seven injections. The number of injections depends upon how big your hand is and if you are comfortable injecting when the plunger is nearly all the way out. When one needs to refill the syringe rather than refilling it with the needle in place, it is best to refill the syringe and put on a fresh needle. Refilling a syringe with a used needle attached increases the risk of a needlestick injury. Refilling the syringe with the used needle in place also can blunt the needle (by sticking it into the side of the tube), which makes injections more difficult.
6. For initial experiments with the method, an injected anesthetic such as avertin or ketamine/xylazine may be preferred as this gives a very much longer period of anesthesia reducing the time pressure or risk that the mouse may come around while being injected. Before starting an experiment, it is highly recommended that some practice is done using mice culled for other purposes.
7. The biosafety cabinet provides an additional shield from the possibility of fluid squirting back out of the needle hole while injecting due to high pressure. The Biosafety cabinet also allows the arms to be held steady by resting against the bottom edge of the glass front of the cabinet.
8. Using the thumb (covered by the rubber page turner) provides a large flat surface to spread the ear for injection. An alternative hand position used by other labs to inject the back of the left ear is to again position the mouse pointing away from the operator and place the left index finger (covered by the rubber page turner) into the inside of the ear. The thumb is then used to pinch the edge of the pinna against the index finger to hold the ear and stretch the pinna over the rubber on index finger.
9. If there are any folds in the skin, put the mouse down and start again. If the mouse is not picked up so that it is suspended by its ear, usually the ear skin will not be taut enough to inject.
10. We usually do one ear per mouse since if one injects both ears on the same mouse, the statistical analysis becomes more complicated since the ears are not entirely independent.

11. To keep the syringe stable during injection, the syringe can be held in a variety of ways. One method that has worked well for us is to lay the barrel of the syringe across the index, middle, and ring fingers (which are held flat), and hold the syringe in place with the thumb and operate the plunger with the little finger.
12. If virus immediately begins to come out from the site of injection, the needle must be inserted further. If the needle has pierced the bottom of the ear, this is seen easily as the color of the rubber page turner becomes clearer and darker as a layer of liquid gets between the ear skin and the rubber. If at first it is not possible to inject liquid, the best approach is to move the needle a little further into the ear to separate the epidermis a little more. As noted above, if too much pressure is created, virus can squirt back out of the needle hole.
13. The size of the dermal bubble can be used as an estimate of the volume injected if a finely graduated glass syringe is not used.
14. At early times after infection, it is possible to also measure the thickness of ears (as has been done classically in models of inflammation). However, at later times, this cannot be done accurately because the scab that forms on lesions can make it very difficult to measure ear thickness with any accuracy.
15. Many strains, for example C57Bl/6 mice, will not tolerate having their ear held in the manner described and will bite. To estimate lesions with this strain, hold their tails as would be done for other methods and use the caliper to gently push the ear back and estimate the lesion size. Alternatively, anesthetize them lightly for making these estimations.
16. It is not practical to score lesion sizes more finely than to the nearest 0.5 mm.
17. The process of deciding what score to give to irregular lesions can seem arbitrary, but with appropriate group sizes and measurements being taken everyday, data produced with this method is generally very clear.
18. The reason that lesions sometimes end up on the inside surface (when the injection was done on the posterior (back) side of the ear) is if the needle goes in a little too deep. This will occur less frequently as the person gets more experienced performing the injections.
19. Sometimes, after larger lesions that involve both sides of the ear resolve, the scab falls off to leave a hole in the ear. In these cases, take care to ensure that the edges around the hole have healed and if not continue to score an unhealed edges as lesions until there is no longer any scabs or broken skin. Do not score the hole as a lesion.

20. It is not possible to completely homogenize all of the tissue as the keratinized layer and the cartilaginous material in the middle of the ear are very tough and elastic. However, the virus is not associated with these parts and so the presence of this material after grinding does not compromise titers. Aim for consistency when grinding.
21. While the sound waves generated during sonication do not damage the virus and reduce titers, the sonication process generates heat and if excessive heat is generated, this can reduce viral titers. To minimize this, the water in the cup of the sonicator should be cooled by adding ice and letting it melt into the water for several minutes (ensure that there is no ice between the sample and the horn that emits the sound waves). Also if samples are thawed in a 37°C water bath, return them to an ice bucket for a few minutes to cool down before sonicating. If samples are refrozen, they need to be sonicated each time they are thawed to break apart virus clumps.
22. The number of dilutions required depends on the virus used, dose, and the time after infection. As a guide, when using VACV strains such as WR, expect to find up to 1×10^7 pfu per ear from days 4 to 8 in ears inoculated with doses from 1×10^2 to 1×10^6 pfu. At later times, titers become highly variable because most of the infectious virus becomes bound up in the scabs, which fall off at different times for different mice. If ears are removed within a few minutes of injection and virus titered, the amount of virus recovered is around tenfold less than the injected dose. This likely reflects inefficiency in the method as well as rapid drainage of much of the injected dose by the lymphatic system.
23. Usual good practice for diluting VACV applies here: (a) always use filtered tips and change tips between dilutions; (b) after taking up virus suspensions, touch the tip to the side of the tube to drain virus on the outside of the tip; (c) when dispensing, do not stick the tip into the medium, but touch the side of the tube and eject; and (d) once virus is ejected, do not use that tip to mix or "washout" the tip.
24. It is possible to plate quite thick homogenates onto cell monolayers without destroying the cells. We have found that even a 1:2 dilution is usually not a problem. Generally, with the amounts of virus expected, dilutions from 1×10^{-2} to 1×10^{-7} cover the range of virus expected at all but late times.
25. To make 500 ml of D2 with 0.4% CMC, dissolve 2 g of CMC in 50 ml PBS in a 500-ml Pyrex bottle by shaking or stirring for at least 24 h. Autoclave this very gelatinous mixture and once cooled, add D2 medium to a total volume of 500 ml and mix well by swirling for several minutes.

References

1. Singh RK, Hosamani M, Balamurugan V, Bhanuprakash V, Rasool TJ, Yadav MP (2007) Buffalopox: an emerging and re-emerging zoonosis. Anim Health Res Rev 8:105–114
2. Ferreira JM, Drumond BP, Guedes MI, Pascoal-Xavier MA, Almeida-Leite CM, Arantes RM, Mota BE, Abrahao JS, Alves PA, Oliveira FM, Ferreira PC, Bonjardim CA, Lobato ZI, Kroon EG (2008) Virulence in murine model shows the existence of two distinct populations of Brazilian vaccinia virus strains. PLoS One 3:e3043
3. Abrahao JS, Guedes MI, Trindade GS, Fonseca FG, Campos RK, Mota BF, Lobato ZI, Silva-Fernandes AT, Rodrigues GO, Lima LS, Ferreira PC, Bonjardim CA, Kroon EG (2009) One more piece in the VACV ecological puzzle: could peridomestic rodents be the link between wildlife and bovine vaccinia outbreaks in Brazil? PLoS One 4:e7428
4. Hill TJ, Field HJ, Blyth WA (1975) Acute and recurrent infection with herpes simplex virus in the mouse: a model for studying latency and recurrent disease. J Gen Virol 28:341–353
5. Nash AA, Field HJ, Quartey-Papafio R (1980) Cell-mediated immunity in herpes simplex virus-infected mice: induction, characterization and antiviral effects of delayed type hypersensitivity. J Gen Virol 48:351–357
6. Moorhead JW, Clayton GH, Smith RL, Schaack J (1999) A replication-incompetent adenovirus vector with the preterminal protein gene deleted efficiently transduces mouse ears. J Virol 73:1046–1053
7. Ikeda S, Tominaga T, Nishimura C (1991) Thy 1+ asialo GM1+ dendritic epidermal cells in skin defense mechanisms of vaccinia virus-infected mice. Arch Virol 117:207–218
8. Tscharke DC, Smith GL (1999) A model for vaccinia virus pathogenesis and immunity based on intradermal injection of mouse ear pinnae. J Gen Virol 80:2751–2755
9. Tscharke DC, Reading PC, Smith GL (2002) Dermal infection with vaccinia virus reveals roles for virus proteins not seen using other inoculation routes. J Gen Virol 83:1977–1986
10. Nelson JB (1938) The behaviour of poxviruses in the respiratory tract. I. The response of mice to the nasal instillation of vaccinia virus. J Exp Med 68:401–412
11. Turner GS (1967) Respiratory infection of mice with vaccinia virus. J Gen Virol 1:399–402

Chapter 10

Measurements of Vaccinia Virus Dissemination Using Whole Body Imaging: Approaches for Predicting of Lethality in Challenge Models and Testing of Vaccines and Antiviral Treatments

Marina Zaitseva, Senta Kapnick, and Hana Golding

Abstract

Preclinical evaluation of novel anti-smallpox vaccines and antiviral treatments often rely on mouse challenge models using pathogenic vaccinia virus, such as Western Reserve (WR) strain or other orthopoxviruses. Traditionally, efficacy of treatment is evaluated using various readouts, such as lethality (rare), measurements of body weight loss, pox lesion scoring, and determination of viral loads in internal organs by enumerating plaques in sensitive cell lines. These methodologies provide valuable information about the contribution of the treatment to protection from infection, yet all have similar limitations: they do not evaluate dissemination of the virus within the same animal and require large numbers of animals. These two problems prompted us to turn to a recently developed whole body imaging technology, where replication of recombinant vaccinia virus expressing luciferase enzyme (WRvFire) is sensed by detecting light emitted by the enzyme in the presence of D-luciferin substrate administered to infected animal. Bioluminescence signals from infected organs in live animals are registered by the charge-coupled device camera in IVIS instrument developed by Caliper, and are converted into numerical values. This chapter describes whole body bioimaging methodology used to determine viral loads in normal live BALB/c mice infected with recombinant WRvFire vaccinia virus. Using Dryvax vaccination as a model, we show how bioluminescence data can be used to determine efficacy of treatment. In addition, we illustrate how bioluminescence and survival outcome can be combined in Receiver Operating Characteristic curve analysis to develop predictive models of lethality that can be applied for testing of new therapeutics and second-generation vaccines.

Key words: Vaccinia virus, Whole body bioimaging, Dryvax vaccine, Receiver Operating Characteristic curve analysis, Predictive models of lethality

Stuart N. Isaacs (ed.), *Vaccinia Virus and Poxvirology: Methods and Protocols*, Methods in Molecular Biology, vol. 890, DOI 10.1007/978-1-61779-876-4_10, © Springer Science+Business Media, LLC 2012

1. Introduction

Bioluminescence imaging techniques have been widely used for studies of microbial and viral infections in cell lines in vitro and in vivo in animal models. Since its isolation from a cDNA library from *Photinus pyralis* in the early 1990s, the firefly luciferase gene was expressed in bacteria, plants, and animals (1, 2). The enzyme coded by luciferase gene is a single polypeptide of 550 amino acids in length (62 kDa) and is active in monomeric form. Firefly luciferase catalyzes the oxidation of D(−)luciferin in the presence of ATP-Mg^{+2} and O_2 to generate oxyluciferin, CO_2, AMP, and light that has a peak emission at 560 nm (3). This light emission can be measured spectrophotometrically (2). The currently available methods allow detection of $\geq 2.4 \times 10^5$ mol of luciferase (3).

The introduction of the luciferase gene into genome of vaccinia virus (VACV) was the first attempt to use measurements of emitted light as an alternative approach to traditional plaque assay to follow replication of the virus (4). VACV is a large DNA-containing virus, which allows for insertion of the luciferase cassette (under a VACV promoter) without affecting viral replication or alteration in pathogenesis in animal model. In addition, the recombinant VACV produces ~10,000 particles per cell and thus a significant pool of luciferase molecules are present in infected organs. Earlier studies have shown that detection of VACV in infected cells in vitro by measuring luciferase expression is 1,000-fold more sensitive than using β-galactosidase reporter system and can detect 0.01 pfu/cell in cell cultures (4). Compared to detection of other fluorescent proteins, measurements of luciferase-generated signal exhibit higher signal-to-background ratio (5). Our own data as well as reports published by other investigators have shown that measurements of photon fluxes correlated in linear fashion with viral loads measured in organs isolated from VACV-infected mice, thus supporting the notion that bioluminescence provides a direct measure of viral dissemination (6, 7).

Detection of VACV in infected animals by registering light emitted from virus-expressing luciferase provides an important advantage over other approaches: it allows for monitoring viral dissemination from the site of challenge to internal organs, and detects virus in otherwise un-accessible anatomical sites. Yet, caution should be taken in interpreting data. The spatial resolution of bioluminescent signal is limited to 2–3 mm due to diffusion of light in the tissue (8). In addition, each 1 cm of tissue causes a tenfold reduction in light intensity due to adsorption. Thus, the signal from deeper organs appears to be lower than that from the organs that are just below or on the surface. Therefore, a direct comparison of recorded bioluminescence between organs would not be accurate without correlation with viral measurements in isolated tissues. Several novel luciferase enzymes have been developed to expand emission spectra and thus permit simultaneous detection of virus and host cells or host proteins. However, care should be taken to account for more limited distribution in vivo and higher susceptibility to oxidation in serum for some of the newly developed luciferases (9).

We have utilized a thymidine kinase-positive (TK+) recombinant WR VACV (WRvFire) expressing luciferase reporter gene under the control of a synthetic immediate-early promoter kindly provided by Dr. Bernard Moss (NIAID, NIH) for lethal challenge of normal BALB/c mice (10). In the chapter, we provide the methods and describe examples of the types of data that can be generated using bioimaging to monitor and quantify WRvFire expression in infected mice. In one example, we used Dryvax-immunized mice that were protected from infection and applied *t*-test to bioluminescence recorded from individual animals to determine whether complete protection correlated with significant reduction in viral loads in internal organs. In this example, we used a cohort of 200 animals that were pretreated with human intravenous vaccinia immunoglobulin (VIGIV) in a dose-ranging study. Using known lethality outcome in 200 animals, we applied Receiver Operating Characteristic (ROC) curve analysis to identify the internal organs and days post infection when bioluminescence provided the most optimal thresholds that discriminated between surviving and non-surviving animals.

In summary, whole body bioimaging helps to greatly reduce the numbers of animals used in the experiments and allows early euthanasia (and reduced suffering). Various treatments can be evaluated using this approach, including vaccines, antiviral drugs, and passive immunotherapy with polyclonal and monoclonal antibodies in pre- and post-challenge modes.

2. Materials

2.1. Anesthesia and Infection of Mice with WRvFire

1. 75% Avertin stock solution: Add 6.5 mL tert-amyl alcohol to a 5-g vial of 2,2,2-tribromoethanol powder (see Notes 1 and 2).
2. 1.5% Avertin solution: 200 μL of 75% Avertin stock solution in 9.8 mL LPS-free PBS, filter sterilize (see Notes 3 and 4).
3. Syringe, 30 mL, and 0.2-μm syringe filter.
4. Balance to weigh mice (e.g., Scout Pro SPE202, Ohaus Corporation).
5. Container to weight mice.
6. Magnifier lamp (e.g., L9087 Full Spectrum Magnifier Desk Lamp, 3×).
7. WRvFire viral stock for infections at 10^7 pfu/mL (see Note 5).

2.2. Bioimaging of Infected Mice

1. IVIS 50 bioimager (Caliper).
2. Living Image 3.02 software (Caliper).
3. D-Luciferin Firefly, potassium salt, 1.0 g/vial (e.g., Caliper cat. no. XR-1001).
4. 15 mg/mL luciferin solution, 15 mg/mL: 1 g of D-luciferin powder dissolved in 66.6 mL of DPBS w/o Ca^{+2} and Mg^{+2} (see Notes 6 and 7).
5. 150-mL vacuum filter system, 0.2 μm.
6. Graduated cylinder.
7. Aluminum foil.
8. 1-mL syringes and 23-G 1.0-in. needles.
9. Nose cones (e.g., VetEquip, Inc. cat. no. 921612).
10. Isoflurane.
11. Charcoal paper (e.g., Art Supplies, 445-109).
12. 70% Ethanol.
13. Tex Pure disinfectant.
14. Sharpie pens.

2.3. Statistical Analysis

1. Excel software.
2. GraphPad Prism V5 software.
3. JMP 7.0 software (SAS Institute Inc., Cary, NC).

3. Methods

The methods described here detail the measurements of bioluminescence of WRvFire in internal organs of infected mice and provide the general guidance for statistical analysis that could be used to establish the efficacy of treatment. We also describe how measurements of bioluminescence could be used as an alternative to lethality or as an end point in studies of anti-smallpox interventions. An example of an experiment that compares WRvFire challenge of groups of vaccinated and unvaccinated mice is shown in Subheading 3.4. An example of an experiment that compares WRvFire challenge of groups of VIG-treated and -untreated mice is shown in Subheading 3.5.

3.1. Intranasal Infection of Mice with WRvFire

1. Weight mice.
2. Anesthetize mice by intraperitoneally (i.p.) injecting the 1.5% Avertin solution into mice at 20 μL/g body weight (approximately 400 μL per 20-g mouse) (see Note 8).

3. Once mice are anesthetized, hold them by scruff and using an illuminated magnifying lamp infect them by slowly dispensing 10 μL of the virus stock into one nostril; be sure that the entire droplet is delivered to nasal cavity (see Notes 9 and 10).
4. Rest mice for 24 h before proceeding with whole body imaging.

3.2. Whole Body Bioimaging of Mice Infected with WRvFire Using IVIS 50 Bioimager

1. Thaw an aliquot of luciferin solution at room temperature.
2. Follow instructions provided by Caliper for turning on flow of O_2 gas on the oxygen tank, for dispensing isoflurane into vaporizer, and for initializing and setting parameters for bioluminescence measurements of the IVIS 50 instrument.
3. Place a sheet of charcoal paper at the stage of the imaging chamber (see Note 11).
4. Label mice 1, 2, 3, 4, 5, 6, etc. using a Sharpie pen (see Note 12).
5. Briefly shake vial of luciferin, load a syringe and holding mouse head at ~30° angle downward, inject 200 μL luciferin i.p. into the first group of three mice (No. 1, 2, 3), and place all three mice in the isoflurane induction chamber for 5–10 min.
6. Once mice are immobile, transfer anesthetized mice into imaging chamber; lay them on their back on the stage protected with charcoal paper. Adjust the heads to allow free delivery of the anesthesia to the nostrils through nose cones inserted in the openings of the tube (see Notes 13 and 14).
7. Cover the mouse heads with another sheet of charcoal paper to avoid saturation of the charge-coupled device (CCD) camera due to high signal from the nasal cavity.
8. Adjust stage to position "D" and acquire black and white photograph of mice, check proper position of animals in the image, and adjust if needed.
9. Switch the instrument to bioluminescence mode and acquire ventral image of mice 1, 2, and 3 by setting exposure time to 2 min (see Notes 15 and 16).
10. Save carefully labeled images of each group of mice (see Note 17).
11. Turn mice on their stomach, cover heads with charcoal paper, and acquire dorsal image (see Note 15).
12. Change the stage to position "B," remove charcoal paper from heads, and acquire dorsal image of heads (see Note 18 and Fig. 1 for an example of the images obtained).
13. Open IVIS chamber, remove animals 1, 2, and 3, and transfer them back into the cage.

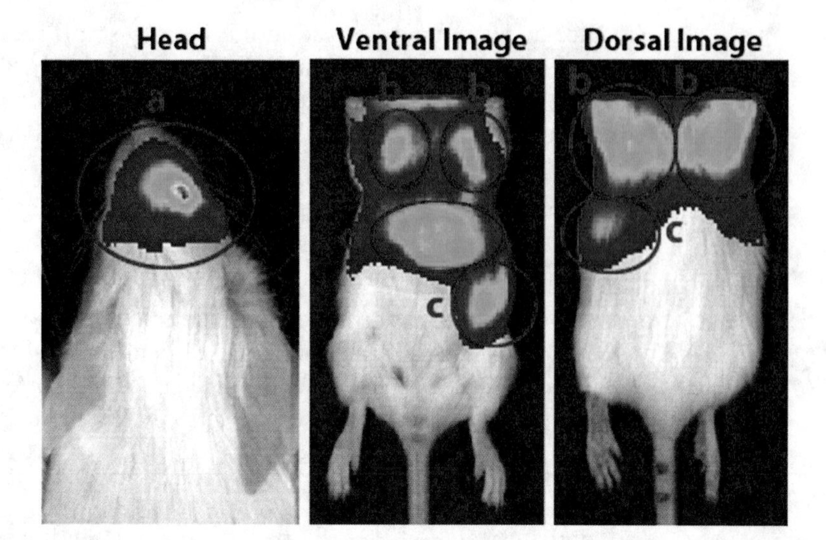

Fig. 1. Images of 5-week-old BALB/c mice infected with WRvFire. Shown are representative black and white images of the heads (*left*), ventral torso (*center*), and dorsal torso (*right*) of mice 3 days after intranasal infection with 10^5 pfu of WRvFire. Strong bioluminescence signals were noted in the nasal cavity (**a**), lungs (**b**), spleen (**c**), and liver (**d**) acquired in the head (*left panel*), ventral position (*center panel*), and dorsal position (*right panel*). *Circles* mark the Regions of Interest (ROIs) used to calculate total fluxes for subsequent analyses.

14. Repeat procedure with the rest of animals keeping the order of imaging and position in the imaging chamber constant every time during the course of the experiment (see Note 19).
15. Mice are injected with luciferin each day of the experiment, ~10 min before imaging.

3.3. Calculation of the Bioluminescence in Infected Mice and Statistical Analysis

1. To prepare data for analysis, the numerical value of the captured light is determined using Living Image software. First, images of mice that belong to the same treatment group and that were recorded on subsequent time points are downloaded using Load as Group option.
2. Using the Region of Interest (ROI) tool in the Living image software, select the area within the organ that exhibits bioluminescent signal to determine the photon/s value for each organ for each day (see Note 20).
3. A selected sequence of images is displayed automatically in a table that can be saved in a *.csv format.
4. Statistical analysis is performed using Excel software by exporting the *.csv table with the ROI data from Living Image into Excel.
5. Using Excel, calculate means and standard deviations (STDEV) of recorded photons/s values for each group of mice.

6. To determine whether the differences in total fluxes between groups of mice are significant, perform two-sample assuming equal variance *t*-test (see Note 21).
7. Record the lethality and generate Kaplan–Meier survival curves of time of death using standard GraphPad Prism V5 software.
8. Add lethality outcome to the Excel table with recorded photons/s values and export it to JMP 7.0 software to calculate area under the curve (AUC) values. (See Notes 22 and 23 and Subheading 3.5 for detailed discussion of the reason why one might need to analyze data in this way).

3.4. Example of Whole Body Bioimaging of WRvFire Infection of Mice Previously Vaccinated with the Dryvax Vaccine

As stated previously, whole body imaging provides a unique opportunity to monitor the levels of viral replication and dissemination within the same host. In this example, nine BALB/c mice per group were either unimmunized (challenge control) or vaccinated by the intraperitoneal route with 10^6 pfu of Dryvax vaccine. Two weeks later, all 18 mice were intranasally challenged with WRvFire (see Note 24). Control mice all succumbed to death within 1 week post challenge, while all Dryvax-vaccinated mice survived (data not shown). Mice were subjected to whole body bioimaging daily for 10 days or until lethality, and the recorded images were used to calculate means and STDEV for total fluxes in the nasal cavity and lungs as shown in Fig. 2. To confirm that differences in the total fluxes between control and vaccinated mice were significant, *t*-test was performed using total fluxes recorded in all animals daily for 7 days. Table 1 shows that there was a significant difference in the luciferase signal in vaccinated mice compared with unvaccinated controls in the nasal cavity at all time points and in the lungs starting from day 3 post challenge (see Note 21). There were no signals

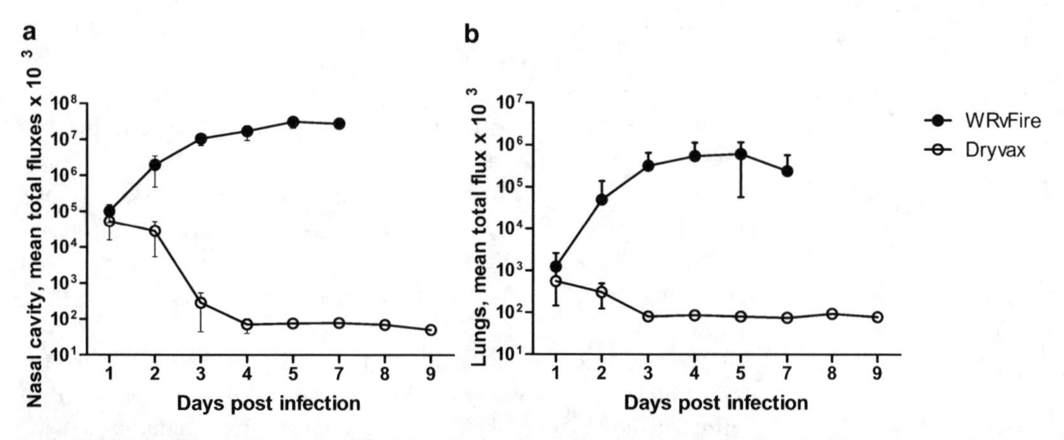

Fig. 2. Bioluminescence in the nasal cavity (**a**) and lungs (**b**) of mice challenged with WRvFire. This figure illustrates the differences in the means of total fluxes between unvaccinated control animals challenged with WRvFire (*solid black circles*, *n* = 9) and mice that were immunized with Dryvax vaccine and then challenged with WRvFire (*open circles*, *n* = 9).

Table 1
t Statistic for the total fluxes in challenged mice that were immunized with Dryvax vaccine ($n = 9$) versus unvaccinated ($n = 9$)

	t Statistic on day post infection[a]				
Location	**1**	**2**	**3**	**4**	**5**
Nasal cavity	2.1[b]	3.7[b]	8.0[c]	6.2[c]	8.4[b]
Lungs	0.3	1.5	2.7[b]	2.6[b]	3.1[b]

[a]Eighteen vaccinated or unvaccinated BALB/c mice were challenged with WRvFire (10^5 pfu). The animals were imaged daily until day 10 post infection or until they succumbed. t-Tests were performed for fluxes in internal organs recorded on days 1–5 between vaccinated and unvaccinated mice. t Critical ≥ 2.12; $n = 18$

[b]$p < 0.05$

[c]$p < 0.000005$

recorded in the spleen or the liver of Dryvax-vaccinated mice at any time point (data not shown). These data confirmed that Dryvax vaccine induces protective immunity against VACV, associated with a significant reduction of luciferase signal in the upper respiratory tract and no dissemination of the virus to spleen and liver.

3.5. Example of Whole Body Bioimaging of WRvFire VACV in Live Mice Following Treatment with Vaccinia Immunoglobulin (VIGIV): Receiver Operating Characteristic curve Analysis

The sample experiment described above shows that when a strong immune response is established that confers 100% protection from lethality, such as following Dryvax vaccination, it can be expected that total fluxes in internal organs of treated animals are significantly different compared with controls. Under these conditions, a direct comparison of means of total fluxes by t-test can be sufficient to support the efficacy of treatment. However, in other cases, the treatment may provide only partial protection (i.e., a certain percentage of animals will survive and the rest will succumb to death). Under this scenario, a direct comparison of means of total fluxes may not be optimal to assess the efficacy of treatment and predict lethality/survival outcome. In addition, it is possible that some treatments, such as treatments with immune globulins, may not prevent viral dissemination to internal organs, even in animals that survive challenge. Therefore, it is important to subject the recorded total fluxes to ROC curve analysis as described in the next example.

Two hundred BALB/c mice in six experiments were intraperitoneally treated with VIGIV at doses ranging from 0.3 to 30 mg/animal (see Note 25) and 2 days later challenged with

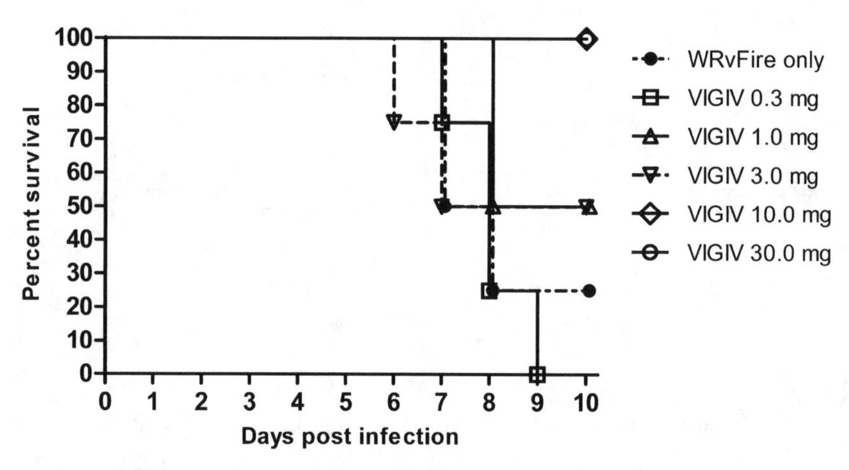

Fig. 3. Survival of WRvFire-challenged mice following treatment with VIGIV. BALB/c mice (six per group) received intraperitoneal injections of VIGIV, at doses shown in the figure, 2 days before intranasal challenge. The numbers of dead mice were recorded each day and were used to calculate the percent survival. At 1- and 3-mg doses, there is about 50% survival. At 10 and 30 mg, there is 100% survival. The data is representative of six experiments. Copyright © American Society for Microbiology (7; 10.1128/JVI.01296-09).

WRvFire. Control mice received PBS. As shown in Fig. 3, treatment with five different doses of VIGIV provided a spectrum of protection from lethal challenge, from 0 to 100%. Mice were subjected to whole body bioimaging daily for 10 days (or until lethality) and total fluxes were recorded in the internal organs. The fluxes of non-surviving ($n = 53$) and surviving ($n = 147$) animals were subjected to statistical analyses. *t*-Test confirmed that total fluxes in the nasal cavities, lungs, spleens, and livers were significantly different between surviving and non-surviving animals (data not shown). Yet, when total fluxes recorded in individual animals were plotted, a clear overlap of signals between surviving and non-surviving individual animals was noted (see Fig. 4).

To generate a lethality prediction model, bioluminescence measured in four internal organs on days 1–5 and the lethality/survival outcome were subjected to ROC analysis to determine whether any particular level of recorded total flux could serve as a threshold between surviving and non-surviving animals (see Note 22). To identify the best predictive organ and optimal time point post infection, the AUC value was calculated by plotting the percent of mice that did not survive (sensitivity) against the percent of mice that survived (1-specificity) for each recorded total flux (see Note 26). Figure 5 provides graphical illustration of the ROC analysis of the bioluminescence recorded in 200 mice in the liver on day 5, where the AUC of 0.91 was obtained.

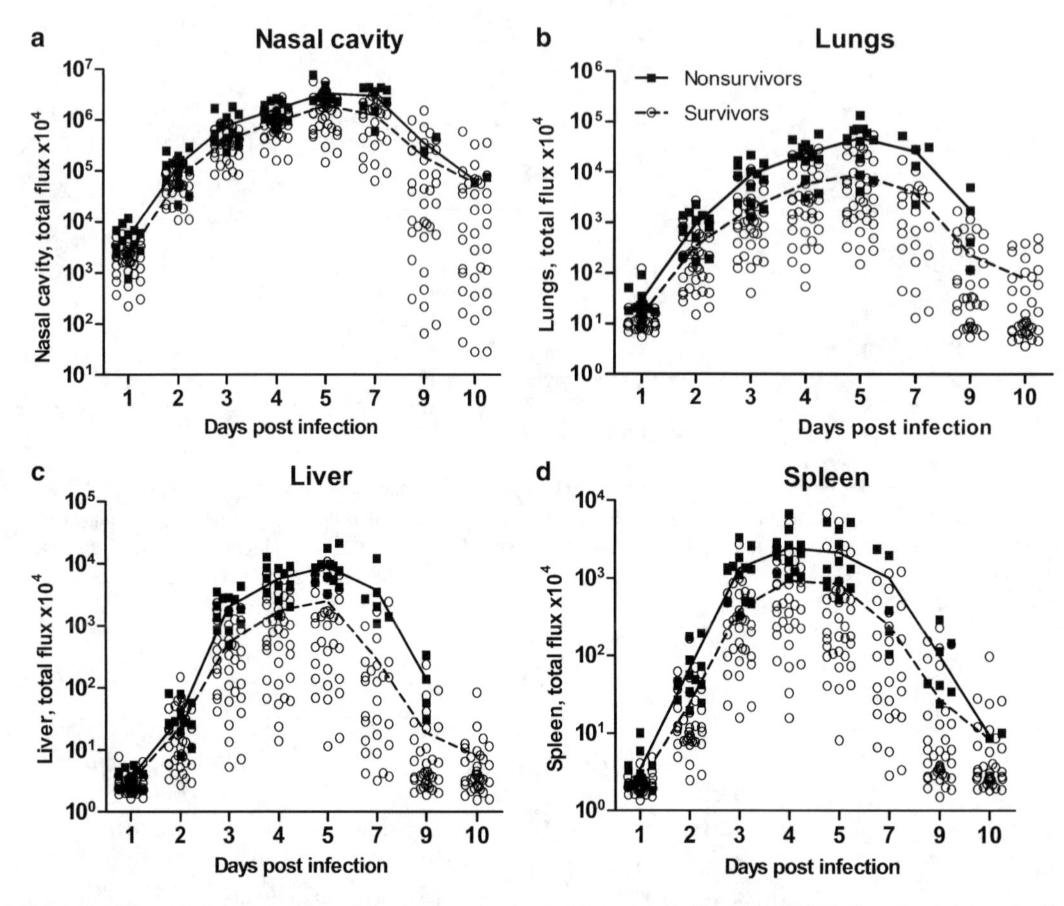

Fig. 4. Total fluxes in internal organs of surviving and non-surviving mice challenged with WRvFire after treatment with VIGIV. This figure illustrates the overlap in the bioluminescent signals between mice that were protected or not protected from the lethal challenge. Two hundred mice, inoculated with VIGIV 2 days prior to challenge, were imaged daily and followed for lethality. Total fluxes in the nasal cavity (**a**), lungs (**b**), liver (**c**), and spleen (**d**) were used to calculate means of total fluxes for surviving (*dashed line*) and non-surviving animals (*solid line*). Total fluxes of 25% of non-surviving (*solid squares*) and 25% surviving animals (*open circles*) at each time point were randomly chosen using Excel's sampling and graphed using GraphPad Prism. Copyright © American Society for Microbiology (7; 10.1128/JVI.01296-09).

Similar analyses of total fluxes were performed for all four organs for five consecutive days and showed that AUC values were >0.85 in the spleen and liver on days 3–5 (see Fig. 6). In parallel, we assessed the predictive power of 25% weight loss to discriminate between surviving and non-surviving animals (see Note 27). The ROC analysis with 25% weight loss as threshold generated AUC values of 0.85 on day 7 and >0.85 on day 9, when

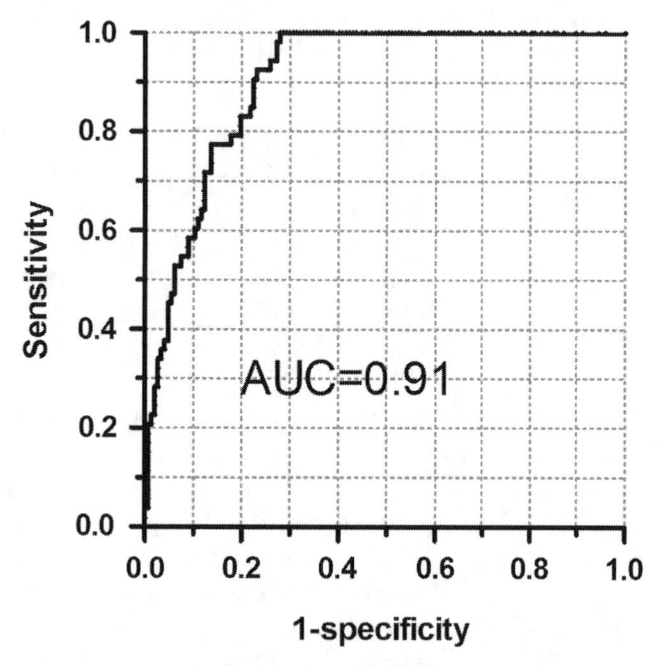

Fig. 5. ROC analysis of total fluxes recorded in livers of 200 mice on day 5 post challenge with WRvFire. This figure illustrates graphic format of ROC analysis. Total fluxes recorded in livers from non-surviving and surviving animals were used to calculate AUC.

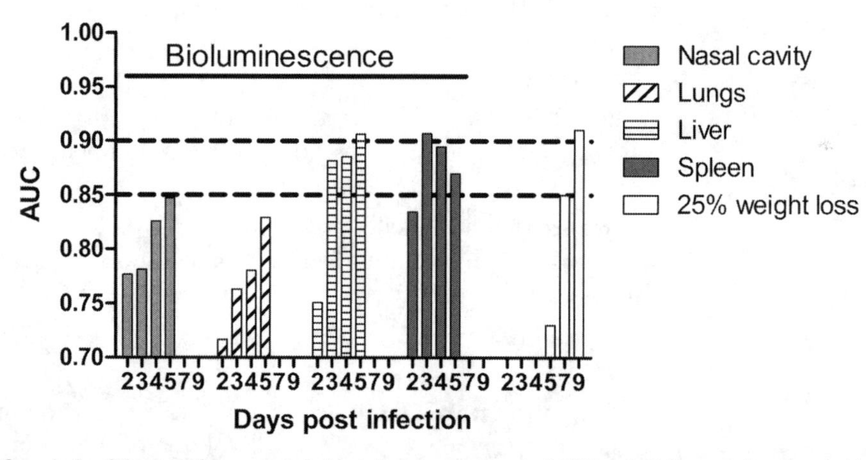

Fig. 6. ROC analysis of the total fluxes recorded in internal organs and of 25% weight loss used as thresholds to predict lethality. This figure illustrates that predictive models of lethality generated using measurements of bioluminescence provided high AUC values (0.85–0.92) and at earlier time (days 3–5) compared with models where 25% weight loss was used as threshold.

the majority of non-surviving animals were already dead. Collectively, the data showed that measurements of bioluminescence in the spleen and liver on days 3–5 post infection can accurately predict survival with higher precision and earlier than the 25% (or 30%) weight loss (see Fig. 6).

4. Notes

1. None of the chemical components used to make Avertin are schedule II drugs and, therefore, one does not need a DEA number.
2. Shake vigorously and protect from light by wrapping vial in foil. Once completely dissolved, this 75% Avertin stock solution can be aliquoted into 1-mL portions and stored at −20°C protected from light. The solution needs to be protected from light (and heat) for the stability of the tribromoethanol, which can generate toxic products upon its degradation.
3. To prepare the working strength 1.5% Avertin solution for injections, first thaw the 75% Avertin stock solution at room temperature and vortex briefly at low speed. Add PBS (if cold, warmed to room temperature), mix thoroughly using the pipette, filter through a 0.2-μm filter into a sterile tube, and protect from light by wrapping with aluminum foil.
4. The ~10 mL volume of 1.5% Avertin is enough to anesthetize approximately 25 mice.
5. Depending on the pathogenicity of the recombinant vaccinia strain, it may be advisable to first determine the dose needed to assure 100% lethality in 6–10 mice/group. Ideally, mice should succumb to death starting from day 5 to 6 and all mice should be dead by day 9–10.
6. The D-luciferin Firefly powder is stored at −20°C. It is advisable to allow powder to warm up at room temperature before dissolving it in DPBS.
7. Filter sterilize the luciferin solution with a 0.2-μm filter flask covered with aluminum foil and then aliquot into smaller volumes (e.g., 1 mL) in cryovials for storage at −20°C.
8. This amount of 1.5% Avertin will result in 30–45 min of anesthesia.
9. It is important to infect mice with WRvFire while animals are under anesthesia and to use small volume of inoculums (e.g., 5–10 μL). We deliver this volume by touching nostril with a pipette tip (easy seen while under the illuminated magnifying glass and slowly deliver the volume into the nostril, which immediately goes in). This small volume reduces animal-to-animal variability in the initial viral loads at the site of challenge (nasal cavity) and spillover to the lungs. We performed preliminary experiments, where we inoculated mice with the same 10^5 pfu of virus but in 10, 15, and 20 μL volumes and found strong bioluminescence signal in the lungs 24 h after injections with 15 and 20 μL volumes, but not 10 μL. According to

Gross et al. (11), the volume of the nasal cavity in a 7-week-old mouse (30 g) is ~30 mm^3. We infect young 5-week-old mice when they are 14–18 g, and thus assume that they have nasal cavities of the smaller volume.

10. As an example of how to prepare 10^7 pfu/mL viral stock, remove one vial containing a predetermined titer of vaccinia from the −80°C freezer, thaw at room temperature, and vortex briefly. Place the vial with viral stock at the bottom of a glass beaker filled with ice and sonicate at 57% power for 15 s using sonicator (e.g., Sonic Dismembrator Model 500, Fisher Scientific). Immediately after sonication, pipette the required volume of the virus into a 4-mL test tube containing a 0.1% BSA in PBS to prepare 10^7 pfu/mL virus stock for infection. The BSA is included as a stabilizer of the diluted virus solution. Mix thoroughly using pipette. Keep prepared viral stock on ice during infections of individual mice and mix by pipetting after 10–12 infections. The residual initial viral stock, the 10^7 pfu stock for infections, and the tips used for virus dilutions and infections of mice should be collected in the biohazard bags autoclaved before disposal into the proper biohazardous waste stream at your institution.
11. The charcoal paper protects the surface of the chamber from contamination and provides a good background for the image.
12. To label mice, we place up to six dots on tails and reapply the dots every other day as they get weaker.
13. As described in the IVIS manual, the induction chamber and IVIS instrument are equipped with isoflurane-absorbing charcoal filters. These filters are weighed before each session and replaced once the weight is increased 50 g over initial weight.
14. Since the IVIS instrument is not housed within a BSL-2 cabinet, it is important that the investigators wear protective clothing (e.g., coat, mask, gloves). Also, all members of our group have been immunized with a licensed smallpox vaccine (now Acambis2000 in the USA). The IVIS instrument in our facility is shared by many investigators. To prevent cross-contamination, we use disposable nose cones from VetEquip that are disposed as biohazard material after each imaging session. Alternatively, Caliper sells individual nose cones that can be autoclaved.
15. If the image is saturated (automatic warning appears on the screen), reduce the exposure time.
16. The intensity of the captured light signal is visualized on the monitor in pseudocolor mode, where the low–medium–strong signal is displayed using blue–green–red color coding, respectively.
17. It is very important to establish a system to classify saved images. IVIS imaging software allows one to add information about each image, such as the name of the experiment, day of

recording image within each experiment, type of treatment, identification number of each mouse, and stage and exposure time. This can be done using one category for each type of information: Series, Experiment, Label, Comment, and Analysis. We also use "D1" and "D2" labels to specify ventral and dorsal images taken at stage D, respectively, and "B2" for dorsal image of the head. This detailed description of each saved image file helps tremendously when analysis of bioluminescence data is performed on a group of mice that were imaged daily and need to be analyzed using an "analyze as sequence" tool provided by Living Image Software.

18. To obtain a clear image of the nasal cavity, remove paper covering mouse heads and move anesthetized animals out of the nose cones and away from the flow of anesthesia gas for no more than 1 min. Since the bioluminescence is very high in the nasal cavity, even at 24 h post infection, a 20–30-s exposure might be sufficient to capture good signals.
19. At the end of the imaging session, it is necessary to perform a thorough decontamination of both the IVIS imaging chamber and the induction chamber using 70% ethanol and TexPure detergent, respectively.
20. Of the options provided by the Living Image analysis software, it is most advisable to display bioluminescence values detected in ROI using photons/s/cm^2/sr mode (total fluxes), where the intensity of the signal is adjusted for time of exposure.
21. *t*-Test calculates the difference between means (averages) for two groups relative to the spread or variability within the group. For significant differences, where p is set at ≤ 0.05, the t value is expected to be equal or above *t critical* which varies depending on the number of animals in the group. The value of *t critical* for each case is shown as part of the table generated by Excel in *t*-test analysis. See Table 1 for an example of data.
22. The need in ROC analysis for generating lethality prediction model was dictated by the observation that unlike clear demarcation after Dryvax vaccination (see Subheading 3.4) a proportion of mice that were treated with VIGIV and survived still exhibited WRvFire dissemination to internal organs (see Subheading 3.5). This is a common situation in the development of diagnostic tests when some subjects are positive in the test yet do not succumb to disease, and it could be resolved using ROC analysis. ROC is a graphical plot of sensitivity vs. 1-specificity which varies for each selected threshold. In our case, the sensitivity or true positive mice were defined as mice that were correctly predicted to die based on a selected total flux threshold and 1-specificity or false positive mice were defined as mice that exhibited the selected total flux but survived.

23. AUC is calculated by plotting the percent of mice that do not survive (sensitivity) against the percent of mice that survive (1-specificity) for each recorded total flux.
24. In our experimental system, 100% lethality was achieved by infecting mice intranasally with 10^5 pfu/mouse (10 μL volume of a 10^7 pfu/mL solution).
25. Vaccinia immunoglobulin (VIGIV; Cangene Corporation, Winnipeg, Canada) was used at 0.3, 1.0, 3.0, 10.0, or 30 mg/animal in 600 μL of PBS.
26. An AUC of 1.0 represents a perfect test, where 100% of lethality (sensitivity) and 100% survival (specificity) are accurately predicted based on a selected threshold. An AUC of 0.5 represents random discrimination. In the medical sciences, AUCs of ≥0.85 and ≥0.9 are considered as strong and perfect models, respectively.
27. Weights of mice were recorded immediately prior to infection (time 0) and daily before imaging. Twenty-five percent reduction in weight compared with weight measured before infection (25% weight loss) is traditionally used by many investigators as indicator for lethality.

Acknowledgments

We express our gratitude to John Scott for help in generating data for this study. This project was funded in part with Federal funds from the National Institute of Health, Department of Health and Human Services, under IAA 224-06-1322.

References

1. de Wet JR, Wood KV, DeLuca M, Helinski DR, Subramani S (1987) Firefly luciferase gene: structure and expression in mammalian cells. Mol Cell Biol 7:725–737
2. Ow DW, Wet DE Jr, Helinski DR, Howell SH, Wood KV, Deluca M (1986) Transient and stable expression of the firefly luciferase gene in plant cells and transgenic plants. Science 234:856–859
3. Gould SJ, Subramani S (1988) Firefly luciferase as a tool in molecular and cell biology. Anal Biochem 175:5–13
4. Rodriguez JF, Rodriguez D, Rodriguez JR, McGowan EB, Esteban M (1988) Expression of the firefly luciferase gene in vaccinia virus: a highly sensitive gene marker to follow virus dissemination in tissues of infected animals. Proc Natl Acad Sci USA 85:1667–1671
5. Lin MZ, McKeown MR, Ng HL, Aguilera TA, Shaner NC, Campbell RE, Adams SR, Gross LA, Ma W, Alber T, Tsien RY (2009) Autofluorescent proteins with excitation in the optical window for intravital imaging in mammals. Chem Biol 16:1169–1179
6. Luker KE, Luker GD (2008) Applications of bioluminescence imaging to antiviral research and therapy: multiple luciferase enzymes and quantitation. Antivir Res 78: 179–187
7. Zaitseva M, Kapnick SM, Scott J, King LR, Manischewitz J, Sirota L, Kodihalli S, Golding H (2009) Application of bioluminescence imaging to the prediction of lethality in vaccinia virus-infected mice. J Virol 83:10437–10447

8. Luker KE, Luker GD (2010) Bioluminescence imaging of reporter mice for studies of infection and inflammation. Antivir Res 86:93–100
9. Pichler A, Prior JL, Piwnica-Worms D (2004) Imaging reversal of multidrug resistance in living mice with bioluminescence: MDR1 P-glycoprotein transports coelenterazine. Proc Natl Acad Sci USA 101:1702–1707
10. Townsley AC, Weisberg AS, Wagenaar TR, Moss B (2006) Vaccinia virus entry into cells via a low-pH-dependent endosomal pathway. J Virol 80:8899–8908
11. Gross EA, Swenberg JA, Fields S, Popp JA (1982) Comparative morphometry of the nasal cavity in rats and mice. J Anat 135:83–88

Chapter 11

Mousepox, A Small Animal Model of Smallpox

David Esteban, Scott Parker, Jill Schriewer, Hollyce Hartzler, and R. Mark Buller

Abstract

Ectromelia virus infections in the laboratory mouse have emerged as a valuable model to investigate human orthopoxvirus infections to understand the progression of disease, to discover and characterize antiviral treatments, and to study the host–pathogen relationship as it relates to pathogenesis and the immune response. Here we describe how to safely work with the virus and protocols for common procedures for the study of ectromelia virus in the laboratory mouse including the preparation of virus stocks, the use of various routes of inoculation, and collection of blood and tissue from infected animals. In addition, several procedures are described for assessing the host response to infection: for example, measurement of virus-specific CD8 T cells and the use of ELISA and neutralization assays to measure orthopoxvirus-specific antibody titers.

Key words: Ectromelia virus, Poxvirus, Animal model, Mouse, Host response, Infection

1. Introduction

Although there have been no natural cases of smallpox since its eradication by vaccination, we are currently faced with the natural emergence of human monkeypox and the potential use of variola virus or monkeypox virus as a biological weapon (1–3). Animal models of orthopoxvirus infections are therefore still important for understanding the progression of disease, as well as discovery and characterization of antiviral treatments. Further, orthopoxviruses, with their large genomes encoding numerous host-response modifiers, serve as valuable tools to study the host–pathogen relationship as it relates to pathogenesis and the immune response.

Ectromelia virus (ECTV) infections in the laboratory mouse have emerged as a valuable model to investigate these questions

Stuart N. Isaacs (ed.), *Vaccinia Virus and Poxvirology: Methods and Protocols*, Methods in Molecular Biology, vol. 890, DOI 10.1007/978-1-61779-876-4_11, © Springer Science+Business Media, LLC 2012

(4–6). The natural host of ECTV is believed to be the mouse or other small rodents such as voles (7). Decades of research have generated an extensive description of the disease in a variety of mouse strains and have led to the identification of both host and viral factors contributing to disease (8–12).

ECTV can infect all laboratory mouse strains and can cause severe disease resulting in death in some strains or mild or unapparent disease in other strains. Different mouse strains can be infected via several different routes, with disease outcomes dependent on both strain and inoculation route; namely, via the footpad, intradermally, intranasally, subcutaneously, or other common routes (13). All mice are susceptible to infection via peripheral routes, but the severity of disease depends upon the mouse strain. For example, following a footpad infection the C57BL/6 and AKR strains are considered resistant to severe disease, while A, BALB/c, DBA, and C3H are considered susceptible to severe disease (14–18). The appropriate route, mouse strain, and dosage must be determined for each investigation.

Here we describe the common approaches for infecting mice with ECTV as well as methods for preparation of the viral inoculum and preparation and analysis of infected tissue. We have previously described inoculation of mice via aerosol in the first edition of this book (19).

2. Materials

2.1. Equipment to Ensure Containment of ECTV in the Vivarium

1. "Care-fresh" bedding.
2. Chlorine-based disinfectants (e.g., 10% Chlorox, Spor-klenz, or Expor solutions).
3. 70% alcohol.
4. Disposable gown.
5. Foot covers.
6. Gloves.
7. Eye protection.

2.2. Propagation of L929 and BSC-1 Cells

1. L929 and BSC-1 cells (frozen in liquid N_2).
2. DMEM-10: Dulbeccos Minimal Essential Media with 10% fetal bovine serum (FBS, e.g., Fetal Clone II from Hyclone).
3. PBS: Tissue culture-grade phosphate buffered saline.
4. Tissue culture flasks.
5. Trypsin solution: 0.05% Trypsin, 0.53 mM EDTA.
6. 37°C water bath.

7. CO_2 incubator: Humidified 37°C incubator with 5% CO_2 atmosphere.
8. Inverted microscope.

2.3. Propagation of the Virus

1. ECTV stock; crude or purified virus (stored below –70°C).
2. DMEM-2: Dulbeccos Minimal Essential Media with 2% FBS.
3. Centrifuge: Beckman centrifuge with GH-3.8 rotor.
4. Sterile 250-mL disposable centrifuge tube.
5. Cup horn sonifier (e.g., Branson Sonifier 250).
6. Recirculating ice water bath.
7. Ear protection.
8. Sterile cell scrapers.

2.4. Purification of Virus

1. Sterile ultracentrifuge tube (e.g., Beckman, polyallomer, Cat # 326823).
2. Sterile Dounce homogenizer; glass, with tight pestle.
3. 36% (w/v) sucrose solution in 10 mM Tris–HCl, pH 9.0 (see Note 1).
4. 40%, 36%, 32%, 28%, and 24% sucrose solutions: Sucrose (w/v) in 1 mM Tris–HCl, pH 9.0 (see Note 1).

2.5. Plaque Assay to Measure Virus Titer

1. PBS-1: PBS with 1% FBS.
2. 24-well tissue culture plate.
3. Sterile tubes.
4. Carboxyl methyl cellulose (CMC) overlay: 10 g CMC, 1 L DMEM, 50 mL FBS, penicillin/streptomycin (see Note 2).
5. Crystal violet stain: 1.3 g of crystal violet, 50 mL of 95% ethanol, 300 mL 37% formaldehyde, bring volume up to 1 L with distilled water (see Note 3).

2.6. Anesthetization

1. Ketamine/xylazine: Ketamine 90 mg/kg and xylazine 10 mg/kg.
2. Syringe with 25-g needle.
3. Anesthetization chamber.
4. CO_2 (80%) and O_2 (20%) mixture (see Note 4).

2.7. Mouse Inoculations and Submandibular Bleeds

1. Intranasal intubation stand (see Note 5 and Fig. 1a).
2. Suture and paper clips.
3. 10-μL pipette.
4. 10-μL barrier pipette tips.
5. Insulin syringe.

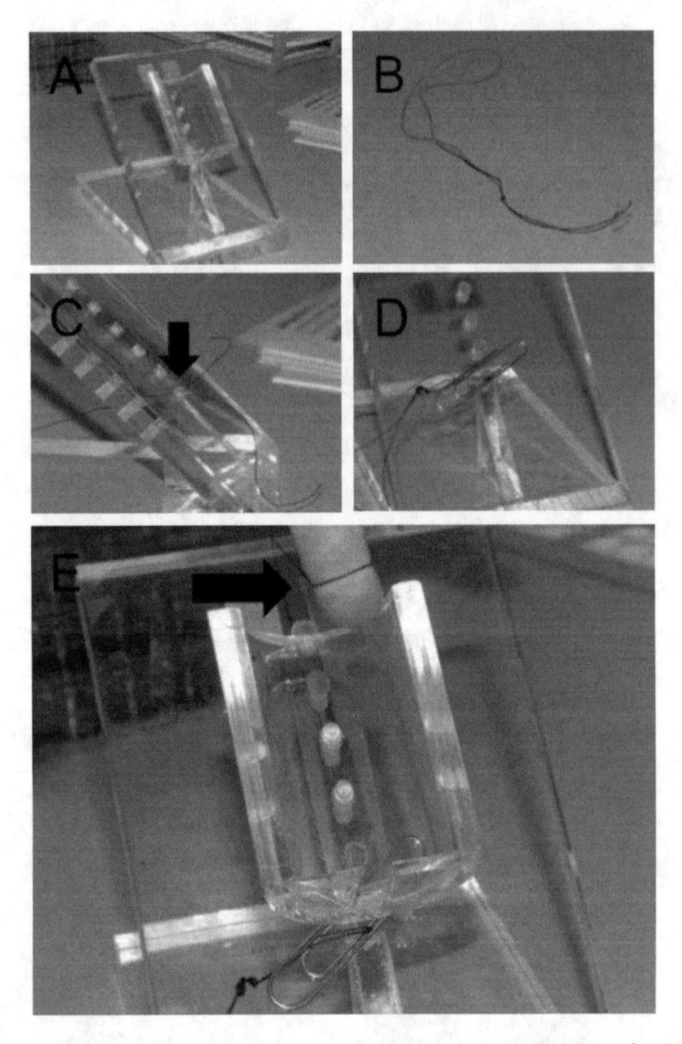

Fig. 1. Set-up for intranasal inoculations. (**a**) An intranasal stand. (**b**) A loop of 2–0-u.s.p surgical suture that will be used to hold the mouse in position. (**c**) Pull the ends of the suture through the hole. (**d**) Tie the suture to a paper clip to secure it onto the stand. (**e**) The suture loop is pulled around the back and over the top of the stand. The mouse will be placed on its back with the head at the top and the suture is placed beneath the front teeth of the mouse to hold the mouse securely into place on the stand.

6. 1-mL syringe with 25-g 5/8 needle.
7. 28-g tuberculin syringe.
8. Rotating tail injector (e.g., Braintree Scientific).
9. 60-W lamp.
10. 18-g needle.
11. Microtainer tubes (see Note 6).

2.8. Euthanasia and Tissue Preparation

1. CO_2 gas.
2. Euthanasia chamber.

3. 70% Ethanol.
4. Scissors, forceps.
5. Pre-weighed tubes (see Note 7).
6. PBS-1: Phosphate Buffered Saline with 1% FBS.
7. Tissue grinders with tight-fitting glass pestles or disposable pestles (e.g., KONTES tissue grinder).
8. Scale.
9. Sterile applicator sticks.

2.9. PCR Analysis of Genome Copies from Blood or Tissue

1. 10 mM dNTP Mix, PCR grade.
2. 50× ROX reference dye (e.g., Invitrogen, Cat#12223-012).
3. 10,000× SYBR Green 1 nucleic acid gel stain (Invitrogen, Cat# S7567) diluted into the appropriate concentration in TAE buffer, pH 7.5.
4. 10× OmniKlenTaq Buffer (e.g., DNA Polymerase Technologies).
5. 2× PEC-2 (e.g., DNA Polymerase Technologies).
6. OmniKlenTaq DNA Polymerase (e.g., DNA Polymerase Technologies).
7. SP028 Primer: 5′- GTA GAA CGA CGC CAG AAT AAG AAT A -3′ (see Note 8).
8. SP029 Primer: 5′- AGA AGA TAT CAG ACG ATC CAC AAT C -3′ (see Note 8).
9. 96-well half skirt PCR plate (e.g., Phenix Research Products MPS-3580).
10. ThermalSeal RT2 Film for qPCR (e.g., Phenix Research Products LMT-RT2).
11. Real-Time PCR System.
12. QIAamp DNA Mini Kit (e.g., Qiagen, Cat# 51306).
13. High Pure PCR Template Preparation Kit for Genomic DNA (e.g., Roche Cat# 11 796 828 001).
14. Standard plasmid DNA (see Note 9).

2.10. ELISA for Orthopoxvirus-Specific Antibody

1. Borate buffered saline: 17 mM sodium borate, 120 mM NaCl, pH 9.
2. Vaccinia virus lysate or BSC-1 soluble antigen in borate buffered saline containing 1% Triton X-100 (see Note 10).
3. Immulon-2 HB plates.
4. Biotinylated Goat anti-mouse IgG gamma.
5. Streptavidin peroxidase.
6. 30% hydrogen peroxide.

7. PBS, pH 7.2, filtered with 0.45 μm HA filter.
8. Normal goat serum (e.g., Vector).
9. PBST: PBS with 0.05% Tween 20.
10. Carbonate buffer: 1.58 g Na_2CO_3, 2.94 g $NaHCO_3$, 11.7 g NaCl (per liter, pH 9.6) filtered with 0.45-μm HA filter.
11. Substrate buffer: 307 mL of 0.1 M citric acid and 193 mL of 0.2 M Na_2HPO_4, pH 4.0 filtered with 0.45-μm HA filter.
12. Blocking buffer: PBS with 0.05% Tween 20, 2% normal goat sera (see Note 11).
13. 3 N HCl.
14. OPD dihydrochloride tablets 5 mg.

2.11. Plaque Neutralization Assay

1. BSC-1 cells.
2. DMEM-2.
3. Purified ECTV.
4. Samples from mouse separated as either plasma or serum.
5. CMC Overlay Media.
6. Crystal Violet Stain.
7. Inverted microscope.

2.12. Measurement of Virus-Specific CD8+ T Cells

1. Splenocytes from mice of the same species as those being measured for $CD8^+$ T cell response.
2. Purified ECTV.
3. Cell irradiator (e.g., XRAD).
4. Antibodies to CD8a (e.g., BD Biosciences clone 53-6.7), CD3 (e.g., BD Biosciences clone 145-2C11), and Interferon gamma IFNγ; (e.g., BD Biosciences clone XMG 1.2).
5. PBS-2: Phosphate Buffered Saline with 2% fetal bovine serum.
6. RPMI-10: RPMI media supplemented with 10% fetal bovine serum.
7. Cytofix/Cytoperm Buffer Kit (e.g., BD Biosciences).
8. 10x Pharmlyse (e.g., BD Biosciences).
9. Golgi Stop or Brefeldin A.
10. CFSE cell labeling kit (e.g., Invitrogen).

3. Methods

3.1. Special Precautions to Ensure Containment of ECTV

Special precautions must be taken when handling ECTV infected mice to prevent an outbreak in the animal handling facility (7). Contaminated bedding can serve as a source of infection, thus the use of excessive bedding or particulate bedding must be avoided (for example "care-fresh" bedding can be used). Effective decontamination requires autoclaving or the use of chlorine-based disinfectants (e.g., 10% Chlorox, Spor-klenz, or Expor solutions). Cages and other equipment should be autoclaved or sprayed or dipped in a container of disinfectant before being removed from the containment area. Personal protective clothing (gown, foot covers, gloves, and eye protection) must be worn by the animal handlers, and disposable items should be discarded in a biohazard waste container in the room, which needs to be autoclaved prior to being put into the biohazard waste stream at your institution. All manipulations should be performed in a laminar flow hood and the surface of the hood and cages should be disinfected prior to return to the rack. Gloves should be kept moist with disinfectant. After the final manipulation, all surfaces of the biosafety cabinet should be sprayed with disinfectant and left wet for 10 min prior to wiping dry (or by spraying with 70% ethanol).

3.2. Propagation of L929 and BSC-1 cells

1. Quickly thaw frozen ampules of cells (stored in Liquid N_2) in 37°C water bath.
2. Transfer thawed cells into a conical tube containing at least 10 mL of DMEM-10 and pellet the cells in a centrifuge at 350×*g* for 5 min at room temperature.
3. Aspirate off the media containing DMSO used to freeze the cells and resuspend the pellet in 5 mL DMEM-10, transfer to a tissue culture flask (T-25), and place in a CO_2 incubator. Allow the cells to grow to confluence.
4. To passage the cells, aspirate cell media and wash cell monolayer with PBS to remove any residual serum containing media. Add 1 mL trypsin solution to cover the monolayer and return to the incubator until cells are no longer attached to the plate (5–10 min) (see Note 12).
5. Add DMEM-10 to the flask and pipette vigorously to remove remaining cells from the plate and to disperse cells into a single cell suspension.
6. Pellet the cells by centrifugation as described in step 2 and then aspirate the trypsin containing media.
7. Resuspend the cells in DMEM-10 and transfer the cells into flasks that are a minimum of three surface areas and a maximum of ten surface areas greater than the previous flask. Record the date and passage number on the flask (see Note 13).

3.3. Propagation of Crude Virus Stock from Cell Lysates

1. Prepare 10 T150 flasks of L929 cells in DMEM-10 and grow until the cells become confluent.
2. Quickly thaw a frozen virus stock in a 37°C water bath then transfer the vial to an ice bucket.

3. Sonicate thawed virus stock for 40 s by immersing the tube into a cup horn water bath on the sonicator being careful not to touch any metal portion of the sonifier. Return the virus to ice for 10 s. Repeat sonification step two more times (see Note 14).
4. Prepare virus dilution in DMEM-2 to achieve a multiplicity of infection (MOI) of 0.2 to 0.5 PFU/cell (5 mL of DMEM-2 per T150 flask).
5. Aspirate off the DMEM-10 in the flasks and overlay cell monolayer with 5 mL of virus dilution per flask. Incubate in CO_2 incubator for 1 h rocking 1–2 times during the hour to ensure an even distribution of the virus.
6. Add 10 mL of fresh DMEM-2 per flask and incubate in CO_2 incubator for 2–3 days.
7. Use a sterile cell scraper to harvest cells when ~75–90% of the cells show cytopathic effect, but are still attached to the flask.
8. Transfer media containing infected cells into a disposable 250-mL centrifuge tube and centrifuge for 5 min at 1,800 × *g* (e.g., for Beckman GH-3.8 rotor: 2,500 rpm) at 5–10°C.
9. Aspirate off media and resuspend cell pellet in 20 mL of DMEM-2 and freeze/thaw in dry ice slurry three times (see Note 15).
10. Sonicate the lysed cells for 1 min in recirculating ice water bath under constant pulse at maximum amplitude and repeat three times (see Note 14).
11. Aliquot the virus into vials labeled with the name of the virus strain, level of purity, date harvested, person preparing the virus, and vial number (see Note 16).
12. Store aliquots below –70°C.

3.4. Preparation of Purified Stock of Virus (Adapted from (20))

3.4.1. Cushion-Purified Virus (See Note 17)

1. Generate a crude stock of virus following steps 1–8 in Subheading 3.3.
2. Resuspend infected cell pellet in 14 mL of 10 mM Tris–HCl, pH 9.0.
3. Keep samples on ice for the remainder of the steps.
4. Homogenize the cell suspension with a glass Dounce homogenizer using 30–40 strokes of a tight pestle (see Note 18).
5. Centrifuge at 1,300 rpm for 5 min at 5–10°C to remove nuclei. Save the supernatant.
6. Resuspend nuclear pellet in 3 mL of 10 mM Tris–HCl, pH 9.0 and centrifuge again at 1,300 rpm for 5 min at 5–10°C.
7. Collect the supernatant and pool with the supernatant from step 5.

8. Divide the pooled supernatants into three ~6-mL aliquots and sonicate each aliquot in an ice water bath under constant pulse at maximum amplitude and repeat three times (see Notes 14 and 19).
9. Vortex the sonicated supernatant and transfer it into a sterile ultracentrifuge tube containing 17 mL of 36% sucrose cushion (see Notes 1 and 20).
10. Centrifuge 80 min at 13,500 rpm in a SW27 rotor.
11. Aspirate and discard the supernatant (see Note 15).
12. Resuspend the viral pellet in 1 mL of 1 mM Tris–HCl, pH 9.0.
13. At this stage the virus is considered a cushion-purified virus. Divide into aliquots labeled with the virus name, date harvested, level of purity, name of person generating virus and number of the aliquot. Samples should be stored at less than –70°C at this point (see Note 21).

3.4.2. Band-Purified Virus (See Note 17)

1. Prepare two sterile 24–40% continuous sucrose gradients per sample in a sterile ultracentrifuge tube the day before purification by carefully adding in a drop-wise fashion 6.8 mL each of 40%, 36%, 32%, 28%, and 24% sucrose solutions sequentially (see Note 1). Cover the top of the tube with parafilm and let sit overnight at 4°C to allow some diffusion of sucrose between layers.
2. Overlay 1 mL of viral pellet from step 12 of Subheading 3.4.1 onto one sucrose gradient (see Note 22).
3. Centrifuge at 26,000×*g* in a SW27 rotor for 50 min (4°C).
4. Collect the virus band (~10 mL volume) with a sterile pipette and place in a sterile 50-mL conical tube (see Note 23).
5. Aspirate the remaining sucrose and discard into a bottle containing 10% bleach to inactivate virus prior to proper disposal (see Note 15).
6. Collect the pellet at the bottom of the tube, which contains aggregated virus, by pipetting it with 1 mL of 1 mM Tris–HCl, pH 9.0.
7. Sonicate the resuspended pellet once for one minute in recirculating ice water bath under constant pulse at maximum amplitude.
8. Re-band this material as in steps 1–4 in the second gradient that was prepared.
9. Collect the band (~10 mL) and pool with the first band from step 4. No need to collect the small pellet at the bottom of this second gradient.

10. Add two volumes (~40 mL) of 1 mM Tris–HCl, pH 9.0 to the pooled bands and mix (see Note 24).
11. Split this volume evenly between two sterile ultracentrifuge tubes and pellet the virus at 33,000×g for 60 min in a SW27 rotor at 4°C. Discard the supernatant (see Note 15).
12. Resuspend the virus pellet in 1 mL of 1 mM Tris–HCl, pH 9.0. Sonicate the band-purified virus for 1 min and divide into aliquots labeled with the virus name, date harvested, level of purity, name of person generating virus, and number of the aliquot (see Note 25).

3.5. Plaque Assay to Measure Virus Titers

1. One day prior to titrating the virus, prepare a 24-well plate of BSC-1 cells to be 90–100% confluent (see Note 26).
2. Sonicate the thawed virus stock at constant pulse amplitude of 75–100 in a cup sonication bath containing ice slurry for three 20 s cycles with a 10 s interval on ice between each sonication cycle (see Note 14).
3. Vortex the virus stock.
4. Make serial tenfold dilutions of the virus stock into PBS-1. First mix 10 μL of virus stock into 1 mL of PBS-1 for a 10^{-2} dilution. Mix. Take 100 μL from this into 1 mL of PBS-1 for a 10^{-3} dilution. Continue tenfold dilutions until the expected range of the virus titer is reached (see Notes 16, 21, and 25).
5. Aspirate all but ~200 μL of media from the wells (two wells for each dilution of virus) and then add 100 μL of viral dilution to each well (in duplicate) (see Note 27).
6. Return the plates to the incubator for 1 h. Warm CMC overlay in 37°C water bath.
7. After 1 h incubation, add 1 mL of CMC overlay to each well (see Note 28).
8. Put plates into incubator and incubate until plaques are visible under a microscope. Plaques will be visible 3–5 days post infection.
9. When the plaques are visible, add 200–250 μL of Crystal Violet stain to each well. Incubate at least 1 h to overnight at room temperature. Wash the plate in a tub of cool running water. Dry inverted. Count the plaques.
10. Calculation of the titer of the virus (in PFU/mL) = [(# plaques per well)/(volume in mL of viral dilution added to the well)]× (1/dilution) (see Note 29).

3.6. Inoculation of Animals

3.6.1. Intranasal Route of Infection

1. Sonicate thawed virus stock and prepare appropriate virus dose to inoculate diluted in PBS without Ca and Mg (and no carrier protein).
2. Anesthetize the animal with ketamine/xylazine intraperitoneally (see Note 30).

3. Secure mouse on intranasal intubation stand using suture and paper clips (see Fig. 1 and Note 5).
4. Load a 10-μL barrier pipette tip with 10 μL of virus suspension.
5. Place the loaded tip next to left mouse nares and release 5 μL from pipette so mouse inhales the inoculum.
6. Dispense the remaining 5 μL to the right nares (see Note 31).
7. Leave mouse on intranasal stand for approximately 1 min before returning mouse to cage.

3.6.2. Subcutaneous Route of Infection

1. Sonicate thawed virus stock and prepare appropriate virus dose to inoculate diluted in PBS without Ca and Mg (and no carrier protein).
2. Restrain animal while tenting skin above shoulder area (see Note 32).
3. Insert needle, bevel up, into cavity created by tented skin. Be careful not to accidentally exit through the skin on the other side of the tent. Check correct positioning by aspirating back. If positive, reinsert needle and aspirate again.
4. Once in proper location, inject slowly.
5. Use one needle/syringe per cage of five animals (see Note 33).

3.6.3. Intradermal Infection

1. Sonicate thawed virus stock and prepare appropriate virus dose to inoculate diluted in PBS without Ca and Mg (and no carrier protein).
2. Anesthetize the animal with ketamine/xylazine intraperitoneally (see Note 30).
3. Locate the injection site according to study design (e.g., ear pinna).
4. Insert the needle into the injection site, bevel up. Bevel needs to be in between skin layers.
5. Inject virus. If needle is placed in the correct spot, the skin should form a bleb (see Notes 33 and 34).

3.6.4. Intravenous Infection

1. Place the mouse in rotating tail injector with tail facing toward you (see Note 35).
2. Once the mouse is in restrained, place the tail under the heat source and stroke the tail in order to dilate the blood vessels.
3. Rotate the mouse so that the lateral tail veins are facing up.
4. Once a vein is visualized, "hold off" the top of the vein using your index and middle finger. Using your ring finger and thumb, hold the tail closer to the end (about 2/3 of the way

down the tail) and bend it down at a slight angle. Make sure there is no slack in the tail.

5. Insert needle bevel up at the bend in the tail. The veins run slightly beneath the skin, so do not insert the needle very deep.
6. Once inserted, do not inject until a flash of blood has been observed in the syringe. Once this happens, you may inject. If any resistance is felt at all, you are no longer in the vein. Remove and try again.

3.7. Bleeding from the Submandibular Vein (21)

1. Restrain the animal by scruffing skin over shoulder blades, making sure to stabilize the head.
2. Firmly poke mouse with the needle at the rear of the jawbone targeting the submandibular vein.
3. Once blood starts to appear, hold the animal over the blood tube for collection (see Note 6). When collection completed, briefly compress puncture site and release hold of the scruff of neck and bleeding should cease.
4. The animal can be bled daily as long as an alternate side of the face is used and does not exceed 10% of the animal's body-weight in 2 weeks.

3.8. CO_2 Euthanasia (See Note 36)

1. Place animals in a chamber and administer CO_2 (100%) gas until breathing has stopped (>20 s).
2. Then perform a cervical dislocation (or other Institutional Animal Care and Use Committee approved procedure) prior to putting the mouse in the "body bag" and into the carcass freezer.

3.9. Processing of Liver, Spleen, and Lung Tissue to Measure Viral Titers in Organs

1. After animal is euthanized, wet hair with ethanol (see Note 36).
2. Make a midline incision.
3. Grasp organ with forceps and cut a sample off for infectivity and place in pre-labeled tube (see Note 37).
4. Weigh the tube containing the tissue and subtract the weight of the tube for the total tissue weight.
5. If glass tissue grinders are used, add 0.1 mL of PBS-1 per 0.1 g of tissue to a tissue grinder. Transfer the tissue to the tissue grinder (a sterile applicator stick may be used here) and pulverize the tissue until a slurry is present. Add no less than 0.5 mL of PBS-1 and no greater than 1 mL to the tissue grinder. Transfer the ground tissue back to the tube.
6. If the disposable pestle is used, add 0.25 mL of PBS-1 to the tube containing the tissue and pulverize the tissue until a slurry is present.

7. Freeze/thaw the sample three times. Samples may be stored below –70°C until they are titrated (using procedures outlined in Subheading 3.5).

3.10. PCR Analysis of Genome Copies Directly from Blood or Tissue (See Note 38)

3.10.1. Whole Blood (Not Clotted) (See Note 6)

1. For real-time PCR on whole blood samples create a master mix containing the recipe of reagents (per reaction) shown in Table 1. Add 1.25 μL of whole mouse blood per reaction (see Note 39).
2. Set up a standard curve using the same master mix in Table 1, but in place of 1 μL of water, use 1 μL of various dilutions of the purified standard plasmid DNA (see Notes 9 and 40).
3. Include 1.25 μL of naïve blood in each reaction for the standard curve (see Note 41).
4. Run real-time PCR using the conditions in Table 2.

3.10.2. Tissue

1. Homogenize tissue samples (lung, spleen, liver, or kidney) as described in Subheading 3.9.
2. Create a master mix containing, per reaction, the recipe of reagents shown in Table 1. Add 1.25 μL of homogenized tissue per reaction.
3. Set up a standard curve as in Subheading 3.10.1, step 2.

Table 1
PCR master mix (volumes/PCR reaction)

Whole blood		Homogenized tissue
10 mM dNTPs	0.50 μL	
10× OmniKlenTaq buffer	2.50 μL	
5 nM SP028 primer	2.00 μL	
5 nM SP029 primer	2.00 μL	
50× ROX reference dye	0.50 μL	
2× PEC	12.50 μL	
SYBR green[a,b]	1.5 μL[a]	0.5 μL[b]
OmniKlenTaq DNA polymerase	0.60 μL	0.10 μL
Water	1.65 μL	3.15 μL
Sample[c]	1.25 μL	1.25 μL
Total volume	25 μL	25 μL

[a]1,000× SYBR green solution used for the whole blood PCR
[b]100× SYBR green solution used for the tissue homogenate and purified DNA PCRs
[c]After aliquoting 23.75 μL of the master mix into individual wells of the PCR plate, add 1.25 μL of whole blood or tissue homogenate

Table 2
PCR reaction conditions

Step	Temperature	Time
1	95°C	10 min
2	95°C	40 s
3	54°C	45 s
4	Plate read	
5	70°C	2 min
6	Repeat steps 2–5 for 40 cycles	
7	Melting cure from 50°C to 95°C, read every 0.2°C, hold 3 s	
8	25°C	Hold

4. Include 1.25 μL of naïve tissue homogenate in each reaction for the standard curve (see Note 41).
5. Run real-time PCR using conditions in Table 2.

3.11. ELISA for Orthopoxvirus-Specific Antibody Titer

1. Dilute antigen 1:2,500 in 1× Carbonate Buffer, add 100 μL/well to an Immulon 2 high-affinity plate and cover with parafilm before incubating overnight at 4°C (see Note 42).
2. Remove plate from 4°C, dump contents, and wash plate three times with 300 μL/well PBS.
3. Add 200 μL/well blocking buffer. Incubate 30 min at room temperature (RT).
4. Dump blocking buffer; no need to wash.
5. To row A add 200 μL PBST, to row B-H add 100 μL PBST. Add 2 μL (1:100) or 4 μL (1:50) of serum sample to appropriate well in Row A (see Note 6). Add positive, negative, and a blank control to each plate (see Note 43).
6. Mix samples in row A by multichannel pipetting up and down 5×. Load the tips with 100 μL of mixed solution and add to row B. Discard the tips and repeat process for rest of rows. Discard 100 μL from row H. The concentration of each subsequent row will be half the row before it, e.g., row A (1:50), row B (1:100), etc. (see Note 44).
7. Incubate 1 h.
8. Dump solution and wash plate three times with 300 μL/well PBS.
9. Add biotintylated anti-Mouse IgG gamma diluted 1:2,500 in PBST at 100 μL/well. Incubate 1 h at RT.

10. Dump solution and wash plate three times with 300 μL/well PBS.
11. Add streptavidin HRP diluted 1:4,000 in PBST at 50 μL/well. Incubate 30 min at RT (see Note 45).
12. Prepare substrate by dissolving one 5 mg OPD tablet in 12.5 mL of substrate buffer (takes 15–30 min).
13. Dump solution and wash plate three times with 300 μL/well PBS.
14. Add 20.6 μL of 30% H_2O_2 to OPD dissolved substrate. Mix well. Add 100 μL/well and incubate 15 min at RT.
15. Add 50 μL/well of 3N HCl. Read on ELISA plate reader at 490 nm.

3.12. Plaque Neutralization Assay

1. One day prior to assay, plate BSC-1 cells in 24-well plates such that there are enough wells for each dilution of the sample being tested (at least six dilutions per sample) as well as controls (immune serum, naïve serum, and untreated virus).
2. Dilute purified ECTV in DMEM-2 to 1,000 PFU/mL. Make serial dilutions of mouse serum/plasma in DMEM-2 similar to the serum dilutions done in step 5 of Subheading 3.11. Mix equal volumes (~60 μL each) of diluted virus and serum including the immune/negative serum as well as samples with no serum diluted (media alone). Incubate 4 h in CO_2 incubator.
3. Aspirate all but ~200 μL of media from the cells and add 100 μL/well of each virus/serum sample. Return cells to incubator for 1 h then overlay with 1 mL pre-warmed CMC Overlay media.
4. Continue incubating until plaques are visible under the inverted microscope. Stain the plates with at least one fourth the overlay volume of Crystal Violet Stain for at least one hour. Destain the plates in a water bath and dry inverted.
5. Count the plaques and determine the dilution at which 50% neutralization occurs by comparing the number of plaques at each dilution to the average number of plaques for the diluted virus incubated with media alone.

3.13. Measurement of IFNγ Secreting CD8+ T Cells

1. One day prior to measuring the virus-specific CD8+ T cells, aseptically harvest the spleen of a naïve mouse of the same genotype and prepare a single cell suspension in RPMI-10 at a final concentration of $1–2 \times 10^7$ cells/mL (see Note 46).
2. Split the cells into separate aliquots with one not to be infected and one aliquot infected with purified ECTV at a multiplicity of infection (MOI) of 0.1 PFU/cell (see Notes 47 and 48).
3. Place the cells in a CO_2 incubator in polypropylene tubes with loosened caps to allow gas exchange overnight.

4. Irradiate the APCs at 2500 RADS (see Note 49).
5. APCs are labeled with CFSE by first washing cells with PBS, pelletting at 350×*g* for 5 min in a Beckman GH-3.8 rotor at temperature greater than 10°C (see Note 48).
6. Cells are resuspended in PBS containing 0.1 μM CFSE and incubated in a CO_2 incubator for 15 min.
7. Cells are washed with PBS and resuspend cells in RPMI-10.
8. Continue incubating for 30 min and check for CFSE incorporation on the flow cytometer compared to unlabeled cells (see Note 48).
9. CD8+ T cells can be analyzed from blood, spleen, or other tissues. Prepare a single cell suspension from the tissue to be analyzed in RPMI-10 at $0.5–1 \times 10^7$ cells/mL. These cells will serve as the responders.
10. Combine $0.5–1 \times 10^6$ responders and an equal number of CFSE-labeled APCs in a 96-well polypropylene round bottom plate. Make one well for each responder mixed with the uninfected APC (MOI = 0) and one well with the responder mixed with the infected APC (MOI = 0.1 PFU/cell). Incubate in a CO_2 incubator overnight.
11. Add 1 μL Golgi Stop per reaction and continue incubating for 4–6 h.
12. Pellet the cells at 350×*g* for five minutes and aspirate media. Wash cells with PBS-2, pellet cells, and aspirate wash.
13. Prepare appropriate amounts of antibodies to CD45, CD3, and CD8a in PBS-2 (see Note 50).
14. Resuspend the cells in this surface stain cocktail, remembering to include cells from the APC preparation that were not loaded with CFSE combined with each antibody individually to serve as single color controls for flow cytometry as well as cells labeled with the fluorophore used to label IFNγ.
15. Incubate at 4°C for 30 min. Wash cells with PBS-2, pellet cells, and aspirate wash.
16. Fix the cells in BD Cytofix/Cytoperm for 30 min at 4°C.
17. Pellet cells and aspirate fix. Wash the cells two times in 1× BD Permwash. Resuspend pellets from each reaction in IFNγ antibody diluted in 1× BD Permwash.
18. Incubate 30 min to overnight at 4°C. Wash cells two times in 1× BD Permwash. Resuspend cells in PBS-2 and analyze on flow cytometer.
19. To ascertain the virus-specific CD8+ T cell response, first determine the number of CFSE negative (responder) cells that are positive for CD45, CD3, and CD8. From this population, compare the number of IFNγ positive cells in the reaction

containing uninfected APCs (MOI = 0 PFU/cell), which serve as the background, to the number of cells in the reaction containing infected APCs (MOI = 0.1 PFU/cell). The difference between these groups is the virus-specific CD8+ T cell response.

4. Notes

1. To make sucrose solutions, weight out sucrose (e.g., 36 g), and then bring volume up to 100 mL with a Tris–HCl solution.
2. CMC is difficult to dissolve. Use the following procedure to prepare CMC overlay. Weigh 10 g of CMC into a 1-L glass bottle with a sterile stir bar in the bottle. Autoclave at least 20 min on dry cycle. Let cool to room temperature. In a biosafety cabinet, add 1 L DMEM to the sterile CMC. Use a sterile 5-mL pipette to dislodge any CMC stuck to the sides of the bottle. Place on a stir plate overnight. The following day, in a biosafety cabinet, add 50 mL fetal calf serum and pen/strep to the media for CMC overlay media with 5% serum, 1× pen/strep. Store at 4°C.
3. To prepare, weigh 1.3 g of crystal violet and add to a 1-L bottle containing a stir bar. Add 50 mL of 95% ethanol to the bottle and stir on a magnetic stir plate for 1 h. Add 300 mL of 37% formaldehyde and distilled water to the 1 L mark on the bottle. Stir overnight at RT, then store at RT until use.
4. Used when one wants to knock the mice down briefly.
5. We had this stand custom built by our shop. It allows one to standardize the protocol and to do a large number of mice at once. Each stand can hold up to five mice. So you put all the mice in place on the stand and then infect them in turn. After you have finished infecting mouse #5 you are ready to start transfer mouse #1 back to the cage. (We house the mice at five per cage.) Other labs hold an individual anesthetized mouse during intranasal and then put the mouse down in a standardized position after each inoculation.
6. EDTA-coated microtainers should be used when plasma or whole blood is needed. Microtainer SST (serum separator tubes) can be used when only serum is needed.
7. Label tubes with study ID, animal's ID, organ, and day of study.
8. Primers amplify 165 base pairs of the EV107 gene. This is the homologous gene to the VACV A4L gene (which codes the A4 core protein).

9. The standard plasmid DNA was generated by cloning the fragment produced by PCR with primers SP028 and SP029 into the pGEM-T vector (Promega).
10. We use BSC-1 cells either infected with vaccinia virus or mock infected (as a control for background binding of serum proteins to the cells alone) at an MOI = 1 until the cytopathic effect is 75% or greater. We wash the cells twice with 3.5 mL/flask of borate buffered saline and then solubilize the cell pellet in 0.5 mL borate buffered saline containing 1% Triton X-100.
11. Add goat serum to PBS, filter through 0.22-μm low protein binding filter, and then add Tween 20.
12. To ensure that cells are no longer attached to the flask, it is best to check flask with an inverted microscope.
13. Cells can be continually passaged up to forty times for reliable results from low passage seed stocks.
14. To prevent the virus from heating up, a recirculating ice water bath can be set up so that ice cold water is being circulated through the cup of the sonifier. The sonifier should be set to constant pulse and maximum amplitude. Put on ear protection and ensure that everyone in the room is also wearing ear protection.
15. All materials that contain virus or have had contact with virus require special handling. Disposable plasticware should be autoclaved prior to disposal in your institutions biohazardous waste stream. Liquids can be inactivated by addition of 10% Chlorox.
16. Crude cell lysates typically contain ~2×10^8 PFU per processed flask. So, the titer of virus from harvested cells from 10 T150 flasks resuspended in 20 mL is ~1×10^8 PFU/mL.
17. Most often cushion-purified virus is used. Band-purified virus is used in instances where one wants to be sure there is no input host debris with the virus. ECTV does band on a sucrose gradient.
18. Check for cell breakage by light microscope.
19. Samples can be stored at −70°C at this point and labeled with the virus strain, harvest date, initials of preparer, and description of the sample. Thaw, re-sonicate, and vortex before continuing with the purification.
20. Slowly add the virus suspension drop wise to the side of the tube just above the surface of the sucrose cushion.
21. Cushioned virus should contain ~2×10^8 PFU per processed flask. So, the titer of cushioned virus from harvested cells from 10 T150 flasks resuspended in 1 mL is ~2×10^9 PFU/mL.

22. If virus was frozen, sonicate and vortex before placing on top of the sucrose cushion.
23. The virus band should be in the middle of the step gradient. To harvest the band, it usually requires pipetting ~10 mL volume from the sucrose gradient.
24. The total volume should be ~60 mL (10 mL from the first banding + 10 mL from the banding of the pellet at the bottom of the first gradient + 40 mL 1 mM Tris–HCl, pH 9.0).
25. Band-purified virus should contain $\sim 2 \times 10^8$ PFU per processed flask. So, the titer of cushioned virus from harvested cells from ten T150 flasks resuspended in 1 mL is $\sim 2 \times 10^9$ PFU/mL.
26. Do not use cells that are over confluent to titer virus.
27. When removing media from wells, one should do these manipulations rapidly to prevent drying out the monolayer. Also, when small volumes cover the monolayer during incubations, it is best to occasionally rock the plates to ensure that the monolayer remains moist.
28. We do not remove the infecting inoculum, but just add CMC directly to the well.
29. Sample titer calculation: If there were 20 plaques in the well from the fourth dilution tube (10^{-5}), titer would be 2×10^7 PFU/mL; $[(20 \text{ plaques})/(0.1 \text{ mL})] \times (1/10^{-5})$.
30. Administer a mixture of ketamine/xylazine by the intraperitoneal route at the dose of 0.1 mL per 10 g of a C57Bl/6, A or SKH-1 mouse. The correct dosage of ketamine/xylazine must be empirically determined for each mouse strain.
31. A 10 μL inoculum volume split between two nares results in an upper respiratory tract infection. To inoculate the whole respiratory tract we use a 50-μL inoculum by administering 25 μL of virus into each nares.
32. Hold the skin fold with forceps to prevent a needle stick. This is best done when the animal is anesthetized with CO_2 (80%)/ O_2 (20%) mix. For administration of CO_2/O_2 anesthesia, place the animals in a chamber and administer CO_2/O_2 gas until desired level of anesthesia is reached, which usually is ~10 s (see Note 4).
33. Dispose of needle and syringe in the appropriate sharps container that after autoclaving goes into your institution's biohazard waste stream.
34. If using the ear pinna, if you insert the needle too far you will go through the ear. In the ear pinna and other sites, you should always see a bleb; if you see the bleb you are correctly between skin layers; if you do not see a bleb you went too deep.
35. The animal does not need to be anesthetized for this procedure.

36. An overdose of ketamine/xylazine may be substituted if tissue from the respiratory tract is going to be used to measure virus infectivity or process tissues for histopathology.
37. When taking a portion of an organ, the key thing to do is to standardize the portion of organ taken and take that same portion throughout the study. For example, if taking a portion of the lung, take the lower left lung quadrant, a section of the largest liver lobe and so forth for the entire study.
38. While the OmniKlenTaq method does not require one to first purify DNA from samples for real-time PCR, some may not want to use the OmniKlenTaq method because the reagents are relatively expensive to use on just a few samples. Thus, one can performing real-time PCR using purified DNA from the blood or tissue and use other real-time PCR kits like the Applied Biosytems kit, which contains the SYBR green, enzyme, etc. This would require one to first purify DNA from a sample using something like Qiagen kit.
39. Remember to include a positive and negative control on every plate, as well as a plasmid template to allow construction of a standard curve. The polymerase and proprietary buffer permit the quantification of DNA directly from blood without purification. For additional information on this procedure see (22).
40. We do serial ten-fold dilutions of the plasmid and use 1 μl of plasmid DNA at concentrations ranging from 8.8×10^8 copies/μL to 8.8×10^0 copies/μL. This therefore gives a standard curve for which one can measure the DNA quantities in the unknown samples.
41. Addition of naïve blood or tissue homogenate per sample, along with the plasmid DNA dilutions is a way to ensure that the reactions for the standard curve have exactly the same components as the samples. We do this because blood or tissue homogenate could contain PCR inhibitors and we want these to remain constant throughout the experiment.
42. Parafilm-covered plates can be stored at 4°C for up to 2 weeks.
43. As a positive control, we use a dilution of a well-characterized pool of sera from a C57Bl/6 mouse that recovered from a footpad infection. The negative control is a 1:50 dilution of a pool of mouse sera from C57Bl/6 mice inoculated with PBS by the footpad route. The blank is PBST.
44. We do the serum dilutions in the coated plates. We tested doing the dilutions in a separate plate and found no difference.
45. You need to titrate each new batch of antigen, goat anti-mouse IgG, and streptavidin HRP to ensure that appropriate signal is obtained at the indicated incubation times.
46. We use the BD Biosciences Pharmlyse protocol.

47. Make sure that there are enough cells for each sample one wants to analyze. Allowing for losses of cell number during successive centrifugation steps, one needs $0.5–1 \times 10^6$ cells at MOI = 0 PFU/cell (uninfected control) and $0.5–1 \times 10^6$ cells at MOI = 0.1 PFU/cell to serve as antigen presenting cells (APCs).
48. Keep a fraction of cells that are not loaded with CFSE to serve as single color controls for flow cytometry, as well as negative controls for CFSE staining.
49. This can be done with an XRAD irradiator or using a blood bank irradiator or other radioactive source. Make sure that the AP function of the cells in not inhibited by the irradiation protocol. We use a T cell hybriboma line that responds to the dominant CD8 T cell epitope in VACV B8R to ensure that irradiated APCs can still present antigen.
50. We titrate each lot of fluorescent antibodies that we use for flow cytometry so the combinations of antibodies used for FACS give expected signals.

Acknowledgments

This work was supported by a subcontract N01-AI-30063 to Southern Research.

References

1. Parker S, Nuara A, Buller RML, Schultz DA (2007) Human monkeypox: an emerging zoonotic disease. Future Microbiol 2:17–34
2. Henderson DA (1999) The looming threat of bioterrorism. Science 283:1279–82
3. Fenner F, Henderson DA, Arita I, Jezek Z, Ladnyi ID (1988) Smallpox and its eradication. World Health Organization, Geneva
4. Saini D, Buller RM, Biris AS, Biswas P (2009) Characterization of a nose-only Inhalation exposure system for ectromelia virus infection of mice. Particulate Sci Technol 27: 152–65
5. Hostetler KY, Beadle JR, Trahan J et al (2007) Oral 1-O-octadecyl-2-O-benzyl-sn-glycero-3-cidofovir targets the lung and is effective against a lethal respiratory challenge with ectromelia virus in mice. Antiviral Res 73:212–8
6. Parker AK, Parker S, Yokoyama WM, Corbett JA, Buller RML (2007) Induction of natural killer cell responses by ectromelia virus controls infection. J Virol 81:4070–9
7. Buller RML, Fenner F (2007) Mousepox. In: Fox JG, Barthold SW, Davisson MT, Newcomer CE, Quimby FW, Smith AL (eds) The mouse in biomedical research. Elsevier, New York, pp 67–92
8. Esteban DJ, Buller RML (2005) Ectromelia virus: the causative agent of mousepox. J Gen Virol 86:2645–59
9. Palumbo GJ, Buller RML (1991) Inhibitors of the lipoxygenase pathway specifically block orthopoxvirus replication. Virology 180: 457–63
10. Brownstein DG, Bhatt PN, Gras L, Jacoby RO (1991) Chromosomal locations and gonadal dependence of genes that mediate resistance to ectromelia (mousepox) virus-induced mortality. J Virol 65:1946–51
11. Delano ML, Brownstein DG (1995) Innate resistance to lethal mousepox is genetically linked to the NK gene complex on chromosome 6 and correlates with early restriction of virus replication by cells with an NK phenotype. J Virol 69:5875–7

12. Brownstein DG, Gras L (1997) Differential pathogenesis of lethal mousepox in congenic DBA/2 mice implicates natural killer cell receptor NKR-P1 in necrotizing hepatitis and the fifth component of complement in recruitment of circulating leukocytes to spleen. Am J Pathol 150:1407–20
13. Schell K (1960) Studies on the innate resistance of mice to infection with mousepox. II. Route of inoculation and resistance, and some observations on the inheritance of resistance. Aust J Exp Biol Med Sci 38:289–99
14. Jacoby RO, Bhatt PN (1987) Mousepox in inbred mice innately resistant or susceptible to lethal infection with ectromelia virus. II. Pathogenesis. Lab Anim Sci 37:16–22
15. Bhatt PN, Jacoby RO (1987) Mousepox in inbred mice innately resistant or susceptible to lethal infection with ectromelia virus. I. Clinical responses. Lab Anim Sci 37:11–5
16. Bhatt PN, Jacoby RO, Gras L (1988) Mousepox in inbred mice innately resistant or susceptible to lethal infection with ectromelia virus. IV. Studies with the Moscow strain. Arch Virol 100:221–30
17. Wallace GD, Buller RM (1985) Kinetics of ectromelia virus (mousepox) transmission and clinical response in C57BL/6j. BALB/cByj and AKR/J inbred mice. Lab Anim Sci 35:41–6
18. Wallace GD, Buller RML, Morse HC III (1985) Genetic determinants of resistance to ectromelia (mousepox) virus-induced mortality. J Virol 55:890–1
19. Schriewer J, Buller RM, Owens G (2004) Mouse models for studying orthopoxvirus respiratory infections. Methods Mol Biol 269: 289–308
20. Earl PL, Moss B, Wyatt LS, Carroll MW (2001) Generation of recombinant vaccinia viruses, Chapter 16. In: Ausubel FM (ed) Current protocols in molecular biology
21. Forbes N, Brayton C, Grindle S, Shepherd S, Tyler B, Guarnieri M (2010) Morbidity and mortality rates associated with serial bleeding from the superficial temporal vein in mice. Lab Anim (NY) 39:236–40
22. Zhang Z, Kermekchiev MB, Barnes WM (2010) Direct DNA amplification from crude clinical samples using a PCR enhancer cocktail and novel mutants of Taq. J Mol Diagn 12:152–61

Chapter 12

Analyzing CD8 T Cells in Mouse Models of Poxvirus Infection

Inge E.A. Flesch, Yik Chun Wong, and David C. Tscharke

Abstract

Mouse models of immunology are frequently used to study host responses to poxviruses or poxvirus-based recombinant vaccines. In this context, the magnitude of CD8⁺ T cell responses is often of interest. Methods to evaluate CD8⁺ T cell responses extend from those that rely on indirect measurement of effector function only, such as cytotoxicity assays, to those that only measure antiviral CD8⁺ T cell numbers and not function, like peptide MHC tetramers. In this chapter, five methods are provided that cover this range: DimerX staining (a variant of peptide-MHC tetramers), intracellular cytokine staining for interferon-γ, CD62L/Granzyme B staining, and in vitro and ex vivo cytotoxicity assays. We also include tables of vaccinia virus peptide epitopes for use in most of these assays.

Key words: CD8 T lymphocyte, Cytotoxic T lymphocytes, CTL, Epitopes, Peptide-MHC tetramers, Mouse model

1. Introduction

Vaccinia virus (VACV) and other orthopoxviruses have been used in mouse models of immunity for more than 50 years (1, 2). The aims of this work range widely from understanding basic principles of immune activation to more applied preclinical testing of recombinant vaccine candidates. For all these purposes, accurate measurement of immunity is required.

CD8⁺ T cells are important antiviral effectors, so are of interest in both a basic sense and also as they are often elicited by recombinant vaccinia vaccines (3). These lymphocytes bear a T cell receptor (TCR) that binds short peptides presented on major histocompatibility complex class I molecules (MHC-I) on the surface of other cells (4). Naïve CD8⁺ T cells that recognize virus-derived peptides are activated during the infection period, and become effector cells. When these activated cells encounter their antigen on an infected

Stuart N. Isaacs (ed.), *Vaccinia Virus and Poxvirology: Methods and Protocols*, Methods in Molecular Biology, vol. 890,
DOI 10.1007/978-1-61779-876-4_12,

cell, they can kill the cell and secrete a range of cytokines, some of which are antiviral and others that modulate immune responses.

Historically, measurement of CD8⁺ T cells has been tied to their ability to kill virus-infected or antigen-loaded cells in vitro. An advantage of this approach is that it clearly shows a relevant function. However, it is indirect as it monitors the killing of targets rather than the CD8⁺ T cells themselves and originally required radioactive ^{51}Cr to track the lysis of target cells. Flow cytometry has enabled nonradioactive cytotoxicity assays and a relatively new twist is to allow the killing to happen in vivo in a mouse rather than in vitro (5, 6). Other current methods also take advantage of flow cytometry. Staining lymphocyte preparations with peptide-MHC tetramers or similar reagents, such as DimerX, allows direct enumeration of CD8⁺ T cells with known antiviral specificities, but not function (7). Alternatively, methods that use antibodies to stain for intracellular cytokines (or surface CD107) after a brief in vitro stimulation with virus peptides allow strict quantification as well as the demonstration of a relevant antiviral function (8–11).

Another option to be considered in experimental design is whether to measure responses to individual virus-derived peptides or to assess the total response to infection. To assess the total response to infection, peptide-MHC tetramers are not useful, but virus-infected cells can be used to stimulate splenocytes prior to intracellular cytokine staining and can be used as targets for cytotoxicity assays in vitro. In addition, because CD8⁺ T cells that are involved in the antiviral response change their expression of a variety of molecules, these activation markers can be used to quantify the total anti-VACV response. Two such markers that are useful in this context are reduced CD62L (L-selectin) on the cell surface and increased intracellular granzyme B (GzmB) (12).

This chapter covers five methods, all of which use flow cytometry and are written assuming familiarity with this technology. They are DimerX staining (a variant of peptide-MHC tetramers), intracellular cytokine staining for interferon-γ (IFN-γ), CD62L/GzmB staining, and in vitro and ex vivo cytotoxicity assays. Table 1 lists the ten most dominant VACV CD8⁺ T cell epitopes mapped thus far for H-2^b (e.g., C57Bl/6) and H-2^d (e.g., BALB/c) haplotype mice, but other less dominant epitopes have also been identified and may be useful in some contexts (11–15). Our work is focused on VACV and so this virus is referenced throughout, but most methods are adaptable to models of cowpox and ectromelia virus infections.

Table 1
Top ten ranked CD8+ T cell epitopes of VACV for H-2b and H-2d haplotype mice

Name[a]	Sequence	MHC[b]	WR Gene	Peptide variations[c]
$B8_{20-27}$	TSYKFESV	H-2K^b	VACWR190	
$A8_{189-196}$	ITYRFYLI	H-2K^b	VACWR127	
$A23_{297-305}$	IGMFNLTFI	H-2D^b	VACWR143	
$A47_{171-180}$	YAHINALEY	H-2D^b	VACWR173	
$A47_{138-146}$	AAFEFINSL	H-2K^b	VACWR173	ECTV: ATFEFINSL
$A3_{270-277}$	KSYNYMLL	H-2K^b	VACWR122	
$K3_{6-15}$	YSLPNAGDVI	H-2D^b	VACWR034	
$B2_{54-62}$	YSQVNKRYI	H-2D^b	VACWR184	Absent: MVA
$L2_{53-61}$	VIYIFTVRL	H-2K^b	VACWR089	ECTV: VIYIFTVHL
$C4_{125-132}$	LNFRFENV	H-2K^b	VACWR024	Absent: ACAM2000, MVA, ECTV
$F2_{26-34}$	SPYAAGYDL	H-2L^d	VACVWR041	Other VACV and CPXV: SPGAAGYDL[d]
$A52_{75-83}$	KYGRLFNEI	H-2K^d	VACVWR178	Absent: MVA
$E3_{140-148}$	VGPSNSPTF	H-2D^d	VACVWR059	ECTV: VGPSNSPIF
$A3_{190-198}$	IYSPSNHHI	H-2K^d	VACVWR122	
$C6_{74-82}$	GFIRSLQTI	H-2K^d	VACVWR022	Other VACV and ECTV: SFIRSLQNI
$I8_{90-98}$	LPNPAFIHI	H-2L^d	VACVWR077	
$I8_{511-519}$	QYIYSEHTI	H-2K^d	VACVWR077	ECTV: QYIYSEYTI
$B2_{49-57}$	KYMWCYSQV	H-2K^d	VACVWR184	Absent: MVA
$D1_{351-359}$	KYEGPFTTT	H-2K^d	VACVWR106	
$J6_{782-790}$	KYAANYTKI	H-2K^d	VACVWR098	

[a]Naming uses VACV Copenhagen nomenclature with amino acid positions in subscript. Order is approximate order of dominance, with the first peptide for each MHC haplotype being the immunodominant epitope after VACV WR infection
[b]MHC restriction; H-2^b haplotype (e.g., C57Bl/6) top half of table, H-2^d haplotype (e.g., BALB/c) bottom half of table
[c]Variations in peptide sequence or predicted expression in VACV strains other than WR (including Copenhagen, Lister, ACAM200, and MVA) and other orthopoxviruses including ectromelia virus (ECTV) and cowpox virus (CPXV)
[d]Unlike SPYAAGYDL in WR, SPGAAGYDL is not usually dominant. ECTV: SNHAAGYDL

2. Materials

2.1. General

1. Eight-week-old female C57BL/6 or BALB/c mice.
2. Tissue culture consumables: Microtiter plates, flasks, pipettes, tubes, filtered tips, etc.
3. 70-μm Sterile cell strainers (e.g., BD Biosciences).

4. Phosphate-buffered saline (PBS).
5. ACK red cell lysis buffer: 7.5 g of NH_4Cl dissolved in 900 ml of water, add 111 ml of 0.17 M Tris–HCl, pH 7.65, adjust the final solution to pH 7.2.
6. D10: Dulbecco's modified Eagle's medium (DMEM) with L-glutamine and 10% fetal bovine serum (FBS).
7. D2: DMEM with L-glutamine and 2% FBS.
8. D0: DMEM with L-glutamine.
9. FACS buffer: PBS with 2% FBS.
10. 80% Ethanol v/v or other disinfectant solution to inactivate virus.
11. Dimethylsulfoxide (DMSO).
12. VACV strains Western Reserve (WR; ATCC, #VR1354) or Modified Vaccinia Ankara (MVA), preferably sucrose cushion purified with a titer greater than 1×10^9 plaque-forming units (pfu)/ml.
13. Cluster tubes (e.g., Corning Life Sciences).
14. Flow cytometer and analysis software.
15. Hemocytometer.
16. Refrigerated bench centrifuge (e.g., Beckman Allegra X-12R or Allegra X-15R with an SX4750 rotor swing-out with maximum radius for buckets of 207.8 mm). The speeds (rpm) given here relate to most standard bench centrifuges that have a similar radius and equate to: 1500 rpm = 525 g and 2100 rpm = 1025 g.

2.2. Synthetic Peptides, Antibodies, and DimerX Reagent

1. Lyophilized peptides: Prepare master stocks in DMSO at 10 mg/ml and store at –70°C (see Note 1).
2. DimerX: Recombinant soluble dimeric mouse H-2D^b:Ig, H-2K^b:Ig, or H-2L^d:Ig fusion proteins (e.g., BD Biosciences).
3. Anti-mouse IgG_1 conjugated with Phycoerythrin (PE) (e.g., clone A85-1 from BD Biosciences).
4. Fc-block: Anti-mouse CD16/CD32 (e.g., clone 2.4 G2 from BD Biosciences or BioLegend).
5. Anti-mouse CD8α conjugated with PE, Allophycocyanin (APC), or APC-Cy7 (e.g., clone 53-6.7 from BD Biosciences or BioLegend).
6. Anti-mouse IFN-γ conjugated with APC (e.g., clone XMG.2 BD Biosciences or BioLegend).
7. Anti-mouse CD62L conjugated with fluorescein isothiocyanate (FITC) (e.g., clone MWL-14 from BioLegend).
8. Anti-human GzmB conjugated with APC (e.g., clone GB12 from Caltag/Invitrogen) or conjugated with AlexaFluor® 647 (e.g., clone GB11 from BioLegend) (see Note 2).

2.3. Intracellular Cytokine Staining and CD62L/GzmB Staining

1. DC2.4 cells: Dendritic-like cells derived from C57BL/6 mouse (16) (see Note 3).
2. P815 cells (ATCC# TIB-64): Mastocytoma cells derived from a DBA/2 (H-2^d haplotype mouse) (see Note 3).
3. Brefeldin A: 5 mg/ml stock in methanol and store at 4°C. Dilute this stock 1:100 in D10 to make a 50 μg/ml solution immediately before use.
4. 16% Paraformaldehyde solution: 1% solution in PBS prepared before use.
5. Saponin: 5% stock solution in water, diluted to 0.5% in FACS buffer before use.

2.4. In Vivo and In Vitro Cytotoxicity Assays

1. CFSE (5(6)-Carboxyfluorescein-diacetate *N*-succinimidyl) stock solution: 1 mg/ml in DMSO stored in small aliquots at –20°C.
2. DiD: Vybrant DiD cell-labeling solution (e.g., Invitrogen).
3. Insulin syringes, 27 G × 1/2 in.
4. Collagenase/DNase I solution: 3 mg/ml collagenase from bovine pancreas, grade II (e.g., Worthington), and 0.03% DNase I (e.g., Roche) in D2, store in small aliquots at –20°C, and dilute 1:3 in D2 before use.
5. EL-4 or RMA cells: Thymoma cell lines from C57BL/6 mice (17).
6. V-bottom 96-well microtiter plates.

3. Methods

3.1. Preparation of Splenocytes from Infected Mice

1. Collect spleens from infected mice in 5 ml D0 or PBS (see Note 4).
2. Pour each spleen and the medium into a cell strainer placed on top of a 50-ml tube (see Note 5).
3. Crush the spleen gently by using the back end (not the rubber end that goes into the syringe) of a plunger from a 1-ml syringe.
4. Rinse the cell strainer with 5 ml D0.
5. Centrifuge cells for 5 min at 1,500 rpm and 4°C, pour off supernatant, and resuspend cell pellets in 5 ml of ACK red cell lysis buffer.
6. Leave at room temperature for 3 min, then add ~25 ml PBS so that the final volume is ~30 ml, mix, and centrifuge again for 5 min at 525 × *g* and 4°C.

7. Resuspend cells in 5 ml D10 and count cells using a hemocytometer.
8. Adjust to 1×10^7 cells/ml for most methods.

3.2. Detection of Peptide-Specific CD8⁺ T Cells Using DimerX Staining

DimerX reagents (MHC class I:Ig dimeric proteins) are a variant of the original peptide-MHC tetramers used to determine the number of epitope-specific $CD8^+$ T cells. In contrast to tetramers, which must be custom made for each peptide, DimerX reagents are loaded with the peptide of choice the day before the assay is performed. This assay is a flexible system for detecting $CD8^+$ T cells with specificity to any antigenic peptide of interest as long as the appropriate MHC allomorph is available.

1. Dilute required peptide stocks to 0.1 mg/ml in PBS.
2. Choose the appropriate DimerX reagent for the peptide. Table 1 lists the MHC restriction for each peptide. Mix the DimerX reagent with peptide in PBS and incubate overnight at 37°C (see Note 6).
3. Mix 0.25–4 μg of peptide-loaded DimerX with 0.25–4 μg of anti-mouse IgG_1 (A85-1)-PE at a ratio of 1:1 or 1:2 of DimerX to A85-1 (see Note 7).
4. Incubate for 1 h at room temperature in the dark.
5. Add 1×10^6 of each splenocyte sample to a well of a round-bottom 96-well microtiter plate, centrifuge plate for 5 min at 1,500 rpm and 4°C, remove supernatant by flicking out (see Note 8), add 50 μl FACS buffer to each well, and resuspend the cells.
6. Add 20 μl Fc-block and incubate for 15 min at 4°C.
7. To wash cells, add 150 μl with FACS buffer, centrifuge plate for 5 min at 1,500 rpm and 4°C, and remove supernatant by flicking. Repeat.
8. Add 50 μl of the peptide-DimerX-A85-1 complex from step 2 with APC-conjugated anti-CD8α antibody (1:200 diluted in FACS buffer) and incubate for 1 h at 4°C in the dark (see Note 9).
9. Wash cells three times with 150 μl FACS buffer; this time, centrifuge for 3 min at $1025 \times g$ and 4°C.
10. Resuspend cells in 50–100 μl FACS buffer and transfer into cluster tubes for flow cytometry.
11. For flow cytometry, set the stopping gate on the $CD8^+$ population and acquire 50,000–100,000 events. Save all events.
12. For analysis, lymphocytes are gated first on a forward scatter (FSC) × side scatter (SSC) plot. Activated lymphocytes can be quite large and granular, so this gate is set broadly. Then, $CD8^+$ cells are gated on a CD8 × SSC plot and followed by $DimerX^+$ cells on a CD8 × DimerX plot (see Note 10 and Fig. 1).

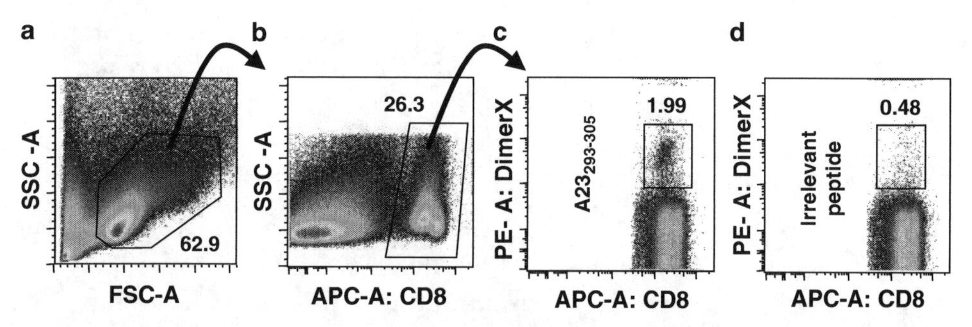

Fig. 1. Flow cytometry gating strategy for detection of VACV peptide-specific CD8+ T cells using DimerX reagents. In the example shown, mice were infected for 7 days with 1×10^6 pfu of VACV WR by the intraperitoneal route and $A23_{297-305}$-specific CD8+ T cells detected in splenocytes using DimerX-D^b loaded with synthetic IGMFNLTFI peptide. (**a**, **b**) Show gating for live lymphocytes and CD8+ T cells, respectively. (**c**) Shows the CD8+, $A23_{297-305}$-DimerX+ population and (**d**) shows the same splenocytes, but stained with DimerX loaded with an irrelevant peptide. *SSC* side scatter, *FSC* forward scatter. See Note 10 for additional discussion of this figure.

3.3. Ex Vivo Stimulation and Intracellular Cytokine Staining for IFN-γ (ICS Assay)

In intracellular cytokine staining (ICS) assays, the number of $CD8^+$ T cells secreting IFN-γ is determined after a short in vitro stimulation of splenocytes from infected mice with relevant peptides (see Note 11). Other cytokines can also be examined (e.g., tumor necrosis factor α or IL-2) if appropriate antibodies are available, but $CD8^+$ T cells expressing these cytokines are a subset of those that can secrete IFN-γ.

1. For stimulations, use either (a) synthetic peptides or (b) virus-infected cells. For (a), dilute 10 mg/ml peptide stocks with D0 to 0.2 μM and add 100 μl to each well of a 96-well round-bottom microtiter plate (see Note 12). For (b), infect DC2.4 cells (H-2^b mice) or P815 cells (H-2^d mice) with VACV at an MOI of 5 pfu/cell for 5 h (see Note 13) and add 100 μl at 2×10^6 cells/ml to each well of a round-bottom microtiter plate.
2. Add 100 μl of each splenocyte sample at 1×10^7 cells/ml in D10 (see Subheading 3.1) to each well.
3. Incubate plate at 37°C with 5% CO_2 for 1 h.
4. After 1 h, add 20 μl of 50 μg/ml brefeldin A to each well (see Note 14) and return to CO_2 incubator for a further 3 h.
5. At the end of this incubation, centrifuge the plate for 5 min at 1,500 rpm and 4°C and discard supernatant by flicking out medium (see Notes 8 and 15).
6. Resuspend cells in 40 μl/well of cold (4°C) anti-mouse CD8α-PE (1:150 diluted in FACS buffer).
7. Incubate plate on ice for 30 min in the dark.
8. Add 150 μl cold FACS buffer to wells, centrifuge plate for 3 min at 2,100 rpm and 4°C, and flick out medium (see Note 8).

9. Resuspend cells in 200 μl cold PBS (no serum or protein), and centrifuge plate for 3 min at 2,100 rpm and 4°C (see Note 16).
10. Resuspend cells in 50 μl/well of 1% paraformaldehyde (see Note 17).
11. Incubate at room temperature for 20 min in the dark.
12. Add 150 μl cold FACS buffer to wells, centrifuge plate for 3 min at 2,100 rpm, and flick out medium.
13. Resuspend cells in 200 μl FACS buffer, centrifuge plate for 3 min at 2,100 rpm, and repeat.
14. Resuspend cells in 40 μl of anti-IFN-γ antibody-APC (1:200 diluted in FACS buffer containing 0.5% saponin).
15. Leave plate in fridge overnight, or on ice for at least 1 h.
16. Wash cells two times with FACS buffer as in steps 12 and 13.
17. Resuspend cells in 50–100 μl of FACS buffer and transfer into cluster tubes for flow cytometry.
18. For flow cytometry, set the stopping gate on the CD8$^+$ population and acquire 50,000–100,000 events, usually as many as possible. Save all events.
19. For analysis, gate on live cells in the FSC × SSC plot. Then, gate the CD8$^+$ population on a CD8$^+$ × SSC plot, and finally gate the IFN-γ$^+$ population on a CD8 × IFN-γ plot (see Note 18 and Fig. 2).

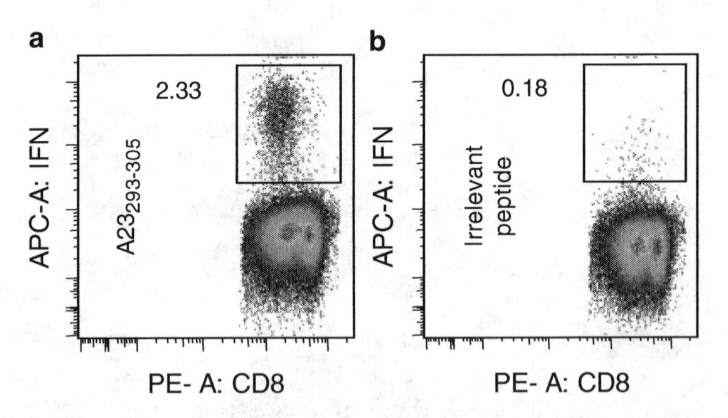

Fig. 2. Flow cytometry results showing detection of VACV peptide-specific CD8$^+$ T cells using ICS. Mice were infected for 7 days with 1×10^6 pfu of VACV WR by the intraperitoneal route and $A23_{297-305}$ (**a**) or an irrelevant peptide (**b**) used to restimulate splenocytes before antibody staining for surface CD8 and intracellular IFN-γ. The plots shown were previously gated for live lymphocytes and the CD8$^+$ events as in Fig. 1. See Note 18 for additional discussion of this figure.

3.4. CD62L/GzmB Staining

Decreased surface CD62L and increased intracellular GzmB expression are markers of activated CD8$^+$ T cells. Using these markers in combination is useful for estimating the total size of the response to VACV at the peak of the acute infection, but is not helpful for studies of memory CD8$^+$ T cell responses. The main advantage of this method is that it does not require in vitro restimulation of cells and therefore is independent of either knowledge of epitopes or access to infected cells. Staining for these markers can be combined with DimerX staining using DimerX loaded with VACV epitopes and in our hands >90% (often >95%) of DimerX$^+$ events will have the activated phenotype (12).

1. Prepare single-cell suspensions of splenocytes (see Subheading 3.1). For this method, the only possible negative control is an uninfected mouse and at least one must be included in each experiment.
2. Add 1×10^6 cells from each sample into wells of a round-bottom 96-well microtiter plate.
3. Centrifuge plate for 5 min at 1,500 rpm and 4°C and discard supernatant by flicking (see Note 8).
4. Resuspend cells in 50 μl of FACS buffer containing anti-CD8α-APC-Cy7 (or CD8alpha-PE) and anti-CD62L-FITC, with each at a final dilution of 1:150.
5. Incubate on ice for 30 min in the dark.
6. Add 150 μl cold FACS buffer to wells, centrifuge plate for 3 min at 2,100 rpm and 4°C, and flick out supernatant.
7. Resuspend cells in 200 μl PBS, centrifuge plate again for 3 min at 2,100 rpm and 4°C, and flick out medium.
8. Resuspend cells in 50 μl/well of 1% paraformaldehyde in PBS and incubate at room temperature for 20 min.
9. Add 150 μl cold FACS buffer to wells, centrifuge plate for 3 min at 2,100 rpm, and flick out supernatant.
10. Resuspend cells in 200 μl FACS buffer, centrifuge plate for 3 min at 2,100 rpm, flick out supernatant, and repeat.
11. Resuspend cells in 50 μl anti-GzmB-APC, diluted 1/200 in FACS buffer with 0.5% saponin.
12. Leave plate in fridge overnight, or for at least 1 h.
13. Wash twice as in steps 9 and 10.
14. Resuspend cells in 50–100 μl of FACS buffer and transfer into cluster tubes for flow cytometry.
15. For flow cytometry, set the stopping gate on the CD8$^+$ population and acquire at least 20,000 events (though usually 50,000–100,000 are easily achievable). Save all events.

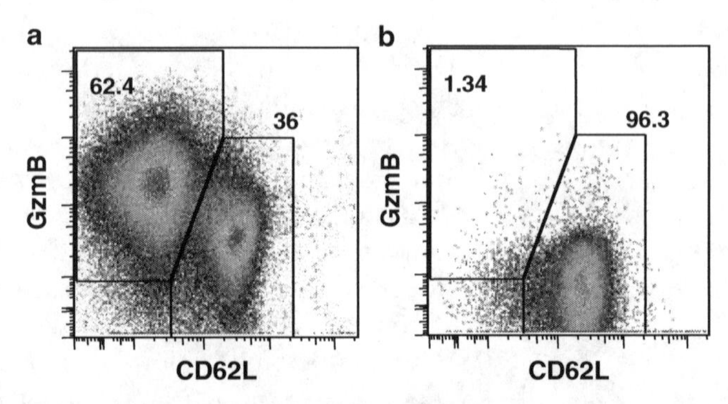

Fig. 3. Flow cytometry results showing detection of the total acute CD8+ T cell response to VACV using CD62L/GzmB staining. Mice were infected for 7 days with 1×10^6 pfu of VACV WR by the intraperitoneal route (**a**) or were uninfected (**b**) and splenocytes were stained with antibodies to detect surface CD8 and CD62L and intracellular GzmB. The plots shown were previously gated for live lymphocytes and CD8+ events as in Fig. 1. Activated (anti-VACV) CD8+ T cells are in the CD62L^{low} and GzmB^{high} gate. Note that neither CD62L nor GzmB alone adequately distinguishes the activated CD8+ T cells, so traditional "quadrant" gating is not appropriate.

16. In analysis, gate on live cells, followed by the CD8+ population on a CD8 × SSC plot, as for DimerX staining and ICS assays (see Fig. 1a, b). The CD8+ events are then displayed on a plot of CD62L × gzmB (see Fig. 3). The activated CD8+ T cells are CD62^{low} and Granzyme B^{high} and the percent in this gate represents the fraction of CD8+ T cells that are responding to the infection. For VACV WR at the peak of acute infection, expect values of 60% of CD8+ T cells or higher.

3.5. In Vivo Cytotoxicity Assay

Killing of cells presenting a relevant antigen is one of the central effector functions of activated CD8+ T cells. This can be determined using an in vivo cytotoxicity assay, which measures the specific lysis of peptide-loaded target cells versus non-loaded cells that are cotransferred into infected mice. These transferred cells are labeled with DiD and two different concentrations of CFSE for cell tracking and discrimination between the two transferred populations. The method is highly sensitive and can be used to detect a CD8+ T cell response as early as 2 days after virus infection if a dominant peptide is used (18).

3.5.1. Preparation of CFSE-Labeled Target Cells

1. Collect spleens from uninfected mice and prepare splenocytes (see Subheading 3.1) (see Note 19).
2. Resuspend splenocytes in 4 ml D0.
3. Add 10 μl Vybrant DiD dye. Add the dye while vortexing at a medium speed to allow uniform labeling of cells.
4. Split cells into two populations, transfer into 15-ml conical tubes, and label 1 and 2.
5. Add 20 μl of 100 μM peptide working stock to Tube 1 (see Note 20). The final peptide concentration in Tube 1 is 1 μM.

6. Incubate both tubes for 1 h at 37°C on a rotator or with occasional shaking to keep cells resuspended.
7. Wash cells in both tubes three times with 5 ml D0 (to remove unbound peptide).
8. Resuspend cells in both tubes in 1 ml D0.
9. Add 28 μl of CFSE working stock solution to Tube 1 (5 μM final concentration; $CFSE^{high}$ in flow cytometric analysis) and 2.8 μl of CFSE working stock solution to Tube 2 (0.5 μM final concentration; $CFSE^{low}$) (see Note 21).
10. Incubate both tubes for 10 min in a 37°C water bath. Mix cells every 2 min.
11. Add 5 ml cold D10 to each tube to quench the labeling reaction.
12. Centrifuge for 5 min at 1,500 rpm and 4°C.
13. Wash cells two more times with 5 ml cold D10.
14. Wash cells with 5 ml PBS.
15. Count cells in both tubes.
16. Combine equal numbers of $CFSE^{high}$ and $CFSE^{low}$ cells.
17. Adjust to $1–2 \times 10^8$ cells/ml PBS.

3.5.2. Injection and Recovery of Targets (Including Lymph Nodes)

1. Inject 200 μl of the CFSE-labeled cell suspension (this is $2–4 \times 10^7$ cells) intravenously into the tail vein of infected/immunized mice and one uninfected control mouse (see Note 22).
2. Four hours after transfer, collect spleens and if required any lymph nodes of interest.
3. Prepare splenocytes as described in Subheading 3.1, up to step 6. Then, resuspend cells in 2 ml of FACS buffer containing 1% paraformaldehyde in PBS, transfer 200 μl to a cluster tube for flow cytometric analysis, and then go to Subheading 3.5.3.
4. If lymph nodes are to be analyzed, remove D2 from the tubes used when harvesting the lymph nodes from the mice and mince nodes with scissors (see Note 23).
5. Add 200 μl Collagenase/DNase I solution and squeeze disrupted lymph nodes with the plunger of a 1-ml syringe to release more cells.
6. Top up tubes to 1 ml with Collagenase/DNase I solution and incubate at room temperature for 20 min.
7. Transfer cells into 15-ml conical tubes containing 3 ml FACS buffer.
8. Centrifuge for 5 min at 1,500 rpm and 4°C.
9. Resuspend cells in 100 μl FACS buffer containing 1% paraformaldehyde and transfer into cluster tubes for flow cytometric analysis.

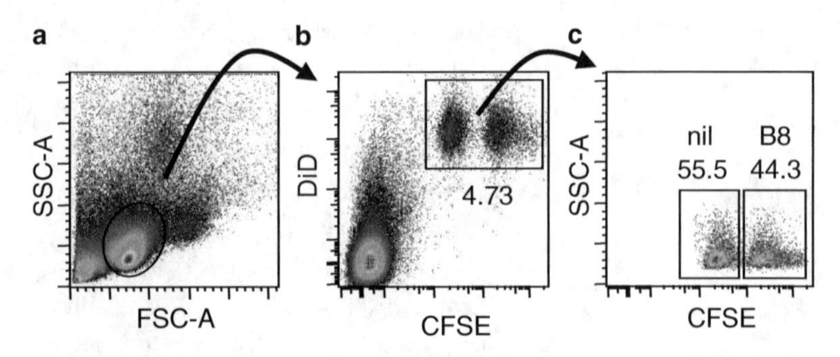

Fig. 4. Flow cytometry gating strategy to detect the remaining labeled target cells recovered from the spleen of an infected mouse used for an in vivo cytotoxicity assay. (**a**, **b**) Show gating for live lymphocytes and DiD$^+$ cells, respectively. (**c**) Shows the gating of CFSE^{high} (loaded with $B8_{20-27}$ peptide) and CFSE^{low} (loaded with an irrelevant peptide or no peptide). In this case, the "ratio test mouse" is 55.5/44.3 = 1.25. To complete the calculation of percent killing, the "ratio uninfected" needs to be determined using the same method, but the labeled and peptide-loaded cells are transferred and recovered from an uninfected mouse. In this example, mice were infected for 2 days with 1×10^6 pfu of VACV WR by the intraperitoneal route. Killing of $B8_{20-27}$ peptide-loaded cells is much higher (can be 90%) by 7 days after infection. See Note 26 for sample calculations.

3.5.3. Flow Cytometric Analysis to Determine Peptide-Specific Cytotoxicity

1. For flow cytometry, set the stopping gate on the DiD$^+$ population and aim to acquire 20,000 events. This will be possible for spleens and large lymph nodes (see Note 24). Save all events.
2. Gate events to exclude dead cells and debris on FSC × SSC, followed by gating on DiD$^+$ events on SSC × DiD plot to identify the transferred cell population. Finally, gate the CFSE^{high} and CFSE^{low} populations from the DiD$^+$ subset and determine the percentage of events in each of these gates (see Note 25 and Fig. 4).
3. To calculate specific lysis, the following formula is used:

$$\text{Specific lysis} = [1 - (\text{ratio uninfected mouse} / \text{ratio test mouse})] \times 100\%,$$

where the ratio in each mouse is percentage CFSE^{low}/percentage CFSE^{high} (see Note 26 and Fig. 4).

3.6. In Vitro Cytotoxicity Assay

This assay is based on similar principles as the in vivo assay in the section above. It also uses different levels of CFSE to identify cell populations loaded with different peptides to determine the loss of cells that are CD8$^+$ T cell targets. Here, non-adherent splenocytes from infected mice are used as effectors and peptide-loaded EL-4 or RMA cells are used as targets. In this assay, DiD$^+$ staining identifies the targets more effectively than in the in vivo assay allowing the use of CFSE-unlabeled targets in addition to CFSE^{low} and CFSE^{high}. In this way, a control and two peptides can be tested in a single well for a direct comparison. The example given here is for H-2^b haplotype (e.g., C57Bl/6) mice, but a similar assay can be done in principle for H-2^d haplotype (e.g., BALB/c) mice, but P815 cells would be used as targets.

3.6.1. Preparation of Effector Cells

1. Collect spleens from infected and uninfected mice and prepare splenocytes (see Subheading 3.1 and Note 27).
2. Resuspend splenocytes from each spleen in 20 ml D10 and transfer to a 75-cm^2 tissue culture flask.
3. Incubate for 1 h at 37°C with 5% CO_2.
4. Harvest non-adherent cells, centrifuge to collect the cells, and resuspend them in D10 (see Note 28).
5. Adjust cell count to 1×10^7 cells/ml in D10 and dilute from here for the appropriate effector-to-target (E:T) ratios (see Subheadhing 3.6.2).

3.6.2. Preparation of Target Cells

1. The assay requires 1×10^4 target cells per well and the number of wells required is usually four wells for each test mouse and for the negative control (splenocytes from an uninfected mouse) for each set of two peptides. The four wells are required to test different ratios of effectors and targets: 100:1, 50:1, 20:1, and 10:1. Once the plate layout is determined, calculate the number of target cells required.
2. Harvest RMA or EL-4 cells (see Note 29), wash once in D0, and count cells. Adjust to 1×10^6 cells/ml and transfer enough volume to provide the number of targets calculated in step 1.
3. Add 1 µl Vybrant DiD dye for every 1×10^6 cells and mix well.
4. Divide equally into three 15-ml conical tubes (2 ml each) and label as 1, 2, and 3.
5. Add peptide #1 working stock (e.g., $B8_{20}$ TSYKFESV) to Tube 1 and peptide #2 working stock (e.g., $A8_{189}$ ITYRFYLI) to Tube 2 to give a final peptide concentration of 0.1 µM for each peptide in each tube.
6. Cells in Tube 3 are loaded with an irrelevant peptide or are left unloaded.
7. Incubate all tubes for 1 h at 37°C on a rotator or with occasional shaking to keep cells resuspended.
8. Wash cells in all tubes three times with 5 ml D0 (to remove unbound peptide).
9. Resuspend cells in each tube in 1 ml D0.
10. Add 14 µl of 100 µl/ml CFSE working stock solution to Tube 1 (2.5 µM final concentration; CFSE^{high} in flow cytometric analysis) and 1.4 µl of 100 µl/ml CFSE working stock solution to Tube 2 (0.25 µM final concentration; CFSE^{low}). Add no CFSE to tube 3.
11. Incubate both tubes for 10 min in a 37°C water bath. Mix cells every 2 min.
12. Add 5 ml cold D10 to each tube to quench the labeling reaction.

13. Centrifuge for 5 min at 1,500 rpm and 4°C.
14. Wash cells two more times with 5 ml cold D10.
15. Resuspend cell in each tube in 2 ml of D10 and combine into a single tube.
16. Count cells and adjust to 1×10^5 cells/ml.

3.6.3. Cytotoxicity Assay

1. Add 100 μl (1×10^4) of target cells into the wells of a V-bottom 96-well microtiter plate according to the desired layout.
2. Add effector cells (from Subheading 3.6.1), including the control "effectors" from an uninfected mouse, to give E:T ratios of 100, 50, 20, and 10 to 1.
3. Top up all wells to 200 μl with D10.
4. Incubate plate for 16 h at 37°C with 5% CO_2.
5. Centrifuge the plate for 5 min at 1,500 rpm and 4°C, and remove supernatant by flicking (see Note 8).
6. Resuspend cells in 200 μl/well FACS buffer.
7. Spin plate for 3 min at 2,100 rpm and 4°C, and flick out medium.
8. Resuspend cells in 60 μl/well of 1% paraformaldehyde and transfer into cluster tubes for flow cytometric analysis.

3.6.4. Flow Cytometric Analysis and Calculation of Percent-Specific Lysis

1. Gate events for live RMA cells on the FSC × SSC, followed by gating on DiD^+ cells. DiD^+ cells are further gated on $CFSE^{high}$, $CFSE^{low}$, and $CFSE^{neg}$ cells (see Fig. 5).
2. To calculate specific lysis, the following formula is used:

$$\text{Specific lysis} = [1 - (\text{ratio negative control} / \text{ratio test effectors}) \times 100\%,$$

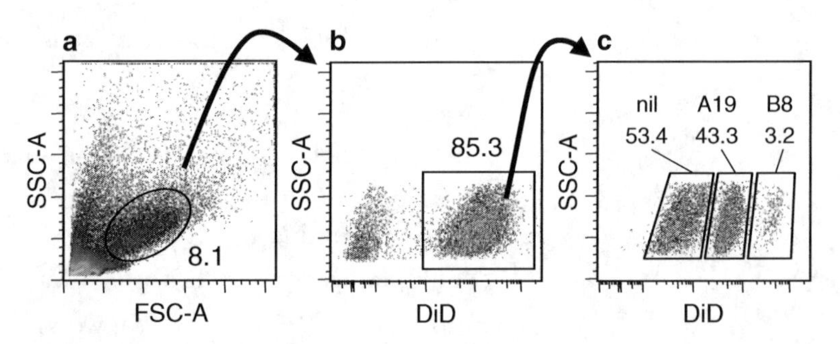

Fig. 5. Flow cytometry gating strategy to detect the remaining labeled target cells recovered from wells after an in vitro cytotoxicity assay. (**a**, **b**) Show gating for live RMA cells and DiD^+ cells, respectively. (**c**) Shows the gating of $CFSE^{high}$ (loaded with $B8_{20-27}$ peptide), $CFSE^{low}$ (loaded with $A19_{47-55}$ peptide), and $CFSE^{neg}$ (loaded with an irrelevant peptide) events. In this case, the "ratio test $B8_{20-27}$" is 53.4/3.2 = 16.69 and "ratio test $A19_{47-55}$" is 53.4/43.3 = 1.23. $A19_{47-55}$ (VSLDYINTM) is a subdominant epitope not listed in Table 1. To complete the calculation of percent killing, the "ratio uninfected" needs to be determined for each peptide using the same method, but with "effector splenocytes" from an uninfected mouse. Mice used to generate the effectors here were infected 7 days previously with 1×10^6 pfu of VACV WR by the intraperitoneal route. See Note 30 for sample calculations.

where the ratios for peptide #1 are the percentage CFSE⁻/percentage CFSE^{high} from negative control and test wells and ratios for peptide #2 are the percentage CFSE⁻/percentage CFSE^{low} from control and test wells (see Note 30).

4. Notes

1. We purchase peptides from Mimotopes, Melbourne, Australia, or Genscript Corp, Piscataway, NJ. Peptides are ordered at a purity of >70% and with both termini unblocked (that is, free amine on N-terminus and free acid on C-terminus). Synthesis and solubility of synthetic peptides can be a problem, but all the peptides listed in Table 1 have been synthesized without issues and all have good solubility in 100% DMSO. Dilute the master stock of peptides in D0 or PBS before use for DimerX loading (see Subheading 3.2) or in vitro stimulation of splenocytes (see Subheading 3.3), respectively. It is important to avoid FBS in media used to dilute peptides because serum contains a range of undefined enzymes, including peptidases.
2. The chosen fluorophores for each of the antibodies in this section work well in the assays described and are a good starting point. However, other combinations of colors may work well, but may require titration and optimization. The anti-human GzmB antibodies both cross-react with mouse GzmB.
3. These cell lines are used as virus-infected antigen-presenting cells.
4. Spleens are taken 7 or 42 days after infection/immunization to measure acute or memory $CD8^+$ T cell responses, respectively.
5. There are many other methods for making single-cell suspensions of splenocytes and the one given here is simple and the one used routinely in our laboratory. The cell strainers are fairly expensive, but can be washed, autoclaved, and reused a few times. An alternative method that does not require the use of these strainers is as follows. Crush the spleens in a petri dish using a plunger from a 10-ml syringe in the 5 ml of medium that they are collected in (use the back end of the plunger, not the rubber end that goes in the syringe). The suspension is then recovered and placed into a 15-ml centrifuge tube (this can be the same tube used to collect the spleen) and 5 ml of fresh medium is used to wash the dish and is added to the centrifuge tube to make a total of 10 ml. The suspension in the tube is then left to rest for 2–3 min so that larger debris falls to the bottom and then the vast majority of the splenocytes are collected by carefully pipetting out the top 9 ml of media.

6. For each peptide, the dose of DimerX per million cells and the molar ratio of peptide to DimerX should be titrated. Amounts that have been used successfully include 2 mg of DimerX fusion protein incubated overnight with a 40 M excess of $B8_{20-27}$ and $A47_{138-146}$ (H-2K^{b}:Ig); $A23_{297-305}$ (H-2D^{d}:Ig) and $F2_{26-34}$ (H-2L^{d}:Ig). This is the best starting point for peptides with high affinity for MHC. An example of a peptide with slightly lower affinity is $K3_{6-15}$ and in this case 4 mg of DimerX (H-2D^{d}:Ig) is loaded with a 40 M excess of peptide. More detailed instructions on optimization are provided with the reagents. The peptide-loaded DimerX can be stored at 4°C for up to a week. It is important to include an irrelevant peptide as a background control for unspecific staining. At present, the only MHC allele available for BALB/c mice is H-2L^{d}.
7. The DimerX to A85-1 mAb ratio is also best determined empirically for each peptide-loaded DimerX complex. This step dimerizes the peptide-loaded DimerX reagent to make a tetravalent reagent with a PE label. Some protocols suggest that cells can be stained with peptide-loaded DimerX first and then with A85-1 mAb, but in our hands this does not work.
8. Always check visually that a pellet of cells can be seen before flicking out supernatants. The flicking is done starting with the plate facing upwards in the palm of the hand and gripping the sides of the plate. The plate is then inverted over a sink or suitable discard tray with a single swift motion that is stopped abruptly so that the supernatant is removed before turning the plate right way up again. This action partly resuspends the cells in remaining medium and so the flicking can only be done once. After supernatant is removed, the cells can be further resuspended by holding the plate against a vortex set to a low speed. Again, look at the cell pellets to ensure resuspension. If cells are well resuspended at this stage, it is not necessary to mix the cells with the next antibody or other reagent that is added.
9. Some fluorochromes are very light sensitive and so it is best to do incubations in the dark and not to expose solutions with antibodies to the light more than necessary. The practice in some laboratories is to do any steps with fluorochome-conjugated antibodies in the dark as much as possible, but a sensible trade-off is required between protecting the fluorochome and being able to see well enough for accurate pipetting.
10. On the FSC×SSC plot (Fig. 1a), dead/dying cells are seen as a population just to the left and centered slightly below the main lymphocyte population and are excluded because they tend to have high nonspecific binding of antibodies. Also on this plot, the gate extends to the right and above this main lymphocyte population because activated T cells can be quite

large and granular. The CD8 × SSC plot (Fig. 1b) shows that there is more than one population of CD8⁺ events (both are included in the gate shown). This is normal, especially after intraperitoneal immunization. CD8⁺ T cells reduce surface levels of CD8 when activated, so ensure that these $CD8^{med}$ events are included in further analysis. Finally, the gate for DimerX staining (Fig. 1c) is drawn with reference to the splenocytes stained with DimerX loaded with irrelevant peptide (background controls, Fig. 1d). The background can be as high as around 0.5% of $CD8^+$ cells.

11. In our hands, most $CD8^+$ T cells that are able to respond to a particular peptide appear to be able to make IFN-γ in this assay (we do not generally see much higher responses when DimerX is used to measure $CD8^+$ T cell responses than when this ICS assay is used). However, the amounts of peptide used in these assays are always superphysiological, usually by several orders of magnitude, so it is not clear that all the $CD8^+$ T cells that are able to make IFN-γ in this in vitro assay would be able to make a cytokine response to an infected cell in vivo.
12. Always include a relevant negative control (i.e., irrelevant peptide or no peptide for peptide stimulations, uninfected cells for cell stimulations). Also take great care not to cross contaminate the peptides. As noted above, the peptides are being used at a concentration that can be 1,000-fold greater than required to trigger a response and this means that submicroliter amounts of contamination can affect results.
13. Before infection, wash cells with D0. Infections are started in 500 μl D0 in round-bottom tubes for 30 min to 1 h at 37°C with shaking. After this time, 10 ml D2 is added and the tubes incubated at 37°C until at least 5 h after the infection was initiated, preferably with slow rotation. Before adding them to the splenocytes, wash cells with D10, count, and adjust to 2×10^6 cells/ml in D10. The ratio of infected cells to splenocytes is 1:5. For best results, more than 80% of cells should be infected. This is most easily tested if a GFP-expressing VACV is available. Note also that P815 from various sources may have different permissivity for VACV infection. The quality of these infections can have a very large impact on the outcome of the assays and so any comparison between mice must be done using cells from the same infection. As a benchmark, for mice acutely infected with VACV strain WR, at least 25–30% of $CD8^+$ T cells in the spleen should be IFN-γ positive when infected cells are used as stimulators in ICS (11, 12, 14).
14. Just dispense the brefeldin A solution without touching the media in the well; there is no need to mix. Take care not to cross contaminate between wells at this stage, especially if using peptides.

15. The next step is done on ice, so it is important that the plate is cooled during this centrifugation. If infected cells are used, flick out medium inside the biosafety cabinet into a tray of disinfectant solution.

16. It is important not to have any protein (BSA or FBS) in second wash and in the paraformaldehyde solution to ensure effective fixation of cells. Paraformaldehyde fixes any proteins, including IFN-γ, within the cells and ensures that cell structure is maintained through saponin treatment.

17. Once cells are fixed, they are dead and any virus should be inactivated. It is also no longer essential to keep washes and centrifuge steps cold, though in general antibody staining steps are usually still done at 4°C.

18. FSC × SSC and CD8 × SSC plots are not shown in Fig. 2, but the gating considerations are the same as for DimerX staining (see Fig. 1a, b and Note 10). The gate for IFN-γ^+, CD8$^+$ events (Fig. 2a) needs to be drawn with reference to a negative control (Fig. 2b), which will be splenocytes from the same mouse that were not stimulated (if infected cells are used, it is important to use uninfected cells for this control). Usual backgrounds for splenocytes from acutely infected mice are around 0.1–0.2% of CD8$^+$ cells (but can be as high as 0.5). For memory, this can be down to tenfold less. In acutely infected mice where very dominant peptides (e.g., $B8_{20-27}$) or infected cells are used, the IFN-γ^-, CD8$^+$ events will appear to be brighter than corresponding cells from the negative control. In these cases, it is important to move the gate for IFN-γ^+ events a little higher (will reduce the % of events) to ensure that all IFN-γ^- events are excluded.

19. For in vivo cytotoxicity assays, targets must be prepared using freshly isolated splenocytes because syngeneic cell lines (e.g., EL-4 or RMA for C57Bl/6 mice) are lost nonspecifically after injection into mice. To increase the yield of these target cells from a single mouse, lymph nodes can be collected in addition to the spleen.

20. To prepare the 100 μM peptide working stock, dilute the 10 mg/ml original peptide stock in D0. An irrelevant peptide can be added to Tube 2, but we have found no differences between loading cells with an irrelevant peptide and leaving them unloaded.

21. To prepare a CFSE working stock solution, dilute the 1 mg/ml stock 1:10 in DMSO. Add CFSE working solution to the cells while vortexing to ensure uniform labeling.

22. The earliest that CD8$^+$ T cell-mediated cytotoxicity can be seen is 2 days after VACV infection and at this time it can only be seen in lymph nodes directly draining the site of infection. Generally, cytotoxicity is robust in the spleen and multiple lymph nodes from 5 days after infection and remains detectable

in memory. The uninfected mouse is used as the control to determine background lysis and to control for any error in mixing of the target cell populations.

23. Lymph nodes should be crushed as fine as possible to allow effective release of lymphocytes from the connective tissues.
24. DiD^+ populations are less than 1% of events and so for small lymph nodes, where it is not possible to collect 20,000 DiD^+ events, just collect as many as possible.
25. Vybrant DiD dye is excited by a 633-nm laser and is detected using the same mirror/filter sets as APC. CFSE dye is excited by a 488-nm laser and is detected by the same mirror/filter sets as FITC.
26. As an example of this calculation, if $CFSE^{low}$ (control) and $CFSE^{high}$ (VACV peptide loaded) populations are recovered in equal proportions from an uninfected mouse (ratio = 1), but in the test mouse 80% of DiD + events are $CFSE^{low}$ and 20% are $CFSE^{high}$ (ratio = 4), the specific lysis is $(1-(1/4)) \times 100 = 75\%$.
27. Splenocytes from the uninfected mouse are required to provide a negative control used to calculate percent killing. This is needed in case there is nonspecific loss of cells in one CFSE-labeled population or there is any error in mixing of the two cell populations.
28. The adherent cells are only loosely attached to the bottom of the flask, so pipette the non-adherent cells out carefully to prevent dislodging the adherent cells.
29. Other cell lines may be useful as targets as long as they are non-adherent. Adherent cells will stick to the bottom of wells and cannot be recovered for flow cytometric analysis.
30. As an example of this calculation, if $CFSE^{neg}$ (control), $CFSE^{high}$ (VACV peptide #1), and $CFSE^{low}$ (VACV peptide #2) populations are recovered in equal proportions (i.e., all are 33.3%) from a well where effectors came from an uninfected mouse, the "ratio negative control" = 1 for peptides #1 and #2. Likewise, if in the test mouse 60% of DiD + events are $CFSE^{neg}$, 10% are $CFSE^{high}$, and 30% are $CFSE^{low}$, the "ratio test" for peptide #1 = 6 and for peptide #2 = 2. The specific lysis for the 2 peptides will be peptide #1 $[1-(1/6)] \times 100 = 83\%$ and peptide #2 $[1-(1/2)] \times 100 = 50\%$.

References

1. Fenner F (1947) Studies in infectious ectromelia of mice. Aust J Exp Biol Med Sci 27:19–30
2. Fenner F (1958) The biological characters of several strains of vaccinia, cowpox and rabbitpox viruses. Virology 5:502–529
3. Drexler I, Staib C, Sutter G (2004) Modified vaccinia virus Ankara as antigen delivery system: how can we best use its potential? Curr Opin Biotechnol 15:506–512
4. Bjorkman PJ, Saper MA, Samraoui B, Bennett WS, Strominger JL, Wiley DC (1987) The foreign antigen binding site and T cell recognition regions of class I histocompatibility antigens. Nature 329:512–518

5. Oehen S, Brduscha-Riem K (1998) Differentiation of naive CTL to effector and memory CTL: correlation of effector function with phenotype and cell division. J Immunol 161:5338–5346
6. Ritchie DS, Hermans IF, Lumsden JM, Scanga CB, Roberts JM, Yang J, Kemp RA, Ronchese F (2000) Dendritic cell elimination as an assay of cytotoxic T lymphocyte activity in vivo. J Immunol Methods 246:109–117
7. Altman JD, Moss PAH, Goulder PJR, Barouch DH, McHeyzer-Williams MG, Bell JI, McMichael AJ, Davis MM (1996) Phenotypic analysis of antigen-specific T lymphocytes. Science 274:94–96
8. Sander B, Andersson J, Andersson U (1991) Assessment of cytokines by immunofluorescence and the paraformaldehyde-saponin procedure. Immunol Rev 119:65–93
9. Jung T, Schauer U, Heusser C, Neumann C, Rieger C (1993) Detection of intracellular cytokines by flow cytometry. J Immunol Methods 159:197–207
10. Betts MR, Brenchley JM, Price DA, De Rosa SC, Douek DC, Roederer M, Koup RA (2003) Sensitive and viable identification of antigen-specific $CD8^+$ T cells by a flow cytometric assay for degranulation. J Immunol Met 281:65–78
11. Tscharke DC, Karupiah G, Zhou J, Palmore T, Irvine KR, Haeryfar SMM, Williams S, Sidney J, Sette A, Bennink JR, Yewdell JW (2005) Identification of poxvirus $CD8^+$ T cell determinants to enable rational design and characterization of smallpox vaccines. J Exp Med 201:95–104
12. Yuen TJ, Flesch IEA, Hollett NA, Dobson BM, Russell TA, Fahrer AM, Tscharke DC (2010) Analysis of A47, an immunoprevalent protein of vaccinia virus, leads to a reevaluation of the total antiviral $CD8^+$ T cell response. J Virol 84:10220–10229
13. Moutaftsi M, Peters B, Pasquetto V, Tscharke DC, Sidney J, Bui HH, Grey H, Sette A (2006) A consensus epitope prediction approach identifies the breadth of murine T_{CD8+}-cell responses to vaccinia virus. Nat Biotechnol 24:817–819
14. Tscharke DC, Woo W-P, Sakala IG, Sidney J, Sette A, Moss DJ, Bennink JR, Karupiah G, Yewdell JW (2006) Poxvirus $CD8^+$ T-cell determinants and cross-reactivity in BALB/c mice. J Virol 80:6318–6323
15. Oseroff C, Peters B, Pasquetto V, Moutaftsi M, Sidney J, Panchanathan V, Tscharke DC, Maillere B, Grey H, Sette A (2008) Dissociation between epitope hierarchy and immunoprevalence in CD8 responses to vaccinia virus Western Reserve. J Immunol 180:7193–7202
16. Shen Z, Reznikoff G, Dranoff G, Rock KL (1997) Cloned dendritic cells can present exogenous antigens on both MHC class I and class II molecules. J Immunol 158:2723–2730
17. van Hall T, van Bergen J, van Veelen PA, Kraakman M, Heukamp LC, Koning F, Melief CJM, Ossendorp F, Offringa R (2000) Identification of a novel tumor-specific CTL epitope presented by RMA, EL-4, and MBL-2 lymphomas reveals their common origin. J Immunol 165:869–877
18. Coles RM, Mueller SN, Heath WR, Carbone FR, Brooks AG (2002) Progression of armed CTL from draining lymph node to spleen shortly after localized infection with Herpes Simplex Virus 1. J Immunol 168:834–838

Chapter 13

Generation and Characterization of Monoclonal Antibodies Specific for Vaccinia Virus

Xiangzhi Meng and Yan Xiang

Abstract

Monoclonal antibodies to specific vaccinia virus (VACV) proteins are valuable reagents in studies of VACV. In this chapter, we describe methods of generating a panel of monoclonal antibodies that recognize a variety of VACV proteins in their native conformation in infected cells. The antibodies thus generated recognize mostly VACV proteins that are involved in virion assembly or/and are major antigens in smallpox vaccine. These antibodies are useful for tracking distinct steps in virion assembly and for studying the B cell epitopes in smallpox vaccine.

Key words: Poxvirus, Vaccinia virus, Hybridoma, Monoclonal antibody

1. Introduction

Vaccinia virus (VACV) expresses nearly 200 proteins in infected cells, approximately half of which participate in various steps of viral replication (1), ultimately resulting in intracellular mature virion (MV) and three different forms of enveloped virion (EV) (2). Different stages of viral replication and different virion forms are often distinguishable by specific viral proteins. Monoclonal antibodies (mAbs) against these viral proteins are thus very useful for dissecting the replication cycle of VACV as well as the function of the viral proteins. A large number of mAbs specific for VACV had been described decades ago (3, 4), but their antigenic targets were not defined and they are no longer available. Ten years ago, a group of VACV-specific mAbs was generated by Alan Schmaljohn's group (5). Although these mAbs were never published, some of them, including a neutralizing antibody against L1, were made available and have been very useful in a number of studies (6).

Stuart N. Isaacs (ed.), *Vaccinia Virus and Poxvirology: Methods and Protocols*, Methods in Molecular Biology, vol. 890,
DOI 10.1007/978-1-61779-876-4_13,

More recently, multiple mAbs against L1 and B5 have been generated with recombinant proteins by Cohen, Eisenberg and coworkers (7, 8), and these mAbs have been used in studies of MV and EV neutralization.

To expand the repertoire of VACV-specific mAbs, we have recently developed a large panel of mAbs from a mouse that was immunized with VACV (9). We describe here our unique combination of protocols, which include immunizations with live VACV and a screen of hybridomas with immunofluorescence analysis of VACV-infected cells. Successful application of the protocols can result in multiple mAbs recognizing a variety of VACV proteins in their native conformations. In our case, more than 60 mAbs were developed from a single immunization, and they are against non-structural proteins (WR148, D13, C3) as well as structural proteins in the virion core (A10) or on the membranes of MV (D8, H3, A14) or EV (B5, A33, F13, A56). Among them, mAbs against A10, A14, and D13 were reported for the first time. The spectrum of the mAbs matched nicely with the profile of polyclonal antibody response after smallpox immunization (10, 11), so they are also useful for studying B cell epitopes in smallpox vaccine and protective effects of the antibodies.

2. Materials

2.1. Immunization

1. Pathogen-free BALB/c mice.
2. WR: Wild type (WT) VACV strain WR (see Note 1).
3. WR.K1L$^-$C7L$^-$ (12) (see Notes 1 and 2).
4. Avertin stock solution: 24 g 2,2,2 tribromoethanol in 15 ml tert-amyl alcohol (see Note 3).
5. Avertin anesthetic working solution (20 mg/ml): 0.5 ml Avertin stock solution in ~39.5 ml PBS; filtered through 0.2-μm filter (see Note 3).
6. Rodent restrainer for tail-vein injection.
7. 27-1/2 gauge needle.
8. Psoralen stock solution: 10 mg/ml in DMSO; sterilized through 0.2-μm filter.
9. Hand-held UV lamps (UVP).

2.2. Hybridoma Generation

1. SP2/0-Ag14 myeloma cells (ATCC #CRL 1581).
2. 50% polyethylene glycol (PEG) (see Note 4).
3. Small surgical scissors for spleen removal.
4. 70 μM nylon cell strainer.

5. DMEM-0: DMEM medium without FBS.
6. DMEM-20: DMEM with 20% FBS.
7. D20/HAT/OPI: DMEM with 20% FBS, HAT Media Supplement, and OPI Media Supplement (see Note 5).
8. D20/HTO: DMEM medium with 20% FBS, HT Media Supplement, and OPI Media Supplement (see Notes 5 and 6).
9. Tissue culture plasticware: 175-cm2 flasks, 96-well flat-bottom microtiter plates, petri dish, and 50-ml conical polypropylene centrifuge tubes.
10. Hemocytometer and trypan blue for counting cells and assessing cell viability (see Note 7).
11. Multichannel pipette.

2.3. Hybridoma Screening

1. HeLa or BHK cells.
2. PBST: PBS with 0.05% Tween-20.
3. 4% paraformaldehyde in PBS.
4. 0.1% Triton X-100 in PBS.
5. Hoechst 33258 dye: 10 mg/ml in PBS and used at 1:2,000 dilution.
6. Cy3-conjugated anti-mouse secondary antibody: Cy3-conjugated AffiniPure Goat anti-mouse IgG (H+L): 0.5 mg/ml and used at 1:200 dilution.
7. FluorSave Reagent.
8. Fluorescence microscope with 100× or 60× oil-immersion objective lens.
9. Microscope slide and cover glass.

2.4. Categorization of the Hybridomas

1. Methionine- and cysteine-free DMEM.
2. ^{35}S-methionine and -cysteine.
3. Lysis buffer: 50 mM Tris–HCl, pH 7.5, 150 mM NaCl, 1% Triton X-100, 0.1% SDS, 0.5% sodium deoxycholate supplemented with protease inhibitor cocktail.
4. Wash buffer: 0.1% Triton X-100, 50 mM Tris, pH 7.4, 300 mM NaCl, 5 mM EDTA.
5. Protein G-sepharose.
6. 2× SDS sample buffer: 100 mM Tris–HCl, pH 6.8, 4% (w/v) SDS, 0.2% (w/v) bromophenol blue, 20% (v/v) glycerol, 200 mM β-mercaptoethanol.
7. Film and cassette.

2.5. Antigen Identification

1. Bis(Sulfosuccinimidyl)-suberate: 50 mM in PBS.
2. 1 M Glycine: dissolved in ddH_2O and adjust the pH to 2.7 with concentrated HCl.

3. Coomassie blue.
4. pGEX6P-1 plasmid (GE Healthcare Life Sciences).
5. BL21(DE3) strain of *E. coli*.
6. Ampicillin stock solution: 100 mg/ml in ddH_2O.
7. LB broth.
8. IPTG: 1 M in H_2O.
9. Sonicator.
10. Molecular biology reagents such as enzymes for restriction digestion and PCR.

3. Methods

3.1. Immunizations

We developed a large panel of VACV-specific mAbs from an immunized BALB/c mouse. The mouse was immunized twice by intranasal infections with live VACV and once with an intravenous injection of UV-inactivated virus (see Note 8). The initial intranasal infection can be done with an attenuated VACV strain or with a sublethal dose (~0.5×10^3 PFU) of WT VACV WR (see Note 2), so the mice survive the infection while developing a strong immune response due to live virus infection. Subsequently, a high dose of WT VACV WR is used to infect the mice to confirm and boost the development of a protective immune response. Finally, 3 days prior to cell fusion to generate the hybridoma, the mice are boosted with an intravenous injection of UV-inactivated VACV to activate VACV-reactive B cells into lymphoblasts, which form hybridomas more efficiently than resting B cells.

1. Anesthetize a 4–6-week-old BALB/c mouse by intraperitoneal injection with Avertin anesthetic working solution at 0.4–0.6 mg/g body weight (see Note 3).
2. Infect the anesthetized mouse intranasally with 10^5 PFU of WR.K1L$^-$C7L$^-$ (12) by applying 10 μl of inoculum into each nostril with a micropipette.
3. After infection, monitor the mouse daily by measuring the body weight and status. (see Note 9).
4. Two weeks after the initial infection, infect the mouse again as described in steps 1 and 2, but with 5×10^6 PFU of WR. Monitor the body weight daily as before (see Note 10).
5. At least 2 weeks after the second infection, and 3 days prior to the planned harvest of mouse spleen, boost the mouse with UV-inactivated WR (see Note 11). The virus is inactivated by limited cross-linking with psoralen and long-wave UV light (13) (see Note 12). To prepare UV-inactivated WR, psoralen is added at a final concentration of 10 μg/ml to ~1 ml of WR

virus (at 10^9 PFU/ml), and the 1 ml mixture is placed in a 35-mm petri dish and exposed to 365-nm UV light emanating from a UV lamp at a distance of 6 cm for 20 min.

6. Inject 50 μl of UV-inactivated WR virus through the tail vein with a 27-1/2-gauge needle.

3.2. Hybridoma Generation

Hybridomas are generated by fusing the spleen cells with SP2/0 cells with a protocol that was modified from a standard protocol (14).

1. At least a week before the fusion, start propagating SP2/0 cells in D20/HT to ensure that the cells grow well and are well adapted to the medium. Split the cells 12–24 h before the fusion.
2. Three days after the final boosting immunizaiton, sacrifice the mouse by cervical dislocation or CO_2 asphyxiation. Sterilely remove the spleen and wash the spleen with DMEM-0 in a 100-mm Petri dish. Perform this and all subsequent steps under sterile condition in a laminar flow hood.
3. Make single cell suspension of the spleen as follows. Use a needle to scratch open the spleen. Put the spleen in a cell strainer on top of a 50-ml conical tube. Use the plunger of a 1-ml syringe to press the cells out of the spleen. Rinse the spleen with DMEM-0 to help cells pass through the screen.
4. Spin down the splenocytes and SP2/0 at $150\times g$ for 5 min. Aspirate off the supernatant and wash the cells with 50 ml DMEM-0. Count the cells with a hemocytometer (see Note 13)
5. Mix 2×10^7 SP2/0 cells with 1×10^8 splenocytes (see Note 14). Divide the cells equally into four portions and perform the fusion separately.
6. Spin down the mixed cells at $150\times g$ for 5 min. Remove the supernatant completely. Loosen the cell pellet by gently tapping the side of the tube. Add 1 ml 50% PEG to the cell pellet drop by drop over the course of 1 min, while continuously swirling the tube in 37°C water bath. Add 10 ml prewarmed D20/HAT/OPI to the tube drop by drop over the course of 1–2 min, while continuously swirling the tube in 37°C water bath.
7. Bring the total volume up to 55–60 ml with D20/HAT/OPI. Plate 200 μl/well to three 96-well plates.
8. Repeat the cell fusion (steps 6 and 7) for the remaining three portions of spleen cells. A total of twelve 96-well plates are prepared for each spleen (see Note 15).
9. Place the plates in a 37°C CO_2 incubator and 7–10 days later inspect the plates for colonies (see Note 16).

3.3. Screen Hybridoma Supernatants with an Immunofluorescence Assay

The hybridomas are screened for their ability to secrete anti-VACV antibodies with an immunofluorescence assay, in which the culture supernatants of the hybridomas are used to stain cells that had been infected with VACV for 8 h at a low multiplicity of infection

(MOI) (see Note 17). Using cells infected at a low MOI allows side-by-side staining of infected and uninfected cells. That is, if the supernatant stains only VACV-infected cells, which display viral DNA factories as areas of cytoplasmic DNA staining, the hybridoma is considered to be specific for VACV. The immunofluorescence patterns are photographed and used later to categorize the hybridomas and corroborate with antigen identification.

1. Seed HeLa or BHK cells at 50% confluency on coverslips placed inside a 24-well plate (see Note 18).
2. Infect the cells with WR at 0.1–0.5 PFU/cell.
3. After 8 h of infection fix cells with 4% paraformaldehyde at room temperature for 20 min.
4. Rinse once with PBST and then permeabilize the cells with 0.1% Triton X-100 for 5 min.
5. Block with DMEM-10 at 37°C for 1 h.
6. Incubate the cells with 100–200 μl of the hybridoma supernatant for 1 h at 37°C.
7. Wash three times with PBST and then incubate with Cy3-conjugated anti-mouse secondary antibody at 37°C for 1 h.
8. Wash three times with PBST and then incubate with Hoechst 33258 dye for 5 min.
9. Wash two times with PBST and once with distilled water.
10. Mount the coverslips on slides with one drop of Fluorsave reagent.
11. Visualize and photograph fluorescent images of the slides by using the highest available magnification on a fluorescence microscope (see Note 19).
12. From wells that are tested positive by immunofluorescence assay, pick up individual colonies of hybridoma by inserting a micropipette tip in the colony and carefully pipetting up 20–50 μl of medium with the cells, and place them into 24-wells containing 1 ml D20/HTO medium per well.
13. When the cells reach close to confluency in 24-wells, rescreen the supernatants with the immunofluorescence assay.
14. Expand cells that are screened positive into a T25 flask. Freeze down several vials of the cells in liquid nitrogen.
15. Collect around 25 ml of the culture supernatant for further analysis.

3.4. Categorization of the Hybridomas

The hybridomas can be further characterized by using their culture supernatants to immunoprecipitate radiolabeled proteins from VACV-infected cells. If the supernatants from several hybridomas precipitate a protein with the same molecular weight

Table 1
Typical immunofluorescence and molecular weight of some common mAbs against VACV

Antigen	Immunofluorescence pattern[a]	MW by SDS-PAGE[b]
D8	Factory; some outside factory; virion-size particles	30
A14	Factory	15, 23
WR148	Entire cytoplasm	87
D13	Factory, aggregates	61
H3	Factory; some outside factory; virion-size particles	32
A56	Cell surface	79
A33	Outside factory	22
C3	Outside factory	27
B5	Outside factory, virion-size particles	40
A10	Factory, particles	93
F13	Outside factory, virion-size particles	38

[a]Factory: staining predominantly of viral DNA factory; outside factory: staining predominantly of areas outside viral DNA factory
[b]Molecular weight (MW) can be calculated from mobility of the protein relative to molecular weight standards on SDS-PAGE by using a program such the Bio-Rad Quantity One

(MW) and show similar immunofluorescence pattern, then the hybridomas can be categorized into one group (Table 1), and only one representative hybridoma from each group need to be focused on initially for antigenic target identification. If the immunoprecipitation and immunofluorescence analysis indicates that a hybridoma is not monospecific, subclone the hybridomas by limited dilution.

1. Infect HeLa cells in 100-mm dishes with WR at 10 PFU/cell.
2. At 8 h later, replace the culture medium with 5 ml of methionine- and cysteine-free DMEM-0 plus 500 μCi of ^{35}S-methionine and -cysteine (see Note 20).
3. After another 8 h, harvest the cells and lyse the cells in 1 ml lysis buffer per dish (see Note 21).
4. Incubate 1 ml of the hybridoma supernatant with 80 μl of the cleared cell lysate for 1 h in a tube rotator.

5. Add 30 μl of 50% (vol/vol) Protein G-sepharose and incubate for another 1 h in the tube rotator.
6. Wash the Protein G-sepharose consecutively with wash buffer and PBS (see Note 21).
7. Resuspend the Protein G-sepharose in 30 μl SDS sample buffer and run 15 μl on a SDS-PAGE gel.
8. Dry the gel and expose it to a film. Measure apparent MW of the proteins in related to MW standards on the autoradiograph (see Note 22).

3.5. Antigen Identification

The most straightforward way for identifying the antigenic target of a mAb is probably analyzing the specific antigen that is precipitated by the mAb with mass spectrometry. This works well for most of the VACV antigens, which are abundantly expressed in infected cells and easily precipitated by antibodies. However, some VACV antigens are not efficiently precipitated by the mAb due to either low expression or low solubility, while some other antigens are very small with few trypsin cleavage sites. In those cases, it may be difficult to identify the antigen with mass spectrometry, but it is often possible to form specific hypothesis about the antigenic target of a mAb by examining the immunofluorescence pattern and the apparent MW of the antigen (see Note 23). The hypothesis could then be tested by applying some common immunological techniques on glutathione-S-transferase (GST) fusion protein of a specific VACV protein or a VACV mutant with a deletion in a specific VACV gene. Alternatively, the antigenic target may be identified by testing the mAb on a microarray consisting of all recombinant VACV proteins (10, 11, 15), as described recently in the identification of a mAb against VACV E3 protein (16).

3.5.1. Identify Antigen by Immunoprecipitation and Mass Spectrometry Analysis

1. Conjugate the mAb to protein G-sepharose by incubating 1 ml of hybridoma supernatant with 30 μl of Protein G-sepharose for 1 h (see Notes 24 and 25).
2. Wash the protein-G sepharose two-times with wash buffer and five times with PBS.
3. Resuspend the protein G-sepharose in 100 μl PBS and add freshly-made Bis(Sulfosuccinimidyl)-suberate to at a final concentration of 5 mM and incubate at room temperature for 60 min.
4. Quench any unused cross-linker by adding 5 μl of 1 M Tris–HCl and elute off any unconjugated antibody by washing the resin three times with 1 M Glycine and two times with PBS.
5. Prepare cell lysate by infecting HeLa cells in 100-mm or 150-mm petri dish with WR at MOI of 10 for 8 h.
6. Harvest the cells and lysed infected cells in 1 ml lysis buffer.
7. Apply the entire 1 ml cell lysates to mAb-conjugated protein G-sepharose and incubate at room temperature for 1 h in the tube rotator.

8. Wash the protein G-sepharose as described above in Subheading 3.4, step 6.
9. Elute proteins from the protein G-sepharose with 50 μl of 1 M glycine.
10. Resolve the eluted proteins by SDS-PAGE.
11. Stain the SDS-PAGE gel with Coomassie blue.
12. Submit Coomassie blue-stained SDS-PAGE gel to an appropriate core facility for mass spectrometry analysis of individual protein bands on the gel.

3.5.2. Test Reactivity of a mAb to a VACV Antigen by Western Blot Analysis of Bacterial Expressed GST Fusion of the VACV Protein (See Note 26)

1. Construct an expression plasmid for a GST fusion of the candidate VACV antigen by PCR-amplifying the viral gene from WR DNA and subcloning the PCR fragment into pGEX6P-1.
2. Transform the plasmid into BL21(DE3) strain of *E. coli.*
3. Inoculate a single colony into 5 ml LB with 100 μg/ml ampicillin and incubate the culture in a shaker overnight at 37°C.
4. Inoculate 1 ml overnight culture into 100 ml fresh LB with 100 μg/ml ampicillin and incubate the culture in a shaker set at 250 rpm and 37°C until OD_{600} reaches 0.6.
5. Set aside 1 ml culture as uninduced sample and add IPTG to the remaining culture at a final concentration of 0.5 mM.
6. Incubate the culture in a shaker at 250 rpm and 25°C for 4 h.
7. Take 1 ml culture as induced sample. (The rest of culture can be harvested and save for other experiments).
8. Spin down bacteria from 1 ml of either uninduced or induced culture sample and add 100 μl SDS sample buffer to the pellet.
9. Sonicate the pellet in a cup sonicator with the maximum power for 2 min.
10. Run 10 μl of the bacterial cell lysate on SDS-PAGE gel and perform separate Coomassie stain and western blot analysis with two identical set of the gels.
11. Use hybridoma supernatant as the primary antibody in the western blot.

4. Notes

1. For the animal work, we use virus purified through a sucrose cushion. This is done following procedures describe in other chapters in this book or according to a standard protocol (17).
2. WR.K1L^{-}C7L^{-} is a highly attenuated WR mutant with specific deletion in the host range genes K1L and C7L (12). It is used

for initial immunization because it does not cause any disease symptom in mice but can elicit a protective immune response. It can be substituted here with another attenuated VACV strain or with a low dose ($<10^4$ PFU) of WT VACV WR. However, when other VACV strains are used, caution should be taken to control the infection dosage so that the infection will not kill or sicken the mouse significantly. WT VACV WR should not be used in exceed of 10^4 PFU, as the intranasal LD_{50} of WT VACV WR in BALB/c mice is approximately 1×10^4 PFU.

3. Avertin (tribromoethanol) degrades in the presence of heat or light to produce toxic byproducts, so the stock solution of avertin should be wrapped with foil and stored in 4°C. A working solution of avertin is made fresh from the stock the day before usage.
4. We purchase ready-to-use PEG solution from Sigma (P7181). The solution contains 50% (w/v) polyethylene glycol (Av. Mol. Wt. 1,450) in DPBS without calcium. It comes filter-sterilized.
5. We reconstitute a vial of HAT (Sigma H0262) with 10 ml sterile cell culture medium. Each vial is used to prepare 500 ml medium. We reconstitute a vial of OPI (Sigma O5003) with 10 ml sterile water. Each vial is used to prepare 1 L of medium.
6. We reconstitute a vial of HT (Sigma H0137) with 10 ml sterile cell culture medium. Each vial is used to prepare 500 ml medium.
7. 100 μl of the cells is incubated with 400 μl PBS and 500 μl trypan blue (0.4% w/v) for 2 min at room temperature. A hemocytometer is then used to count live (clear) and dead (blue) cells.
8. Because mice are infected with live virus, this immunization protocol requires no adjuvant and appears to elicit an antibody response similar to the one that results from smallpox vaccination (9). We found that the majority of the hybridomas that are generated from this immunization protocol are specific for VACV. This is in contrast to the relatively low percentage of antigen-specific hybridomas that usually result from immunization with a protein antigen. Most of the hybridomas that are not reactive to VACV appear to recognize cytoskeleton, similar to that described by a previous report (18).
9. A BALB/c mouse that has been infected with WR.K1L$^-$C7L$^-$ usually does not lose any body weight. However, if another strain of VACV is used and the mouse should be monitored and humanely sacrificed if it meets endpoint criteria approved by your IACUC. In our studies, losses of 30% of the original body weight required mice to be euthanized.
10. Mouse that has been previously infected with WR.K1L$^-$C7L$^-$ may lose up to 15% of its original body weight when it is challenged with 5×10^6 PFU of WT WR. The survival of the mouse

with such a high dose VACV WR challenge confirms that the initial immunization is successful and serves as a boost to the initial immunization. It is thus unnecessary to evaluate serum antibody titer for VACV before cell fusion.

11. The final boost is added to follow the standard immunization protocol for mAb development. It was meant to activate resting B lymphocytes into actively dividing blastoid cells, which are the best cells for fusion with SP2/0 cells. It is thus done exactly 3 days before the harvest of the spleen for cell fusion. We have not tested whether this step is necessary for mAb development after live VACV infection.
12. Exposing VACV to long-wave UV light for 6 min in the presence of 5 μg/ml psoralen was shown to block more than 95% of VACV early gene translation (19). However, this treatment does not abolish the ability of the virus to enter into cells.
13. A spleen from a mouse usually has 1×10^8 to 2×10^8 B cells.
14. SP2/0 cells are added to spleen cells at the ratio of 1:5 to 1:10.
15. The cells are plated into twelve 96-well plates so that the majority of the wells that have colonies only have one colony, which is usually monospecific for a VACV antigen. This reduces the need for subcloning the hybridoma cells by limited dilution.
16. The medium with HAT is used for selecting hybridomas. After the hybridoma is established, the selection is removed by omitting aminopterin from the medium. (Aminopterin inhibits dihydrofolate reductase and forces the cells to use HGPRT and HT). The cells are first adapted to grow in HT-containing medium and then eventually in regular DMEM medium.
17. We found that for identifying hybridomas specific for VACV, immunofluorescence analysis is more specific and more sensitive than other commonly used screening methods such as ELISA. ELISA of infected cell lysate has a high percentage of false positive, while ELISA of purified virus appears to only detect antibodies against major virion proteins. In contrast, immunofluorescence analysis of infected and uninfected cells unambiguously identified hybridomas that are specific for VACV. In addition, the immunofluorescence patterns help in identification of the antigenic targets of the hybridomas.
18. HeLa and BHK cells are relatively big and flat with large cytoplasms, so they are ideal for immunofluorescence analysis of VACV proteins and structures in the cytoplasm.
19. We typically use a 100×/1.4 N.A. oil-immersion super-apochromat objective lens, as it shows more clearly the immunofluorescence pattern of the antigen inside the cells, which is critical for categorizing the hybridomas and identifying their antigenic targets. For example, structural proteins of MV usually localize

with viral DNA factories, while proteins of EV membrane predominantly localize in areas outside the viral DNA factories. Antibodies against proteins that are on the virion surface often stain virion-size particles.

20. To avoid contaminating the CO_2 incubator with radioactivity, we put the dishes inside a Nalge Nunc Beta Storage Box placed inside the CO_2 incubator.
21. All Solid and liquid ^{35}S waste must be disposed into appropriate and properly labeled waste containers. Since steps include exposure of infected cells to a detergent, which inactivates virus, the radioactive waste is noninfectious.
22. The apparent MW of the proteins relative to MW standards can be measured with a program such as Bio-Rad Quantity One.
23. Although the immunization protocol described here differs from the practice of smallpox vaccination, the spectrum of mAbs that are generated usually falls within the profile of polyclonal antibody response to smallpox vaccine (10, 11). Therefore, the antigenic targets of the mAbs should be first considered to be among 25 VACV proteins that are consistently the target of antibodies after vaccination with smallpox vaccine (11).
24. The antibody is conjugated to protein G-sepharose prior to the immunoprecipitation to prevent the antibody from being eluted along with the precipitated protein, which may complicate protein identification by mass spectrometry. The procedure we described here will let the antibody bind to protein G-sepharose via the Fc domain before they are covalently linked together with a cross-linker, so the orientation of the antibody is maintained for optimal antigen binding. This cross-linking step can be omitted if the precipitated antigen can be easily distinguished from the immunoglobulin on the SDS-PAGE gel or if preconjugating the antibody to sepharose significantly decreases the ability of the antibody in precipitating the antigen.
25. For all initial characterization of the hybridomas including immunoprecipitation for mass spectrometry, supernatants of hybridomas grown in D20/HTO should be adequate. We usually choose a batch of FBS with a low concentration of bovine immunoglobulin for hybridoma work, so the immunoprecipitation is not affected by the presence of 20% FBS. The growth medium for the hybridomas can be gradually switched first to DMEM-20 with HT then to DMEM-20 only. The hybridomas can also be grown in serum-free medium for hybridoma culture (Invitrogen) to produce mouse antibody devoid of bovine immunoglobulin.

26. The advantage of this approach is that the GST-fusion protein can be easily overexpressed in *E. coli* and requires no purification before being analyzed with western blot. In fact, other proteins in whole bacterial cell lysates serve as a nice negative control for the specificity of the antibody in western blot. Specific antibodies often recognize only the GST fusion protein among the whole bacterial cell lysates, even for some antibodies that are later found to recognize a conformational epitope, probably because a small but sufficient amount of the overexpressed GST fusion proteins undergo some refolding during the western blot. However, there are also antibodies that do not recognize the antigen when it is part of a GST fusion expressed in prokaryotic cells.

Acknowledgments

Y.X is supported by NIH grant AI079217 and contract 272200900048C-0-0-1.

References

1. Upton C, Slack S, Hunter AL, Ehlers A, Roper RL (2003) Poxvirus orthologous clusters: toward defining the minimum essential poxvirus genome. J Virol 77:7590–7600
2. Moss B (2007) Poxviridae: the viruses and their replication. In: Knipe DM, Howley PM (eds) Fields virology. Lippincott Williams & Wilkins, Philadelphia, pp 2905–2946
3. Ichihashi Y, Oie M (1988) Epitope mosaic on the surface proteins of orthopoxviruses. Virology 163:133–144
4. Wilton S, Gordon J, Dales S (1986) Identification of antigenic determinants by polyclonal and hybridoma antibodies induced during the course of infection by vaccinia virus. Virology 148:84–96
5. Hooper JW, Custer DM, Schmaljohn CS, Schmaljohn AL (2000) DNA vaccination with vaccinia virus L1R and A33R genes protects mice against a lethal poxvirus challenge. Virology 266:329–339
6. Lustig S, Fogg C, Whitbeck JC, Eisenberg RJ, Cohen GH, Moss B (2005) Combinations of polyclonal or monoclonal antibodies to proteins of the outer membranes of the two infectious forms of vaccinia virus protect mice against a lethal respiratory challenge. J Virol 79:13454–13462
7. Aldaz-Carroll L, Whitbeck JC, Ponce de Leon M, Lou H, Hirao L, Isaacs SN, Moss B, Eisenberg RJ, Cohen GH (2005) Epitope-mapping studies define two major neutralization sites on the vaccinia virus extracellular enveloped virus glycoprotein B5R. J Virol 79:6260–6271
8. Aldaz-Carroll L, Whitbeck JC, Ponce de Leon M, Lou H, Pannell LK, Lebowitz J, Fogg C, White CL, Moss B, Cohen GH, Eisenberg RJ (2005) Physical and immunological characterization of a recombinant secreted form of the membrane protein encoded by the vaccinia virus L1R gene. Virology 341:59–71
9. Meng X, Zhong Y, Embry A, Yan B, Lu S, Zhong G, Xiang Y (2011) Generation and characterization of a large panel of murine monoclonal antibodies against vaccinia virus. Virology 409:271–279
10. Davies DH, Liang X, Hernandez JE, Randall A, Hirst S, Mu Y, Romero KM, Nguyen TT, Kalantari-Dehaghi M, Crotty S, Baldi P, Villarreal LP, Felgner PL (2005) Profiling the humoral immune response to infection by using proteome microarrays: high-throughput vaccine and diagnostic antigen discovery. Proc Natl Acad Sci USA 102:547–552
11. Davies DH, Molina DM, Wrammert J, Miller J, Hirst S, Mu Y, Pablo J, Unal B,

Nakajima-Sasaki R, Liang X, Crotty S, Karem KL, Damon IK, Ahmed R, Villarreal L, Felgner PL (2007) Proteome-wide analysis of the serological response to vaccinia and smallpox. Proteomics 7:1678–1686

12. Meng X, Chao J, Xiang Y (2008) Identification from diverse mammalian poxviruses of host-range regulatory genes functioning equivalently to vaccinia virus C7L. Virology 372: 372–383
13. Tsung K, Yim JH, Marti W, Buller RM, Norton JA (1996) Gene expression and cytopathic effect of vaccinia virus inactivated by psoralen and long-wave UV light. J Virol 70:165–171
14. Yokoyama WM, Christensen M, Santos GD, Miller D (2006) Production of monoclonal antibodies. Curr Protoc Immunol Chapter 2: Unit 2.5
15. Davies DH, Wyatt LS, Newman FK, Earl PL, Chun S, Hernandez JE, Molina DM, Hirst S, Moss B, Frey SE, Felgner PL (2008) Antibody profiling by proteome microarray reveals the immunogenicity of the attenuated smallpox vaccine modified vaccinia virus ankara is comparable to that of Dryvax. J Virol 82:652–663
16. Weaver JR, Shamim M, Alexander E, Davies DH, Felgner PL, Isaacs SN (2007) The identification and characterization of a monoclonal antibody to the vaccinia virus E3 protein. Virus Res 130:269–274
17. Earl PL, Moss B, Wyatt LS, Carroll MW (2001) Generation of recombinant vaccinia viruses. Curr Protoc Mol Biol Chapter 16: Unit 16.17
18. Dales S, Fujinami RS, Oldstone MB (1983) Infection with vaccinia favors the selection of hybridomas synthesizing autoantibodies against intermediate filaments, one of them cross-reacting with the virus hemagglutinin. J Immunol 131: 1546–1553
19. Ramsey-Ewing A, Moss B (1998) Apoptosis induced by a postbinding step of vaccinia virus entry into Chinese hamster ovary cells. Virology 242:138–149

Chapter 14

Bioinformatics for Analysis of Poxvirus Genomes

Melissa Da Silva and Chris Upton

Abstract

In recent years, there have been numerous unprecedented technological advances in the field of molecular biology; these include DNA sequencing, mass spectrometry of proteins, and microarray analysis of mRNA transcripts. Perhaps, however, it is the area of genomics, which has now generated the complete genome sequences of more than 100 poxviruses, that has had the greatest impact on the average virology researcher because the DNA sequence data is in constant use in many different ways by almost all molecular virologists. As this data resource grows, so does the importance of the availability of databases and software tools to enable the bench virologist to work with and make use of this (valuable/expensive) DNA sequence information. Thus, providing researchers with intuitive software to first select and reformat genomics data from large databases, second, to compare/analyze genomics data, and third, to view and interpret large and complex sets of results has become pivotal in enabling progress to be made in modern virology. This chapter is directed at the *bench virologist* and describes the software required for a number of common bioinformatics techniques that are useful for comparing and analyzing poxvirus genomes. In a number of examples, we also highlight the Viral Orthologous Clusters database system and integrated tools that we developed for the management and analysis of complete viral genomes.

Key words: Poxvirus, Vaccinia virus, Smallpox, Bioinformatics, Genomics, Dotplot, Multiple sequence alignment, VOCs, VGO, BLAST, JDotter, MSA

1. Introduction

Since the first complete poxvirus genome was announced in 1990 (1), more than 110 genomes have been sequenced (see Note 1). There are now both many diverse and very closely related genomes (e.g., the series of variola virus genomes) in the data collection; this leads to the need for different types of analysis. As the new research field of bioinformatics has developed during the last 10 years, many new bioinformatics tools have been designed and implemented. These provide a variety of analysis options to wet-lab researchers.

Stuart N. Isaacs (ed.), *Vaccinia Virus and Poxvirology: Methods and Protocols*, Methods in Molecular Biology, vol. 890,
DOI 10.1007/978-1-61779-876-4_14,

However, for the average virologist, even the process of organizing, manipulating, and analyzing the data contained in poxvirus genomes that may be >200 kilobases (kB) is not trivial. It is often very time consuming and tedious to perform; what, at the outset, seems a simple analysis (e.g., collect 20 diverse poxvirus DNA polymerase proteins and perform a multiple alignment) may in fact take all day to perform. To address this significant problem, over the last 10+ years, our group has developed the Viral Orthologous Clusters (VOCs) database for managing genomics data from a variety of virus families, including poxviruses. It not only provides graphical user interfaces (GUIs) to allow users to easily access, process, display, and retrieve the data, but also integrates a variety of easy-to-use bioinformatics software tools. This gives the virologist a suite of tools to help with the most commonly performed analyses that molecular virology demands.

For this chapter, the authors have selected software tools that perform common bioinformatics tasks and that are relatively straightforward to use. Once researchers can efficiently access and retrieve genomic sequences in a data format suitable for most bioinformatics programs, then they have many more opportunities for performing bioinformatics analyses in a reasonably painless and timely manner. Therefore, we spend considerable time demonstrating the power and utility of the VOCs database system. Our primary goal is to provide readers with easy-to-follow demonstrations of (1) data retrieval and management and (2) several common bioinformatics methods for analyzing poxvirus genomes. The latter include (1) comparison of the entire poxvirus genomes; (2) comparison of poxvirus genes and proteins within orthologous families; (3) similarity searching of poxvirus proteins against databases for prediction of function; and (4) searching of poxvirus proteins for motifs and functional domains. Although it is not possible to go into great detail for all the software mentioned, help files and manuals are available and detailed instructions will be provided on how to access the software that is used through the Virology.ca site (see Note 2).

Finally, it should be noted that much of this work falls under the topic of *Comparative Genomics* and that our tools have been specifically designed to manage and analyze multiple genomes in comparative processes. Also, although we only discuss the analysis of poxviruses here, the database and tools can be used for other viruses, including intron-containing viruses such as herpesviruses and baculoviruses, and can function with much larger DNA sequences, e.g., bacterial genomes.

2. Materials

1. To work on typical bioinformatics projects, researchers need a reasonably up-to-date desktop computer and operating system that can support the installation of Java SE 6 (see Note 3).
2. In developing our own tools, we have preferred the JAVA client-server format (see Note 3), where a copy of the software is automatically loaded onto the user's computer. This is preferred because of the greater flexibility and functionality it allows in the client software and the use of JAVA also overcomes the problem of supporting a large variety of WWW browsers and different operating systems.
3. Key software:
   (a) VOCs: An SQL database (see Note 4) of information derived from all complete annotated poxvirus genomes. A powerful, but easy-to-use, interface has been built to query the database. It can be accessed via the Virology.ca Web site or directly at http://athena.bioc.uvic.ca/tools/VOCS/Poxviridae.
   (b) *Viral Genome Organizer* (VGO): Software for visualizing and comparing complete poxvirus genomes. The program allows custom retrieval of protein/DNA sequences from a genome and can display preprocessed result files from a variety of analyses alongside the genomic data. A variety of searches can be performed on the genome-sized DNA sequences. It can be accessed via the Virology.ca Web site or directly at http://athena.bioc.uvic.ca/tools/VGO/Poxviridae.
   (c) *DOTTER*: Runs on UNIX and windows systems (2). We have developed a JAVA client–server version of DOTTER called JDOTTER (3) that can use DNA sequences in FASTA files or import genomes/sequences from the VOCs database (http://virology.ca/tools/JDotter/Poxviridae). JDOTTER can also produce dotplots for protein sequences, and multiple DNA or protein sequences.
   (d) *Base-By-Base* (BBB): Software for generating, visualizing, and editing multiple sequence alignments. The program can use DNA and protein sequences in FASTA files or import genomes/sequences from the VOCs database. It uses standard alignment algorithms, such as ClustalW (4), T-Coffee (5), and MUSCLE (6). It has the ability to display annotations from GenBank files and allows users to add their own annotations, including primer-binding sites.

It can be accessed via the Virology.ca Web site or directly at http://virology.ca/tools/BaseByBase/Poxviridae.

(e) *Genome Annotation Transfer Utility* (GATU): Software for annotating genomes using a reference genome; writes GenBank files. Provides for interactive annotation with researcher. It can be accessed via the Virology.ca Web site or directly at http://virology.ca/tools/GATU/Poxviridae.

(f) *ARTEMIS*: A JAVA (see Note 5) program for helping with DNA sequence annotation (7). It runs on multiple platforms and is available from http://www.sanger.ac.uk/Software/Artemis/.

(g) *HHPRED*: A Web-based interface to a search tool that uses Hidden Markov Model (HMM) profile matching to identify distantly related proteins to the query protein. It is available at http://toolkit.tuebingen.mpg.de/hhpred.

(h) *Sequence Searcher*: A JAVA (see Note 5) program for searching multiple DNA and protein sequences for user-specified sequence motifs (8).

(i) *NAP*: A global nucleotide to amino acid alignment program (9). This program runs on UNIX machines, but we have recently developed a JAVA client–server version of the program and incorporated it into VOCs and VGO, as well as have developed a stand-alone client that is available at http://virology.ca/tools/NAP/.

3. Methods

3.1. Connecting to the Virology.ca Web Site

The Virology.ca Web site is reached by pointing an Internet Browser to http://www.virology.ca. It is well worth spending a little time browsing this site, but most of the software discussed here can be found as follows.

1. From the *Organisms* menu/submenu, choose *dsDNA/Poxviridae*.
2. Select tool from the *VBRC Tools* menu.
3. Click on the *Click to Launch* button.
4. Some programs require JAVA-Web Start to be installed on the local computer. This is built into the Macintosh OS X operating system; on other operating systems, you will be alerted if you need to install JAVA-Web Start (see Note 5).

3.2. Organizing Poxvirus Genome Data

Although a 200–300 kB genome is small by current sequencing capacities, analysis of the information contained within a single poxvirus genome, never mind 100+ such genomes, is much more efficient with specialized bioinformatics software. Among the various programs we have developed over the last 10 years for characterization of the poxvirus genomes, the two described below have the specific goals of (1) providing a dedicated SQL database to organize and store poxvirus genomic sequence data together with various types of sequence annotation and (2) giving virologists intuitive common interfaces to this database and the other analysis software, thereby eliminating, as far as possible, the necessity to connect to and learn the use of multiple Web sites or programs. These two tools are described first because several other sections of this review refer to analyses that they perform.

3.3. Viral Orthologous Clusters

VOCs (10) is a JAVA client–server application that accesses a large, up-to-date, MySQL database containing all complete and fully annotated poxvirus genomes. The database is administered by the corresponding author's laboratory and is available to all researchers via the Internet at the Virology.ca Web site or directly at http://virology.ca/tools/VOCS/Poxviridae. The VOCs database stores complete genome sequences, which are parsed from GenBank files, along with the DNA and protein sequences from the individual genes, promoter region sequences, annotations, molecular weights, predicted isoelectric points, calculated nucleotide (nt) and amino acid (aa) frequencies, and calculated codon usages. Thus, the researcher can interact with the data at multiple levels, for example: (1) Complete genomic DNAs, e.g., genomes can be searched for subsequences; (2) gene and protein sequences, e.g., these can be retrieved from the database and aligned or searched; and (3) comparatively, e.g., determine which genes are common to all poxviruses.

Some of the poxvirus GenBank sequence files contain errors in annotations and DNA sequences; we update our database as soon as we become aware of any corrections to these files, and similarly we annotate genomes with new information from the literature; therefore, VOCs is the most current source of poxvirus genomic information. In fact, due to the number of errors in the Vaccinia virus (Tian Tan strain) genome sequence, we have excluded it from the VOCs database (10).

One of the key features of VOCs is that each poxvirus gene/protein is assigned to an orthologous gene family based on BLASTP (protein query search against a protein database) similarity scores. Currently (see Note 1), the database contains 114 complete poxvirus genomes with >22,000 predicted genes or gene fragments. These predicted genes have been grouped into 489 ortholog families (requiring a representative from two or more different viruses) that have been given names based on function (if known). Among the applications integrated into VOCs are Position-Specific Iterative BLAST (PSI-BLAST), BLASTP, BLASTX (DNA query translated and searched against a protein database), TBLASTN (protein query search against a DNA database translated to six reading frames on the fly) (11); VGO (12); BBB (13); JDotter (3); and NAP (for comparing a protein against a DNA sequence (9). Importantly, the BLAST searches can be restricted to run against only the sequences in the VOCs database (which greatly simplifies the results by limiting them to poxvirus sequences) or the complete protein database at NCBI can be used.

Each time a user opens the VOCs program, JAVA-Web Start automatically checks the version of the client on the user's machine and downloads an updated version if required. VOCs was specifically designed to be used by molecular virologists and its interface makes it simple for researchers to make a variety of otherwise complex SQL database queries easily and rapidly. Three windows are used to navigate through the VOCs program: the Sequence Filter, Ortholog Group Filter, and Genome Filter windows. The Sequence Filter window allows users to search the database for genes based on such parameters as gene name, size, *pI*, NCBI protein ID, as well as nucleotide and amino acid sequences and constraints. The Ortholog Group Filter window allows users to search the database based on family name, ID, numbers of genes per family, and specific annotations. Although the bottom part of all three VOCs interface windows allows the user to select or exclude particular genomes for subsequent queries, the Genome Filter window allows users to restrict the set of genomes that are used and actually appear in the lower section of the windows; for example, the Genome Filter window can be used to limit the view to the 8 Capripoxvirus genomes instead of showing all 114 genomes. Table 1 shows some example queries that can easily be made within VOCs. Once a database query has been selected, the results of the query can be seen as a Gene Count or Ortholog Group Count in which case only the number of hits, of that particular type, in the database is displayed; or the list of genes and families can be viewed in a new window by clicking on the appropriate *View* button. The *count* function is used to check the query and prevent the accidental download of large numbers of genes, which would waste time. For an additional safeguard, VOCs asks users to confirm that they want to display more than 1,000 items. The Gene and Ortholog Results Tables can be sorted by clicking on a column header; for example, *sort by A + T%,* which reveals that there is considerable variation among the genes. Within these tables, the genes and Ortholog Groups are also available for display or further analysis using the aforementioned tools. It is quite simple to select an Ortholog Group and then view all genes within that group. The predicted protein sequences, DNA, or promoter region sequences can be selected for input into a tool to generate a multiple alignment, or particular sequences can be used to search the VOCs database using any of the BLAST programs.

Table 1
Examples of queries that can be performed with VOCs

Window	Query
SF	Find genes with names that match *A10*
SF	Find genes with DNA sequence containing a given sequence of nucleotides (ACGATCGATT)
SF	Find proteins with $4.5 < pI < 6.5$
SF	Find proteins with 25,000 < MW < 30,000 kDa
SF	Find proteins with *serine* content greater than *13%*
SF	Find proteins with *leu + ile + val + ala > 40%*
SF	Draw a map of genes in ECTV genome
SF	Create a gene comparison table for multiple viruses compared to a reference genome (Table 2)
OGF	Find the gene family containing ectromelia virus gene 108
OGF	Find gene families containing a single (unique) gene
OGF	Find gene families that contain a gene from all Orthopoxviruses
OGF	Find gene families present in myxoma virus but absent from SFV

SF Sequence filter window, *OGF* Ortholog group filter window

The most user-requested feature in VOCS is the ability to create a gene content (ortholog) comparison table for multiple viruses compared to a reference genome (Table 2). Here, multiple poxvirus genomes are compared to a reference genome to highlight the presence/absence of genes among multiple viruses. The table of results is also useful for determining the names of the orthologs in different viruses.

For users wishing to set up their own database, there is also an administrative client program (VOCs-Admin) that offers a variety of additional functions over the client program, allowing management of the database. These functions include the following: add a genome and all its genes into the database from a given GenBank format file; delete or modify genomes and/or genes; assign genes to Ortholog Groups; and edit, delete, or modify user notes.

3.3.1. VOCs Use Example #1: Alignment of Poxvirus Uracil DNA Glycosylase Proteins

1. To select the gene family, Open VOCs as described in Subheading 3.1.
2. Click on *Ortholog Group Filter* at the top of the window.
3. (a) If you know words in the group name, in the *Ortholog Group Selector* section, select checkbox at the left of *Ortholog Group Selector,* use option *contains*, and type the keyword uracil.

Table 2
Part of the table generated from VOCs Tools: *Genome Comparison*

CPXV-BR gene No.	Ortholog group	CPXV-BR gene No.	CMLV-CMS gene No.	ECTV-Mos gene number
CPXV-BR-119	CPV-B-116	CPXV-BR-119		
CPXV-BR-120	Unknown (Cop-H7R)	CPXV-BR-120	CMLV-CMS-121	ECTV-Mos-091
CPXV-BR-121	Large capping enzyme	CPXV-BR-121	CMLV-CMS-122	ECTV-Mos-092
CPXV-BR-122	Virion Core (Cop-D2L)	CPXV-BR-122	CMLV-CMS-124	ECTV-Mos-093
CPXV-BR-123	VV_Cop-D ORF B	CPXV-BR-123	CMLV-CMS-125	
CPXV-BR-124	Virion core (Cop-D3R)	CPXV-BR-124	CMLV-CMS-126	ECTV-Mos-094
CPXV-BR-125	Uracil-DNA glycosylase	CPXV-BR-125	CMLV-CMS-127	ECTV-Mos-095
CPXV-BR-126	NTPase, DNA replication	CPXV-BR-126	CMLV-CMS-129	ECTV-Mos-096
CPXV-BR-127	VETF-s (early transcription factor small)	CPXV-BR-127	CMLV-CMS-132	ECTV-Mos-097
CPXV-BR-128	RNA pol 18(RPO18)	CPXV-BR-128	CMLV-CMS-133	ECTV-Mos-098
CPXV-BR-129	Carbonic anhydrase/virion	CPXV-BR-129	CMLV-CMS-134	ECTV-Mos-099
CPXV-BR-130	mutT motif/NTP-PPH	CPXV-BR-130	CMLV-CMS-135	ECTV-Mos-100
CPXV-BR-131	mutT motif/NPH-PPH/ RNA levels regulator	CPXV-BR-131	CMLV-CMS-136	ECTV-Mos-101
CPXV-BR-132	NPH-I/Helicase, virion	CPXV-BR-132	CMLV-CMS-137	ECTV-Mos-102
CPXV-BR-133	Small capping enzyme	CPXV-BR-133	CMLV-CMS-141	ECTV-Mos-103

CPXV-Brighton was chosen as the reference genome for comparison to Camelpox-CMS and ECTV-Moscow; orthologous genes are displayed

(b) If you are not sure of how the group is named, use the *Orthologs Groups section* of this window (selecting checkbox for *Select ONLY Ortholog Groups*) to scroll through the alphabetized group names to find *Uracil-DNA glycosylase*. (c) If you know the gene number in any virus in VOCs, this could be selected directly or from the list of genes for that virus (see Note 6).

4. To make an alignment of all proteins in the group (currently 114), click on the *OrtGrpView* (Ortholog Group View) button, select the gene family by clicking on the group in the table, click on the *Align* button at the bottom of the window, and select alignment program of your choice.
5. Or, if you want to make an alignment of a subset of the Ortholog Group, click on the *GeneView* button, and select the proteins to be aligned by clicking on the rows of the table

using shift or command key to select several genes. From *Alignment* menu, choose *Protein Sequence Alignment* and the type of alignment software (use MUSCLE for more distantly related proteins).

6. The alignment is returned in a BBB multiple sequence alignment editor window (see Note 6).

3.3.2. VOCs Use Example #2: Find Ortholog Groups that Have a Protein Member Present in Myxoma Virus but Not in Rabbit Fibroma Virus

1. Open VOCs as described in Subheading 3.1 or if the tool is already open, go to the *select* menu and *clear all* to remove any previous search parameters.
2. Click on *Ortholog Group Filter* tab at the top of the window.
3. Scroll down to the *Ortholog Group Query* section, and select checkbox for *"Select Ortholog Groups that…"*.
4. Select *Contain* radio button; select *MYXV-Lau from* menu; and click *Add Criteria* button below this.
5. Select *AND* operator radio button at the right side; and click *Add Operator* button.
6. Select *Do NOT Contain* radio button; select *RFV-Kas* from menu; and click *Add Criteria* button. Your query will be shown in the window.
7. Click on *OrtGrpCnt or* OrtGrpView button. The search requires only a few seconds.
8. The result reveals that myxoma virus has only four genes that are absent from Rabbit Fibroma Virus.

3.3.3. VOCs Use Example #3: Find Ortholog Groups that Have Gene Present in All Poxvirus Genomes (i.e., the Ortholog Group Will Have the Maximum Number of Viruses)

1. Open VOCs as described in Subheading 3.1 or if the tool is already open, go to the *select* menu and *clear all* to remove any previous search parameters.
2. Click on *Ortholog Group Filter* tab at the top of the window.
3. Scroll down to the *Ortholog Group Size* section, select checkbox for *"#viruses,"* and enter the number of genomes in the database into the left box (this number is shown in the right box).
4. Click on *OrtGrpCnt or* OrtGrpView button. The search requires only a few seconds.

3.4. The Viral Genome Organizer

VGO (12) was originally based on Genotator (14), an annotation workbench designed for the analysis of eukaryotic genomic sequences, but has been extensively redesigned and now uses a JAVA interface. VGO provides a simple-to-use graphical interface to complete poxvirus genomic sequences. Because VGO also understands and accesses the VOCs database, it manages large amounts of information including the complete genome sequence, all gene and protein sequences, and can display (1) coding regions designated in the genome GenBank file; (2) computer-predicted open reading frames (ORFs) of any user-selected size; (3) all start and stop codons; (4) search results for restriction sites or any other subsequence defined in a regular expression (e.g., TTTTTNT); (5) graphs of nucleotide composition; and (6) results from a user-defined input file (see Note 7; Fig. 1). VGO also permits the user to quickly display gene and protein sequences by clicking on their graphical representation in its genome map; these sequences are then available in windows that allow the user to copy/paste sequences into other analysis tools. A particularly useful feature of VGO is that it allows a user to select and retrieve any region (DNA sequence) of a genome by simple interactions with the graphical genome map.

VGO was designed with *comparative genomics* in mind, and can display multiple genomes in a single window (limited only by screen "real-estate"). As a *stand-alone* tool, VGO can read GenBank files; however, additional powerful features are available if genome files are imported from the VOCs database. For example, the *Auto-highlight Related Genes* feature is then able to highlight orthologs of any selected gene in the other viral genomes being displayed. When comparing multiple genomes, it is convenient to label the genes by their VOCs *Family Number* (Ortholog Group) since

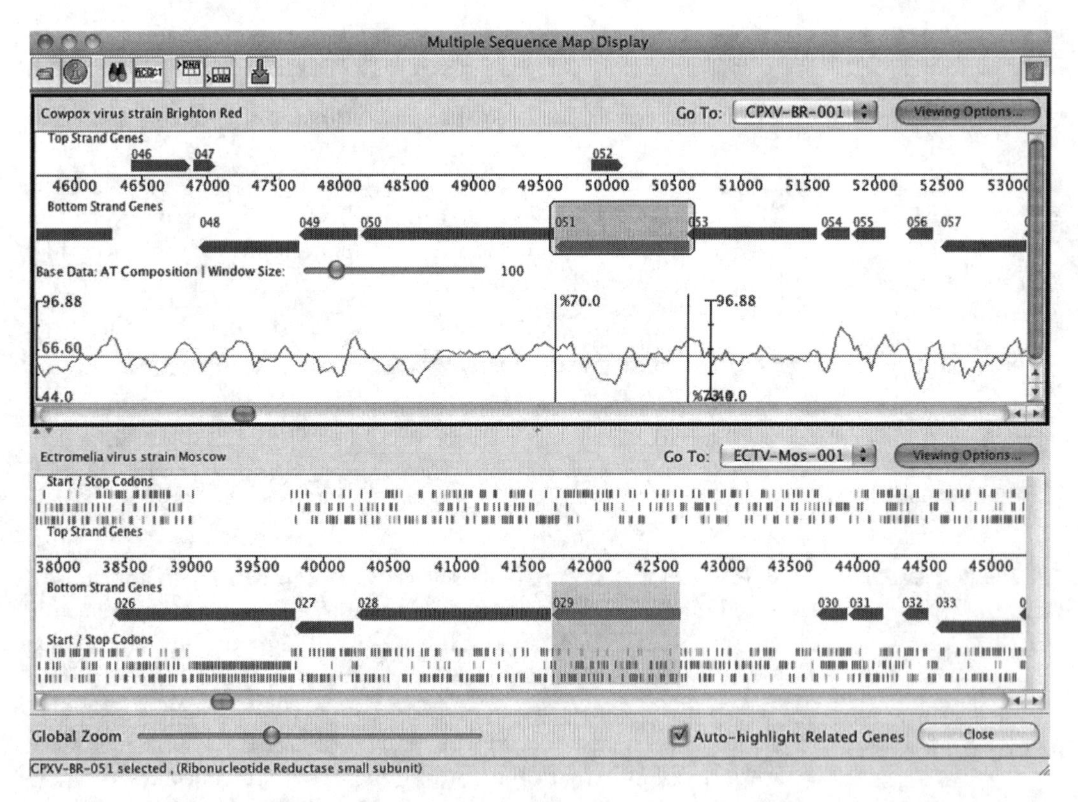

Fig. 1. Viral Genome Organizes (VGO). Depiction of analysis from Subheading 3.4.1. *Shaded* genes are orthologs. On the computer screen (and the electonic version of this chapter), Start/Stop codons are green and red, respectively. In this example, the cowpox genome is also displaying the A+T% graph.

orthologous genes in the multiple viruses will then appear with the same number. For example, this solves the problem that the DNA ligase gene is called A50R, J4R, K4R, 168R, 148, 171, and 188 in a series of Orthopoxviruses, but they are all in the same VOCs Ortholog Group, #292, which is also named DNA ligase.

3.4.1. VGO Use Example #1: Determine if a Large (1 kB) Noncoding Region in Ectromelia virus (ECTV-Moscow) Corresponds to a Gene-Coding Region of Cowpox Virus (CPXV-Brighton) (See Note 8)

1. Open VGO as described in Subheading 3.1.
2. From *File* menu, choose *Open*; select both *ECTV-Moscow* and *CPXV-Brighton* (see Note 9); and click OK.
3. From the *Organism List* (pink bars in VGO window), click on the two viruses for analysis (see Note 9).
4. From *View* menu, choose *Sequence Map*.
5. If stop/start codons are not visible by default, use the *Viewing Options* button to show them for both viruses, also selecting display *GenBank name*. Stretch the windows vertically to see all frames for the two viruses (see Note 10).
6. In the *ECTV-Moscow* window, use the *Global Zoom scroll* bar to zoom until genes can be viewed as a reasonable size and gene names appear (see Notes 10 and 11).
7. In the *ECTV-Moscow* window, use the *Go To menu* to move to gene 028.
8. Select the *Auto-highlight Related Genes* box at the bottom of the window.
9. Click on the *ECTV-Moscow* gene 029 to select it (color changes to orange and the area is shaded).
10. Scroll along the cowpox genome until the orthologous highlighted gene (orange) appears (in this example, it is gene 051).
11. Manually align the two genes in the CPXV and ECTV windows using scroll bars.
12. Click on *ECTV-Moscow* gene 030 and VGO shows that this is equivalent to CPXV 054. This suggests that the ECTV 1-kB region between 029 and 030 may be homologous to the cowpox 052 gene (Fig. 1). To test this hypothesis, continue with the following instructions.
13. Open NAP (Nucleotide-amino acid alignment (9)) from the Virology.ca *Tools* menu. This program aligns, with gaps, a protein sequence with a translated DNA sequence; thus, it is easy to test if a region of DNA has small deletions/insertions that interrupt ORFs.
14. Double click on the CPXV 053 gene to display the protein sequence and copy it into the *Protein Sequence* section of the *NAP window* using standard *copy/paste* commands.
15. Select *Genome Subsequence* from the VGO *View* menu. A new window appears.

16. Use your cursor to drag a box over the region of the ECTV-Moscow genome between gene 029 and 030 (this fills in the correct coordinates into the *Genome Subsequence* window).
17. Click the *Display button* in the *Subsequence Grabber* window and copy the DNA sequence into the *DNA Sequence* section of the *NAP window*.
18. Click the *Submit button*.
19. Scroll through the NAP alignment of the DNA and protein sequences. Aligned positions (codon and amino acid) are shown with *colons* between them (partial conservation uses *periods*), and gaps are indicated with dashes.
20. This *NAP* result clearly demonstrates that the two regions of the genomes are homologous but that a small number of nucleotide deletions cause a series of frameshifts in the ECTV-Moscow gene, which destroy its functionality. This is can also be seen by looking at the start/stop codons depicted in the six-frame genome translation in VGO. Also note that the ORF labeled 052 in Cowpox has not been annotated in the ectromelia virus genome and a repeat region in ectromelia gene 026 is absent from the cowpox ortholog.

3.4.2. VGO Use Example #2: Search ECTV-Moscow Genome for All TTTTTNT Sequences (See Note 12)

1. Open VGO and the ECTV-Moscow genome as described in Subheading 3.4.1.
2. From *Analysis* menu, select *Search Selected Sequence/Reg. Expression search* (nucleotide pattern is represented as a regular expression; http://en.wikipedia.org/wiki/Regular_expression).
3. Type *TTTTTNT or TTTTT.T* into the box, and click OK.
4. Results are displayed above (forward strand) and below (reverse strand) the gene representations.

3.5. Comparison of Poxvirus Genomic Sequences

One of the indispensable tools for the comparison of large DNA sequences in pairwise fashion is the dotplot. Each nucleotide, or small window of nucleotides, of one sequence is compared to every nucleotide of another and the results are visually displayed in an easy-to-understand plot. For analysis of poxvirus genomes, the software must be able to handle DNA sequences in excess of 300 kB. We have found that DOTTER (2) is an extremely effective tool because *after* the plot is calculated, the user can change scoring parameters in real time while viewing the plot. It is also possible to zoom into particular regions of the dotplot by selecting an area with the cursor. This results in recalculation of the plot in that smaller region. Fig. 2 shows a dotplot comparison of the genomes of variola virus (strain Bangladesh) and Monkeypox virus (strain Zaire). The regions of high similarity, as well as gaps in the alignment, are immediately obvious. The *greyramp tool* (Fig. 2; inset)

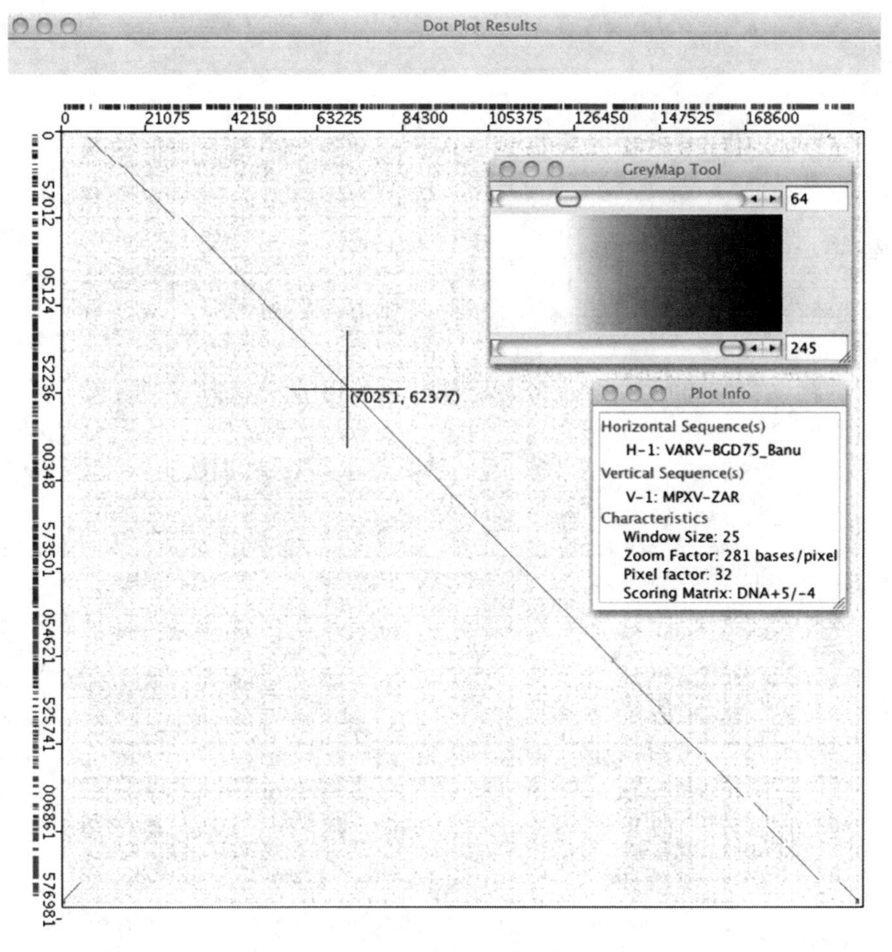

Fig. 2. Dotplot of complete genomes (DNA) of variola virus (strain Bangladesh) vs monkeypox virus (strain Zaire). *Insets* show (1) *Greyramp Tool* for changing scoring parameters and (2) Plot information. *Small blocks* along axes are usually colored to represent genes with transcription orientation.

is used to rapidly change scoring parameters without the need for recalculation of the complete dotplot. An alignment tool is also available (not shown) which displays a continuously scrollable window showing the alignment of the two DNA sequences at any point in the plot chosen by the user. Dotplots are particularly useful for detecting direct and inverted repeats; part of the poxviral Inverted Terminal Repeats is visible in the top right of the dotplot in Fig. 2. Users of dotplots should be aware that the resolution of the plot is low (related to the number of nucleotides/screen pixel) when large sequences are compared; therefore, to get a good sense of fine sequence similarity from a Dotplot, it is necessary to "*zoom-in.*" We have created a JAVA interface for DOTTER (JDOTTER, (3)) that permits it to be used as a graphical display in the VOCs interface with gene or protein sequences or genome segments selected by the user.

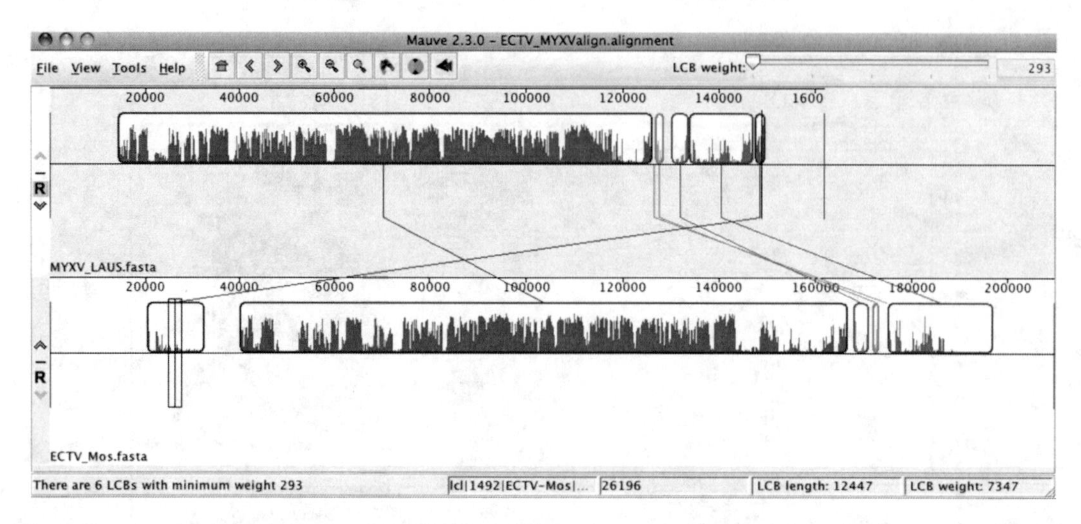

Fig. 3. Mauve comparison of Myxoma virus (MYXV) and Ectromelia virus (ECTV) genomes. MAUVE color codes regions that are aligned to each other, for example the large region spanning from nucleotide 18,000 to 125,000 on the MYXV genome corresponds to the large region spanning from nucleotide 40,000 to 165,000 on the ECTV genome.

For genome sequence comparisons using pairs or multiple genomes, the MAUVE aligner (15, 16) can create global alignments of sequences that contain rearrangements, including inversions, of DNA segments, which are very useful for comparing poxvirus genomes. MAUVE alignments are fast and particularly useful because of the visual overview of the alignment (Fig. 3) that shows the user all blocks that align, including translocations and inversions. To perform an alignment in MAUVE, either FASTA-formatted or GenBank files of the genomes are required; annotations in GenBank files are more useful because the information helps navigate through the genomes. The "*Align Sequences*" option provides a quick alignment of the genomes, whereas the "*Progressive Mauve*" alignment is slower and more sensitive. Once the alignment has been performed, the visual overview of the alignment is displayed (Fig. 3). Importantly, the user can zoom into the alignment allowing for visualization of SNPs at the nucleotide level. However, if the user needs to manually edit an alignment created by MAUVE, a separate alignment tool must be used since MAUVE does not provide editing functions.

Another useful way to display similarities between poxvirus genomes is with Synteny plots. These are scatter diagrams of protein orthologs shared between pairs of virus genomes (17).

3.6. Annotation of Poxvirus Genomes

Having determined the DNA sequence of a poxvirus genome, there remains the sometimes challenging, but always time-consuming, problem of annotating the genome. The level of difficulty is inversely proportional to the similarity of the genome to other previously properly annotated sequenced genomes. Although poxvirus genes do not contain the introns that complicate the prediction

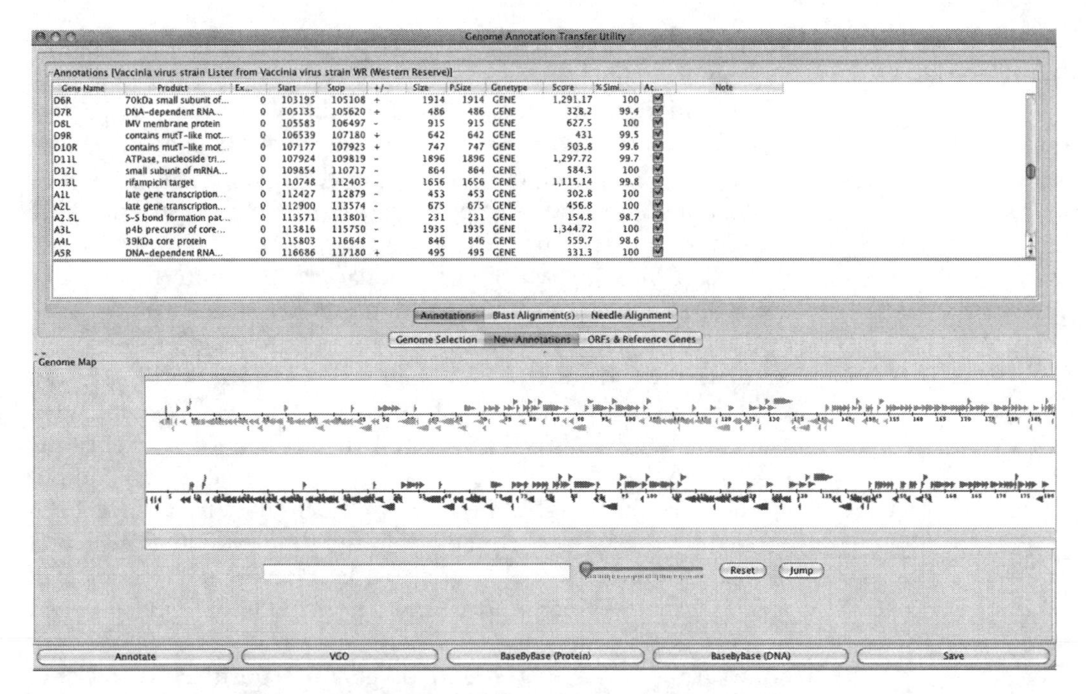

Fig. 4. Gene Annotation Transfer Utility (GATU). *Top panel* shows list of genes predicted to be located in the genome requiring annotation. *Bottom panel* shows a genome map of the reference genome (*top line*) and predicted genes of the genome to be annotated (*bottom line*). On the computer screen, gene symbols are colored to provide clarity.

of many eukaryotic genes, one of the trickier problems (which is exacerbated in GC-rich genomes because of the low frequency of stop-codons in noncoding sequence) is to decide which ORFs should be designated as a gene and annotated in the GenBank file.

Many genes are simple to annotate because they are conserved in a number of diverse poxviruses. To take advantage of the growing resource of annotated poxvirus genomes, we have developed GATU (18) (Fig. 4), a tool that uses a reference genome to transfer as many annotations as possible to a new target genome. Depending on the similarity between the reference and target genomes, 70–100% of the genes may be annotated by GATU with essentially no effort by the user. An important feature of GATU is that the tool leaves final control of the annotation process in the hands of the user (Fig. 4). Although obvious annotations suggested by GATU are preselected in checkboxes for acceptance, the user can reject these if required. Similarly, novel ORFs or those that significantly vary from orthologs in the reference genome are provided to the user as *suggested ORFs*, leaving the final decision in the hands of the expert. When the user is happy with the final annotations, these can be written out as a GenBank file for submission to a public database.

However, small ORFs that have no obvious ortholog within the reference or other poxvirus genomes are especially problematic for annotators. The VGO genome display tool, which can be used to examine colinearity between the target and reference genomes, can help in these situations; it also provides A + T% plots for searching for potential AT-rich promoter regions and a convenient display of start/stop codons in the six possible coding frames to allow the annotator to look for potential sequencing errors that have broken otherwise complete orthologous genes. We have also found that ARTEMIS (7), which has a similar graphical interface, can be a very useful tool in helping to annotate genomes. In addition to displaying information such as ORFs, start–stop codons, six-frame translations, and nucleotide composition graphs, ARTEMIS can provide links to BLAST search programs and features semiautomatic naming of ORFs. However, it cannot take advantage of the information present in previously annotated reference genomes in the way that GATU does. A number of criteria have been applied to poxvirus gene prediction, the simplest being ORF size; others include potential overlap with other genes, presence of promoter-like elements, isoelectric point, amino acid composition (19), and codon usage. For vaccinia virus, and other AT-rich poxviruses, the coding strand of genes is positively correlated with purine content (20).

In the area of genome annotation, less is probably more; it is easier to add annotations than take them away once a GenBank file has been submitted. For example, vaccinia virus (strain Copenhagen, (1)) was initially annotated with major ORFs (genes) and 65 minor ORFs; most of the latter substantially overlap with larger genes, usually on the opposite DNA strand. These extra ORFs were named "X-ORF-Y," where X represents the *HindIII* genome fragment and Y is a letter representing the rank of the ORF from left to right. It is unlikely that any of these minor ORFs are functional genes, but to the inexperienced eye it appears that this virus has a large series of unique genes. It would, therefore, be useful if annotation systems could include the option for *probably nonfunctional ORF* descriptions since relatively minor mutations can easily destroy gene function without changing the ORF very much. An example of this is the introduction of a point mutation that creates a stop codon or a small frame-shifting indel near the 5′ end of a gene's coding sequence; in these situations, one would expect normal mRNAs to be produced, but translation to lead to a severely truncated protein. It is, therefore, grossly inaccurate to label such ORFs as functional genes, even though they might be 95% as long as the original gene.

Another useful tool for the characterization of new genomes is NAP (9), a global nucleotide versus amino acid alignment program. We have made an easy-to-use JAVA client–server interface for NAP; it is available as a stand-alone tool at http://virology.ca/tools/NAP. The utility of this program lies in the fact that it not only generates alignments between protein and DNA sequences but that it also produces a gapped-global alignment. Therefore, it becomes very easy to locate and evaluate potential sequencing errors that may break genes into fragments by the introduction of frame-shifting mutations or stop codons.

3.7. Searching for Distantly Related Proteins

One of the most common questions in bioinformatics is: "What is my protein (or DNA) sequence similar to?" Similarity searches in which a protein or DNA sequence is searched against a database of all known sequences are most often performed with one of the BLAST programs (11). There are various search algorithm strategies, and an important design factor is how they balance search sensitivity and search speed. In this regard, it is important to note that frequently BLAST search parameters are not set to their most sensitive; for example, *WORD SIZE*, the match length that triggers an extended alignment of sequence regions, should be adjusted to the minimum possible. For "routine" protein database searches, BLASTP is sufficient to find database matches at >30% identity. However, a more sensitive database search program is PSI-BLAST which automatically constructs a new position-specific scoring matrix (PSSM), using a multiple alignment of the highest scoring database matches, to be used in the next round of an iterative series of BLAST searches. Thus, on each round of searching, the program uses a modified scoring matrix that reflects the most conserved residues in sequences that have already been identified as similar to the query. The author has found, for example, that PSI-BLAST is superior to BLASTP in the identification of members of the poxvirus uracil DNA glycosylase family which share very little similarity with other uracil DNA glycosylases (21). A recent addition to the NCBI PSI-BLAST utility is the Constraint-based Multiple Alignment Tool which is enhanced for creating multiple alignments of distantly related proteins (COBALT (22)).

Another method to identify distantly related proteins involves using profile-based searches. In general, profile-based searches rely on comparison of different profiles of a set of related sequences rather than a single sequence to identify homologs in a database and can be more sensitive than sequence-based approaches, like PSI-BLAST (23). One example of a profile-based search is HHpred (http://toolkit.lmb.uni-muenchen.de/hhpred) which uses a profile based on an HMM to perform the search (23). HHpred begins by taking the query protein sequence and running several iterations of PSI-BLAST in order to create a multiple alignment of the query plus related protein sequences. This alignment is then turned into an HMM which is then used to search HMMs created from the protein databank (PDB), structural classification of proteins (SCOP), and CATH databases. The HMMs of both the query and the proteins in each of the databases are created based on secondary structure either from a prediction of secondary structure using PSI-PRED or from the tertiary structure of the protein if a crystal structure is in the database. Results (hits) are displayed based on the probability that the query sequence is a true match to the hit sequence. HHpred results list both an E-value and a probability score that are interpreted much like BLAST results, where an E-value that is close to zero means that the probability of the hit being random is close to zero. Given that there are still several VACV proteins that have an unidentified function, we regularly use HHpred to try and identify distant relatives to these "unknown" proteins. One such scan of the VACV G5R protein revealed that it was similar to an archaeal flap-endonuclease (FEN-1) protein (24) and a scan of the VACV G8R protein revealed that it was similar to the proliferating cell nuclear antigen (PCNA) protein in yeast (25).

Often, once it has been established that the protein of interest has a distant relative to another protein in the database, it is very useful to model its tertiary structure and use the results to help formulate hypotheses and develop biochemical experiments on the protein of interest. We use two different tools to model tertiary structure: the Protein Homology/analogY Recognition Engine (PHYRE (26)) produces results relatively quickly (within an hour) and models only the subsection of the protein which aligns with a hit in the database, whereas Robetta (27), which can take considerably longer to give results (over 1 month in some cases), will model difficult regions using ab initio approaches. PHYRE works by first creating a profile of the query sequence aligned with several related sequences that are identified in an initial PSI-BLAST run. This profile is then compared to the profiles of other sequences in the PDB database and the top hit is then used as the template to build a model of the region that aligned with the hit. Modeling involves a replacement algorithm, where the residues in the backbone of the template that aligned with the query sequence are replaced by residues seen in the query sequence. The backbone of the template structure does not get changed unless the query contains insertions or deletions relative to the template. Because PHYRE also employs a profile–profile-based search in its initial stages, it can be used to confirm HHpred results.

Since PHYRE only models the region of the query protein that aligns sufficiently with the template sequence, it is often advantageous to also use a more comprehensive modeling program such as Robetta to view the putative tertiary structure of the protein of interest. Robetta is a fully automated structure prediction server that uses comparative modeling methods to model protein structure if a related structure exists in the PDB database or ab initio prediction methods if no structure exists. When the query sequence is initially submitted to the Robetta servers, the algorithm begins to break the query sequence down into different domains using the Ginzu protocol (28). The Ginzu protocol involves scanning the protein sequence for domains using PDB-BLAST, then HHsearch, PFAM comparisons, and finally PSI-BLAST looking for sequences that could be used as a template for each identified domain. Each of these domains is modeled using comparative modeling if the domain is found to have a related structure in the database, or ab initio modeling if it does not. We used Robetta as a second step in the *in silico* characterization of the VACV G5R protein and it identified an archaeal flap-endonuclease protein as the best template to use in the subsequent comparative modeling step. The resulting model was subsequently compared to the crystal structure of the human FEN-1 protein and was found to be highly conserved in both secondary and tertiary structures and with three of the five main features of the FEN-1 protein including the active site, suggesting that the G5R protein should be classified as a flap-endonuclease protein. Subsequently, this protein was shown to be required for genome processing in infected cells and to be involved in repair of double-strand breaks by recombination in transfected cells (29).

3.8. Motif Searches

A protein motif can be defined as a string of amino acids that is characteristic of a functional/structural unit or protein domain of a series of proteins. A protein domain is often defined as an "independently folding structural unit"; thus, a motif is usually considerably smaller than a domain. One of the most commonly used motif databases is PROSITE ((30); www.expasy.ch/prosite/). In September 2009, PROSITE contained more than 1,500 different entries. Motif searches are frequently useful in identifying domains within a protein when no other large areas of similarity exist with other proteins.

For example, [KR]-[LIVA]-[LIVC]-[LIVM]-x-G-[QI]-D-P-Y is the PROSITE (PS00130) motif for uracil DNA glycosylases (UNG). It is important that PROSITE motifs are updated over time, as new members of a protein family are recognized. For example, since the first edition of this book, the number of acceptable amino acids at the second position of this motif was increased to four with the inclusion of alanine. Similarly, one needs to appreciate that although this motif detects over 400 UNG proteins, there are also 4 UNGs that are not found by searching with this particular motif; these are designated *false negatives* in the PROSITE documentation. These *false negatives* are a consequence of maintaining this strict motif that does not generate any false positive hits. This particular PROSITE motif is written as a regular expression (a *pattern* in PROSITE), a format that allows mismatches and variability in spacing between residues in the motif. Software tools within VOCs allow users to search sequences with any regular expressions; at www.expasy.ch/prosite/, software can be used to search (1) a protein sequence against all motifs or (2) a PROSITE or user-created motif against all protein sequences. Advantages of performing the search through VOCs are the speed and the fact that only poxvirus sequences are searched. It should be noted, however, that the PROSITE (PS00130) motif only matches 104 of the114 UNGs present in VOCs, whereas a modified motif [KLNR]-[LIV]-[LIVC]-[LIVM]-x-G-[QIY]-[RD]-[SP]-[YF] will find 113 of 114 UNGs in VOCs but matches several *false negative* proteins in SWISS-PROT. Some PROSITE motifs (*profiles* in PROSITE) are scored using an amino acid match scoring matrix, but are included in the ScanProsite search at www.expasy.ch/prosite.

In addition to PROSITE, there are several systems that integrate multiple motif/profile searching tools. The ProfileScan Server (http://hits.isb-sib.ch/cgi-bin/PFSCAN) searches user-supplied protein sequences for Pfam (*P*rotein *Fam*ilies database of alignments and HMMs) and PROSITE motifs. The Pfam Web site (http://pfam.sanger.ac.uk/) lists families for many of the poxvirus proteins that are common to all poxviruses and is an excellent source of information, although the output can be rather overwhelming and difficult to interpret. InterPro (http://www.ebi.ac.uk/interpro/) is another comprehensive assembly of motif and sequence databases that is in turn connected to a variety of other databases. It can be useful for looking at conserved motifs or domains in protein families, but for the most part, it is overkill for simple motif searches.

3.9. Multiple Sequence Alignments

The generation of multiple sequence alignments (MSAs) is a staple technique in comparative molecular biology. Determining which amino acids are conserved at particular locations in a group of proteins may help predict which residues are likely to have important roles in the structure or biochemistry of the proteins. These conserved residues are also used to define the motif(s) characteristic of protein families and MSAs are the basis of many types of phylogenetic analyses. A variety of computer programs are available to align DNA and protein sequences, attempting to maximize a score from matching amino acids or nucleotides and also minimize penalty scores due to insertions and deletions (indels). One of the best known alignment tools is CLUSTALW (4), but there are newer algorithms such as T-COFFEE (5) and MUSCLE (6) that have been reported to produce better alignments when using distantly related sequences (<30% identity). DNA alignments of short- and medium-length sequences, such as promoters and genes, respectively, can be achieved with the same software, but sequences the length of poxvirus genomes require specialized tools such as DIALIGN2 (31) and MAUVE (16).

However, whatever software is used, MSAs usually need final *hand-editing*, especially around gaps, using a sequence alignment editor to produce final accurate alignments. A final hurdle is the production of a publication quality figure, if required. We developed the BBB tool for creating, editing, and viewing not only MSAs of proteins and short DNA sequences, but also complete poxvirus genomes. BBB was developed by our group as a JAVA MSA editor to interface with VOCs, but it also functions as a stand-alone tool and can save files to be used at later times, in the same way that word processors do. Over the years, BBB has been enhanced with several unique features, for example: (1) easy display of differences between adjacent sequences; (2) display of three-frame translation of DNA sequences; (3) display of *top* or *bottom* strand for a DNA sequence; (4) ability to read in GenBank files including annotation on DNA sequences; (5) ability to read sequences from VOCs database; (6) ability to search for internal sequences using regular expressions or fuzzy (allowing mismatches) motifs; (7) ability to add *user-comments* to sequence regions; (8) ability to realign internal regions of an MSA with several algorithms and import results to the existing alignment; (9) generate %identity table for sequences in an alignment; (10) map primer sequences to a DNA sequence; and (11) save alignment pictures for publication. Finally, one of the most powerful features of BBB is its ability to summarize differences between genomes in an MSA. When provided with two or more aligned and annotated genomes from closely related viruses (e.g., isolates of variola virus), BBB can detect all nucleotide differences and present this information with an analysis of the consequences of the differences (truncation of ORF, silent mutation, coding change, in predicted promoter region).

Another powerful JAVA, and therefore platform independent, MSA editor is JalView (http://www.jalview.org); some of the features unique to JalView include the display of predicted secondary structure (calculated using Jpred 3 (32)) and the linking of a protein sequence in the MSA window to its structure (if available) in a Jmol structure viewing window (an open-source Java viewer for chemical structures in 3D).

3.10. Display of RNA Sequencing Data

An example of the power of Next-Generation (nucleotide) Sequencing (NGS) and its potential contribution to the analysis of the viral life cycle was recently provided by sequencing RNA libraries made at several times post infection with tissue culture cells infected with vaccinia virus (33). The final result, after much computational processing, is a measure of how many times a particular genome nucleotide is found in the short sequences from the RNA library. The many sequences obtained from a single sequencing run together with the short length of the sequences allow for mapping of the 5′ ends of transcripts, timing of transcription, and to some extent relative quantification of the transcripts. Although comprehensive programs are available to analyze such data, their use usually has a very steep learning curve and the many features of these tools are not required by the typical molecular virologist. Therefore, we have built a simple-to-use viewer for this kind of data into BBB. After loading an annotated genome into BBB, which displays the genome sequence (top or bottom strand) with

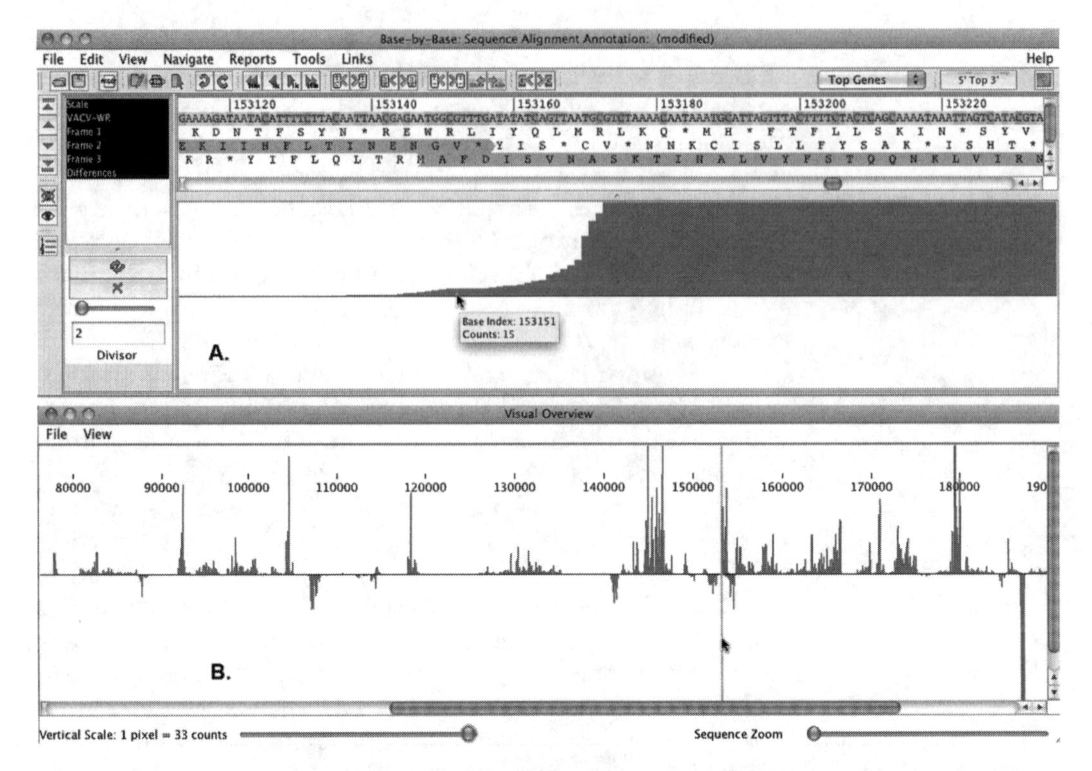

Fig. 5. Display of RNA sequencing data in Base-By-Base. (**a**) VACV-WR genome displayed with three-frame translations (top strand) with histogram of early RNA sequences that map to each nucleotide. Counts for each nucleotide can be viewed by moving the cursor over the different parts of the histogram. (**b**) Overview of data shown in (**a**); this view shows sequences mapping to top and bottom strands, representing transcription to the *right* and *left*, respectively. RNA sequencing data was kindly provided by Dr. Bernard Moss.

three-frame translations and positions of genes, the RNA sequencing summary data from MochiView (34) is read into the program and viewed together with the sequence, essentially as a histogram (Fig. 5a). BBB also provides a genome-scale overview of the data (Fig. 5b).

3.11. Future Work

It is difficult to predict where the explosion in genomic and proteomic information will take us. It is, however, obvious that genome sequencing is getting easier and cheaper by the year and our need for bioinformatics databases and tools grows correspondingly. As more and more poxvirus genomes are sequenced, the variation is less and new questions can be asked (e.g., the sequencing of complete genomes of variola virus). However, with this will come the need to develop novel software analysis tools.

Another problem that must be addressed is that not only are the databases growing rapidly with new sequence information, but genome annotations quickly become out of date as new functions are discovered for viral proteins and occasionally new genes are discovered in genomes sequenced many years ago. What is needed is a notification system to provide users with updates of relevant new information in the databases. Although most journals can e-mail titles of new publications to users based on keyword information, the reporting of new hits in biological sequence databases is a far more complex problem because of the volume of data in both the public protein databases and the poxvirus database and difficulties arising from small changes to the accession numbers or sequences themselves in the NCBI databases.

Finally, as more protein structures are solved for poxvirus proteins, good bioinformatics tools to interface between structural data and the large volume of sequence data are still needed. This will maximize the benefit of high-resolution structures and should enhance our understanding of the poxvirus life cycle and development of therapeutics.

4. Notes

1. A review of this topic has been submitted.
2. Virology.ca (http://www.virology.ca) was formerly part of the Poxvirus Bioinformatics Resource Center (PBRC; http://www.poxvirus.org) and the Viral Bioinformatics Resource Center (VBRC; http://www.biovirus.org).
3. Current desktop computers have more than sufficient "horsepower" to run basic bioinformatics tools; it is more useful to buy extra RAM and a second computer screen (for displaying data and results). Most of the software described in this chapter is available in either (a) a simple WWW-form (e.g., BLAST at NCBI) interface that sends requests to a remote computer and displays the results in a WWW browser or (b) a JAVA client–server format in which the local client program is far more sophisticated than a WWW interface and only connects to the remote server to download information from a database or to off-load computationally intensive calculations. JAVA is a multiplatform, platform-independent, object-oriented programming language.
4. SQL stands for *sequence query language* database (note that the word *sequence* does not refer to DNA or protein sequence).
5. Downloading JAVA-Web Start is straightforward and takes only a few minutes even for a novice. When JAVA-Web Start downloads one of the JAVA clients of VOCs, VGO, or JDotter to the user's local machine for the first time, it displays a warning window by default, which informs the user that permitting this program to be installed on his or her computer is potentially dangerous. This is because the software is like most other programs on your computer and can write to your hard drive.

The default warning message also includes information about the origin of the software and developers to allow the user to determine if the software comes from a trustworthy site. We recommend that you accept the software. The Virology.ca site is secure behind a firewall to prevent potential hackers from tampering with these files. Our group uses this software daily in exactly the same way as external users; therefore, we would quickly detect any problems with the software.

6. If searching for a vaccinia gene in the *Gene Family Analyzer* returns no sequences, check that the virus selector has not been used to restrict the search to another virus or that no other search criteria were inadvertently selected. The *Clear All* feature of the *Select* menu can be used to remove all queries.
7. An example of the results from a user-defined input file is a simple format text file defining position, length, and color of boxes to be drawn on the VGO sequence map. We frequently use it to graphically display results of otherwise obtuse text files generated by promoter prediction programs.
8. Between ECTV genes 028 and 029, there is approximately 1 kB without any annotated gene. It is unusual to have non-coding regions in poxvirus genomes and, therefore, it would be useful to determine how this region in ECTV compares to the CPXV genome. The hypothesis to be tested is: Does this region of ECTV contain a *fragmented gene* equivalent to a complete gene in CPXV?
9. Use control or command (Macintosh apple) key for multiple selections.
10. Scrolling is faster with fewer items displayed on the screen. To maximize speed, do not display start/stop codons unless needed and zoom in before scrolling.
11. The annotated virus genes are shown in dark blue above (transcribed to the right) and below (transcribed to the left) the scale bar.
12. The TTTTTNT (T_5NT) sequence is a transcriptional stop signal that signals the polymerase transcribing EARLY poxvirus genes to fall off the template resulting in relatively homogeneous transcripts from early genes. This signal is not recognized by the viral polymerases that transcribe intermediate and late genes.

Acknowledgments

The authors thank the many programmers, researchers, and students who have been involved in the development and testing of this software. This work has been supported by NIH/NIAID (Grant

AI48653-02 and Contract HHSN266200400036C) and funds from the Natural Sciences Engineering Research Council of Canada. Drs. C. Upton, R. M. L. Buller, and. E. J. Lefkowitz were the original developers of the Poxvirus Bioinformatics Resource.

References

1. Goebel SJ, Johnson GP, Perkus ME, Davis SW, Winslow JP, Paoletti E (1990) The complete DNA sequence of vaccinia virus. Virology 179:247–266
2. Sonnhammer EL, Durbin R (1995) A dot-matrix program with dynamic threshold control suited for genomic DNA and protein sequence analysis. Gene 167:GC1-10
3. Brodie R, Roper RL, Upton C (2004) JDotter: a Java interface to multiple dotplots generated by dotter. Bioinformatics 20:279–281
4. Thompson JD, Higgins DG, Gibson TJ (1994) CLUSTAL W: improving the sensitivity of progressive multiple sequence alignment through sequence weighting, position-specific gap penalties and weight matrix choice. Nucleic Acids Res 22:4673–4680
5. Notredame C, Higgins DG, Heringa J (2000) T-Coffee: A novel method for fast and accurate multiple sequence alignment. J Mol Biol 302: 205–217
6. Edgar RC (2004) MUSCLE: multiple sequence alignment with high accuracy and high throughput. Nucleic Acids Res 32: 1792–1797
7. Mural RJ (2000) ARTEMIS: a tool for displaying and annotating DNA sequence. Brief Bioinform 1:199–200
8. Marass F, Upton C (2009) Sequence searcher: a Java tool to perform regular expression and fuzzy searches of multiple DNA and protein sequences. BMC Res Notes 2:14
9. Huang X, Zhang J (1996) Methods for comparing a DNA sequence with a protein sequence. Comput Appl Biosci 12:497–506
10. Upton C, Slack S, Hunter AL, Ehlers A, Roper RL (2003) Poxvirus orthologous clusters: toward defining the minimum essential poxvirus genome. J Virol 77:7590–7600
11. Altschul SF, Madden TL, Schäffer AA, Zhang J, Zhang Z, Miller W, Lipman DJ (1997) Gapped BLAST and PSI-BLAST: a new generation of protein database search programs. Nucleic Acids Res 25:3389–3402
12. Upton C, Hogg D, Perrin D, Boone M, Harris NL (2000) Viral genome organizer: a system for analyzing complete viral genomes. Virus Res 70:55–64
13. Brodie R, Smith AJ, Roper RL, Tcherepanov V, Upton C (2004) Base-by-base: single nucleotide-level analysis of whole viral genome alignments. BMC Bioinformatics 5:96
14. Harris NL (1997) Genotator: a workbench for sequence annotation. Genome Res 7:754–762
15. Rissman AI, Mau B, Biehl BS, Darling AE, Glasner JD, Perna NT (2009) Reordering contigs of draft genomes using the Mauve aligner. Bioinformatics 25:2071–2073
16. Darling AC, Mau B, Blattner FR, Perna NT (2004) Mauve: multiple alignment of conserved genomic sequence with rearrangements. Genome Res 14:1394–1403
17. Lefkowitz EJ, Upton C, Changayil SS, Buck C, Traktman P, Buller RM (2005) Poxvirus Bioinformatics Resource Center: a comprehensive Poxviridae informational and analytical resource. Nucleic Acids Res 33:D311–316
18. Tcherepanov V, Ehlers A, Upton C (2006) Genome Annotation Transfer Utility (GATU): rapid annotation of viral genomes using a closely related reference genome. BMC Genomics 7:150
19. Upton C (2000) Screening predicted coding regions in poxvirus genomes. Virus Genes 20:159–164
20. Da Silva M, Upton C (2005) Using purine skews to predict genes in AT-rich poxviruses. BMC Genomics 6:22
21. Li W, Pio F, Pawlowski K, Godzik A (2000) Saturated BLAST: an automated multiple intermediate sequence search used to detect distant homology. Bioinformatics 16:1105–1110
22. Papadopoulos JS, Agarwala R (2007) COBALT: constraint-based alignment tool for multiple protein sequences. Bioinformatics 23:1073–1079
23. Söding J, Biegert A, Lupas AN (2005) The HHpred interactive server for protein homology detection and structure prediction. Nucleic Acids Res 33(Web Server issue):W244–8
24. Da Silva M, Shen L, Tcherepanov V, Watson C, Upton C (2006) Predicted function of the vaccinia virus G5R protein. Bioinformatics 22:2846–2850
25. Da Silva M, Upton C (2009) Vaccinia virus G8R protein: a structural ortholog of proliferating

cell nuclear antigen (PCNA). PLoS One 4:e5479

26. Kelley LA, Sternberg MJ (2009) Protein structure prediction on the Web: a case study using the Phyre server. Nat Protoc 4:363–371
27. Kim DE, Chivian D, Baker D (2004) Protein structure prediction and analysis using the Robetta server. Nucleic Acids Res 32:W526–531
28. Kim DE, Chivian D, Malmstrom L, Baker D (2005) Automated prediction of domain boundaries in CASP6 targets using Ginzu and RosettaDOM. Proteins 61(Suppl 7):193–200
29. Senkevich TG, Koonin EV, Moss B (2009) Predicted poxvirus FEN1-like nuclease required for homologous recombination, double-strand break repair and full-size genome formation. Proc Natl Acad Sci USA 106: 17921–17926
30. Falquet L, Pagni M, Bucher P, Hulo N, Sigrist CJ, Hofmann K, Bairoch A (2002) The PROSITE database, its status in 2002. Nucleic Acids Res 30:235–238
31. Morgenstern B (1999) DIALIGN 2: improvement of the segment-to-segment approach to multiple sequence alignment. Bioinformatics 15:211–218
32. Cole C, Barber JD, Barton GJ (2008) The Jpred 3 secondary structure prediction server. Nucleic Acids Res 36:W197–201
33. Yang Z, Bruno DP, Martens CA, Porcella SF, Moss B (2010) Simultaneous high-resolution analysis of vaccinia virus and host cell transcriptomes by deep RNA sequencing. Proc Natl Acad Sci USA 107:11513–11518
34. Homann O, Johnson A (2010) MochiView: versatile software for genome browsing and DNA motif analysis. BMC Biol 8:49

Chapter 15

Antigen Presentation Assays to Investigate Uncharacterized Immunoregulatory Genes

Rachel L. Roper

Abstract

Antigen presentation to T lymphocytes is the seminal triggering event of the specific immune response, and poxviruses encode immunomodulatory genes that disrupt this process. Discovery of viral proteins that interfere with steps in the antigen presentation process requires a robust, easily manipulated antigen-presenting and T lymphocyte response system. Use of fresh primary antigen-presenting cells (APC) is preferable because cell lines that can present antigen in vitro are often not representative of APC in vivo and are typically weak stimulators. To study immunomodulatory poxvirus genes, we have used infected primary rat macrophages to present a model antigen, the myelin basic protein peptide, to a cognate CD4+ RsL11 T cell clone. Using this system, viruses can be assessed for difference in immunomodulation, and viral gene functions may also be assayed by comparing effects of wild type virus and mutant viruses (e.g., a deletion in the putative immunomodulatory gene). While antigen presentation can be thought of as a single event, it can also be considered as a larger process comprising multiple steps including: antigen acquisition, antigen processing, peptide loading onto MHC molecules, transport to the surface, MHC binding to T cell receptor, interaction of costimulatory molecules, cell signaling, cytokine synthesis by both cells, and proliferation of antigen specific T lymphocytes. This system allows for the initial determination of whether there is a phenotype and then also allows the stepwise deconstruction of the system to analyze this process at several points to focus in on the mechanism of immunomodulation. We have used this model system to elucidate the function of a highly conserved but previously uncharacterized poxvirus gene that we showed was important for virulence in rodents. The experimental system developed should be broadly applicable to analyzing viral effects on immunity.

Key words: Antigen presentation, CD4, Immunomodulation, T lymphocyte, Cytokine, Vaccinia, Poxvirus, MHC

1. Introduction

Viruses have an impressive number of strategies for survival, replication, and spread in the host organism. Orthopoxvirus genomes encode approximately 200 genes, many with functions that remain

Stuart N. Isaacs (ed.), *Vaccinia Virus and Poxvirology: Methods and Protocols*, Methods in Molecular Biology, vol. 890,
DOI 10.1007/978-1-61779-876-4_15, © Springer Science+Business Media, LLC 2012

to be elucidated (1). For example, viral virulence genes encode the ability to (1) grow in otherwise restrictive cell types (host range genes) (2–5), (2) block inflammation and immune responses (6–9), (3) inhibit cellular apoptosis (10–12), (4) enhance spread of virus particles using host proteins (13, 14), (5) interfere with cell signaling (15, 16), and (6) globally regulate cellular gene expression (17, 18). Viruses control immune responses to facilitate their survival; including blocking antigen processing and presentation, MHC expression, cytokine and chemokine production, antibody, and cytotoxic T-cell-mediated killing (8, 9).

When a new virulence factor is identified in a virus, it is often desired to determine its effects on immunity. Immune modifiers can be broadly categorized as secreted or cell associated proteins. Secreted factors may also affect neighboring uninfected cells and are likely to fall into the category of growth factors, soluble receptors that bind and inactivate immune components, or soluble factors that bind cell surface receptors, thus delivering an immune-dampening signal or blocking an immunostimulatory signal. Viral proteins that remain cell associated can act intracellularly or intercellularly in a myriad of ways including to inhibit signaling, antigen processing, MHC trafficking, Fas/ligand interactions, and protection from complement-mediated lysis. This chapter describes techniques our laboratory has developed to rapidly assay viral effects on antigen presentation and subsequent macrophage and T lymphocyte activation.

Numerous studies have shown the importance of various aspects of innate and adaptive immunity to defense against viral pathogens. CD4+ helper T lymphocytes are important in stimulating and shaping the immune response because they provide "help" to both B and T lymphocyte effector cells. CD4+ T cells secrete cytokines that stimulate antibody production and induce immunoglobulin isotype switching in B lymphocytes and also provide help that simulates the development of CD8+ cytotoxic T lymphocytes that kill virally infected cells. While there is evidence that multiple branches of the immune system provide protection from poxvirus infections, the particular importance of CD4+ T lymphocytes in survival and recovery from poxvirus infections has been shown in studies using knockout mice, where it was found that the absence of MHC class I molecules or CD8+ T cell responses did not diminish protection, but that decreases in CD4+ or MHC class II expression caused a loss of protective immunity (19). Therefore, we have explored the effects of poxvirus infection of antigen-presenting cells (APC) and their interactions with CD4+ T lymphocytes (8). We expanded this research to include the study of the vaccinia virus A35R virulence gene, which we showed was not required for viral replication in tissue culture but which dramatically decreased virulence in a rodent intranasal challenge model (9, 20).

We wanted to develop an assay that would capture an immune response effect in any of multiple steps in the process of antigen presentation: antigen uptake by the APC, presentation in the context of MHC, macrophage-T cell interaction and mutual stimulation resulting in changes in surface activation proteins, and soluble effector and cytokine synthesis by both cells. We describe here our use of an antigen presentation cell assay using primary rat APC recruited in vivo to a site of inflammation. These cells are an excellent model, since it is often difficult to culture lines that truly represent APC in vivo. Rats are a good model animal as they are natural hosts of orthopoxvirus infection, they can transmit the viruses to primates (21), their APCs are able to be infected by vaccinia virus (22), and they are large enough to provide sufficient cells for numerous assays. The rat APC are infected with different viruses or virus mutants, pulsed with a model antigen, and then assayed for their ability to stimulate the model antigen specific CD4+ T cell line (see Fig. 1) (8, 9, 22). Various cytokines, chemokines, and bioactive mediators may be measured in the supernatants from these antigen presentation assays to determine if there is a decreased response or if the virus or viral gene has altered the character of the immune response. For example, we have used Luminex fluorescent bead technology to measure 23 different cytokines in one 50 μl aliquot of the harvested supernatants (8, 22). We detected MIP1α, IL-1β, GMCSF, IL-1α, IL-2, IL-6, IFN-α, IL-17, IL-18, GROKC, RANTES, MCP1, and TNFα. Significant quantities of eotaxin, G-CSF, leptin, IL-4, IL-5, IL-9, IL-13, IP-10, and VEGF were not detected. Macrophage and T lymphocyte responses may be easily measured with assays for nitric oxide (NO) and interleukin-2 (IL-2) production, respectively.

We have used this system, or parts thereof, to assess, viral killing of APC, viral replication, induction of apoptosis, and MHC

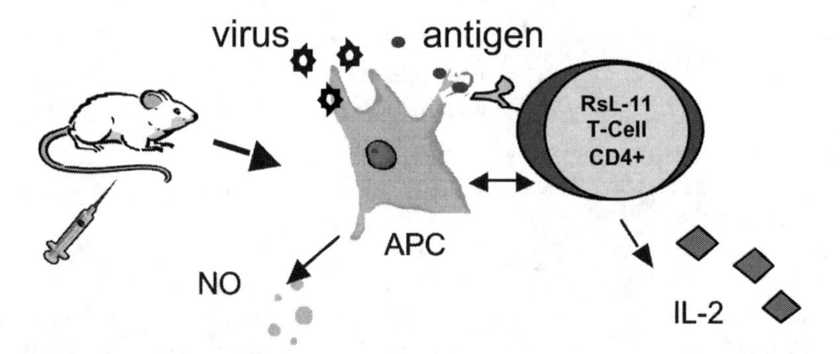

Fig. 1. Antigen-presenting cells (APC) are elicited by injection of killed *P. acnes*, and peritoneal exudates cells are harvested 2–3 days later. APC are infected with wild type or mutant virus for 3 h, pulsed with antigen, and added to CD4+ T cell clone RsL11. The APC present antigen to the RsL11, and the activated RsL11 produce IL-2 and other cytokines and stimulate the APC to make nitric oxide and cytokines. These bioactive compounds are measured to determine the effects of virus and viral genes on antigen presentation efficiency.

class II and costimulatory protein surface expression (8, 9, 22). This system has the advantage that it is possible to run one assay and determine the end point to see if there is any effect. If there is an effect, the system can then be broken down into steps to determine the location of the block and the changes in the immune response (8, 22).

2. Materials

1. Lewis rats, 3 months to 1 year old.
2. Inactivated *Propionibacterium acnes* (see Note 1).
3. RPMI/10% FBS: RPMI containing 10% FBS, 2 mM glutamine, 100 μg/ml streptomycin, 100 U/ml penicillin, 50 μM 2-ME (beta mercaptoethanol) (see Notes 2 and 3).
4. Hanks' balanced salt solution (HBSS).
5. Guinea pig myelin basic whole protein (GPMBP) or GPMBP peptide fragment 68–82 (PQKSQRSQDENPV).
6. CD4+ RsL11 T-cell clones (8, 22, 23).
7. 0.1 M sodium nitrite.
8. Griess reagent: 1% sulfanilamide, 0.1% N-[1-naphthy] ethylenediamine in 2.5% phosphoric acid (see Note 4).
9. CTLL-2 cells (ATCC # TIB-214).
10. Interleukin 2.
11. MTS,3-(4,5-dimethylthiazol-2-yl)-5-(3-carboxymethoxyphenyl)-2-(4-sulphophenyl)-2H-tetrazolium, inner salt: 2.0 mg/ml dissolved in PBS (light sensitive) (see Note 5).
12. PMS, phenazine methosulfate: 0.1 mg/ml dissolved in PBS (light sensitive).
13. MTS/PMS: 100 μl PMS per 2.0 ml of MTS, made fresh prior to addition to the culture plate containing cells.

3. Methods

3.1. Peritoneal Macrophage (APC) Isolation

1. Lewis rats (3 months to 1 year old) are injected intraperitoneally using a 25-gauge needle with 200 μg of inactivated *P. acnes* diluted in 5 ml of HBSS.
2. Two to three days later, the rats are sacrificed and the peritoneum aseptically opened.
3. Peritoneal exudate cells (APC) are harvested by washing the peritoneal cavity and internal organs three times with 13 ml

each time of cold HBSS and aspirating fluids. Use the pipette to gently agitate organs to collect the APC. Be careful not to perforate any blood vessels or organs. You should be able to remove a total of ~40 ml from the peritoneum in a 50-ml conical tube. The cells should be kept on ice from harvest until the time of infection.

4. Pellet the cells by centrifugation at 800 × *g* for 10 min.
5. Discard media and resuspend the cells in 20 ml RPMI/10% FBS (see Notes 2 and 3).
6. Centrifuge the cells at 800 × *g* for 10 min.
7. Resuspend the cells in 10 ml RPMI/10% FBS.
8. Count the cells (2–4 × 10^7 cells can be expected per rat).

3.2. Virus Infection of Rat Peritoneal Macrophages

1. Separate required numbers of cells for each experimental group (e.g. wild type, deletion mutants, or uninfected as control) into 15-ml conical tubes (see Notes 6 and 7).
2. Bring the volume in each tube to approximately 1 ml RPMI/10% FBS. As infection rates are dependent on virus and cell concentration, keep the volumes the same between comparison groups.
3. Calculate the desired multiplicity of infection (MOI, the number of virus per cell, usually 3–10), and add virus (wild type, deletion mutants, or uninfected as control) to the conical tubes containing APC, and mix by flicking the tube. Do not vortex the cells, as they may be damaged. Remember to keep volumes constant in all tubes (see Notes 8 and 9).
4. Place conical tubes in 37°C incubator with 5% CO_2. Loosen the tube cap to allow for gas exchange and lay the tube in a nearly horizontal position by resting it on another empty conical tube or pipette to keep liquid from leaking out (see Note 10).
5. During the 2-h incubation with virus, resuspend cells by gentle finger flicking every 20 min, using care to keep the cap in place to maintain sterility. While virus infection should be complete by 2 h, you may wish to leave the infection for longer to allow for adequate protein expression and processing. We have seen good immunoinhibitory activity with 3–5 h infection times (22).

3.3. Loading of APC with Model Antigen

1. After the cells are infected, add 5 ml of warm (37°C) RPMI/10% FBS to the APC in 15-ml conical tubes and pellet the cells by centrifuging at 800 × *g* for 5 min.
2. Aspirate out media containing unbound virus (see Notes 11 and 12).
3. Resuspend cells in 1 ml warm RPMI/10% FBS and add the antigen, guinea pig myelin basic protein or peptide, at a 50 nM final concentration (see Note 13).

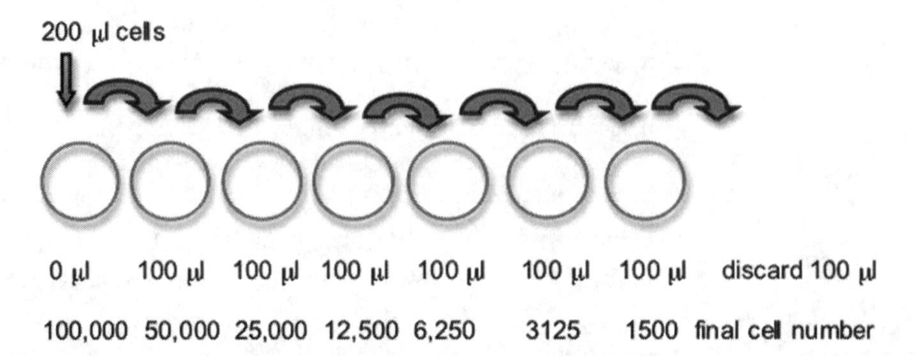

Fig. 2. Make a 1:2 dilution series by putting 100 μl of media in wells 2–7 of a 96 well plate. We recommend to do this in triplicate to allow for statistical analysis. Add 200 μl of cells to the first well (no media), and take 100 μl from the first well and add it to the second well and mix. Take 100 μl from the second well and add it to the third well and mix. Repeat for the next four wells. When you get to the last well, add the 100 μl, mix, and then remove and discard 100 μl, so that all wells will have a 100 μl final volume. If you begin with 200,000 cells in 200 μl in the first well, you will create a final titration curve with the number of APC per well shown in the figure.

4. Incubate for 30 min, with caps loose and tubes placed on the diagonal as described above in a 37°C incubator to allow antigen uptake (see Note 10).
5. Add 5 ml of warm (37°C) media to the APC in 15-ml conical tubes and centrifuge cells at 800 × *g* for 5 min to wash out antigen (see Note 14).
6. Resuspend 600,000 cells in 600 μl RPMI/10% FBS to yield 200,000 cells/200 μl.
7. Set up 1:2 serial dilutions in triplicate in a flat-bottom 96-well tissue culture plate with 100 μl of RPMI/10% FBS per well and transferring 100 μl volume each time (see Fig. 2).
8. Put 200 μl of the cell suspension in the first well and then dilute by moving 100 μl from one subsequent well to the next with mixing.
9. Discard the final 100 μl from the last well so that there will be a volume of 100 μl in each well after dilutions have been made.
10. Remember to include controls to monitor the proper functioning of the assay. Control wells should include media only, antigen but no APC, and APC and no antigen (see Note 15).
11. Add 100 μl of RPMI/10% FBS containing 25,000 Lewis rat CD4+ RsL11 cells (24) to each well (see Note 16).
12. Incubate the cells for 15–72 h at 37°C in 5% CO_2 to allow for stimulation of both the cells and cytokine secretion.
13. After the desired incubation period, remove the plates from incubator and check media color to note if cells have overgrown (yellow indicates acid production and cell crowding). To take supernatant at a specific time point, tilt plates by propping

up one edge in a tissue culture hood and carefully and slowly collect 40–50 μl of supernatants from the top of the deep side of the well (a multichannel pipette works well). Avoid aspiration of cells, since supernatants are desired for analysis of soluble effectors released as the result of antigen presentation (see Notes 17 and 18).

14. Transfer supernatants from the desired time points directly to an empty 96-well tissue culture plate (or microfuge tubes if needed for another assay format) and assay immediately or freeze for later analysis using the assays described below.
15. Put the plates back in the incubator to incubate for collection of supernatants for later time points. 24 and 48 h are useful time points, but we have detected differences at 12 h and up to 96 h (see Note 19).

3.4. Analysis of Supernatants

Various cytokines, chemokines, and bioactive mediators may be measured as a result of antigen presentation as described in Subheading 1. We describe here two assays for measuring bioactive mediators released as a result of antigen presentation, one for the measurement of IL-2 secreted by T lymphocytes (22, 24), and one for measurement of nitric oxide released by APC after stimulation by the activated T lymphocytes (8, 22).

3.4.1. Measuring the APC Response, Nitric Oxide Measurement

In order to measure the response of the APC to the antigen presentation stimulation from the activated T lymphocytes, supernatants may be assayed for the presence of nitric oxide (22, 25).

1. Supernatants should be in 96 well plates, either fresh or thawed and warmed to room temperature after storage.
2. A solution of 0.1 M sodium nitrite can be used as a positive control and to make a standard curve for sample comparison. To generate a standard curve, make a dilution series from 100 μM to 0 μM sodium nitrite on the plate (see Note 20).
3. Add 50 μl of Griess Reagent to 40–50 μl of the harvested supernatants (and the wells with the sodium nitrite controls) in a 96-well plate.
4. Wait for 5 min and read the absorbance on a plate reader at 540 nm.
5. NO production should only be detected in supernatants from cultures that contained the model antigen and both cell types. If NO is detected in the absence of any of the three essential components of antigen presentation, there is a contaminant affecting the assay.

3.4.2. Measuring the T Lymphocyte Response, IL-2 Bioassay

The T lymphocyte response to antigen stimulation can be monitored by measuring the IL-2 production. The amount of IL-2 can be easily measured using CTLL cells. CTLL are an IL-2 dependent

T cell line that responds to mouse and rat IL-2 by proliferating. Thus, they can be used as an IL-2 bioassay where the absorbance will correlate with CTLL proliferation, IL-2 concentration in the media, and antigen presentation efficiency.

1. CTLL are continuously grown in culture in IL-2 containing media (e.g., RPMI/10%) (see Note 21).
2. For this assay, the cells are used when they are IL-2 deprived and almost dying because this will keep the background low. CTLL should be used when the flask is confluent and it appears that ~50% of the CTLL are dead (see Note 22).
3. You will want 10,000 live CTLL/well, so plan to have at least this many cells. Remove the cells from the flask and place in a 50 ml conical tube.
4. Add the same volume of RPMI (with no IL-2) to the cells and centrifuge at 800 × *g* for 10 min to wash out residual IL-2.
5. Aspirate supernatant and resuspend CTLL cells in 10 ml RPMI (with no IL-2) and centrifuge at 800 × *g* for 10 min.
6. Resuspend CTLL in 5 ml of RPMI (with no IL-2) and count live cells (e.g. by trypan blue exclusion, see Note 23).
7. Dilute the cells to the desired number of cells required for the test samples at a concentration of 10,000 CTLL/well (usually in 150 μl per well).
8. Add CTLL to 96-well flat-bottom tissue culture treated plate containing supernatants (40–50 μl) to be tested for IL-2. A multichannel pipette is useful for large numbers of wells or plates.
9. Mix the cells during addition and make sure you add the same number of cells to each well because this will affect your readout.
10. Important control groups are described in Notes 24 and 25.
11. Check the plates using a microscope to ensure that there are approximately the same numbers of CTLL per well.
12. Incubate the plates at 37°C, 5% CO_2 overnight.
13. Check the plates using a microscope. Wait until the cells in the positive control (IL-2 containing wells) appear large and bright and the cells in the negative wells (no IL-2) appear almost dead. Typically, this is about 15–30 h after plating.
14. Add 10 μl of MTS/PMS to each well. MTS/PMS is light sensitive, so all of the following steps should be done in low light.
15. Once there is a visible color change due to the MTS/PMS addition (usually 6–12 h after adding MTS/PMS), read the absorbance in the plates at 492 nm (and 690 nm reference). We usually read plates twice a day until a color/absorbance plateau begins to be reached.

16. The plates should be observed under the microscope each day for possible contaminations and to ensure that the plate reader and assay are working properly, i.e., wells with the most proliferation have the highest absorbance readings (see Note 26).

3.5. Analysis of Results

If antigen presentation is inhibited by the virus or viral gene, there is expected to be a reduction in the amount of both IL-2 and NO produced, since both of these compounds are made as the result of antigen presentation (8, 22). However, the particular effect on the system will give clues as to the mechanism of the viral gene because the APC first stimulates the T lymphocyte and then the activated T lymphocytes activate the APC. For example, if IL-2 production were normal and NO were reduced, it might suggest that the APC were able to present antigen, and the T lymphocytes were able to respond in terms of IL-2, but that the either the subsequent response by the APC to T cell mediators was blocked or the T cell secreted a different profile of cytokines after antigen presentation, thus differentially affecting the APC. A number of scenarios are possible, since there is a sequence of events and the virus may block at a particular point.

If an interesting phenotype is discovered, it is next possible to deconstruct this assay system to further elucidate the mechanism of action of the virus or viral gene. The APC can be isolated and assayed in the absence of T lymphocytes by infecting the APC and then stimulating NO production with IFNγ and/or LPS (8). If APC are found to be affected, they can be further assessed for viral effects. We have found that while vaccinia virus does not replicate well in these APC, it does affect metabolism, apoptosis induction, the expression of surface costimulatory markers, MHC, and antigenic peptide, (8, 9, 22, 26). Similarly, the T lymphocyte can be isolated and studied for viral effects by incubating the T lymphocytes with virus and then stimulating the T cells with phorbol myristate acetate, ionomycin, or concanavalin A (8). However, we and others have found that T lymphocytes are not easily affected by vaccinia virus infection (8, 22, 27) and most viral effects may be focused in the APC.

4. Notes

1. *P. acnes* was formerly called *Corynebacterium parvum*. Inactivated *P. acnes* is commercially available as a veterinary product called EqStim (Neogen). This can also be prepared in-house.
2. Some lots of FBS are better at supporting the survival and function of immune cells. Unfortunately, no particular supplier

of FBS or no known ingredients routinely allow guaranteed immune responses. Therefore, it is best to test serum lots before setting up large experiments. Once you find a FBS lot that works well, save that lot of FBS for your in vitro immunologic experiments.

3. Immune cells often require the presence of 5×10^{-5} M beta-mercaptoethanol. We strongly recommend its inclusion in media.
4. A Griess reagent assay kit is also available from Promega.
5. MTS is also available as part of a kit from Promega (cell Titer 96).
6. It is important to use polypropylene conical tubes for infection because the APC will not adhere to them. If tissue culture treated plastics are used for the infection, the APC may adhere and be lost, as they will not be in suspension for addition to the T lymphocytes during the assay.
7. We typically do a 1:2 titration of APC from 100,000–1,500 APC per well in triplicate, thus requiring approximately 600,000 cells per group. We recommend to make a titration of APC in order to see a range of responses and maximize the possibility of capturing differences in efficiency of presentation.
8. The titer of the viruses is very important, as changes in titer may affect the efficiency of antigen presentation. We recommend the use of virus stocks that have been repeatedly titered with high replicate number in order to ensure confidence of titers when comparing two or more viruses. In addition the titer can be reconfirmed on the day of infection.
9. Purified virus preparations may be employed for infections in order to remove the possibility of measuring effects of trace cytokines or other bioactive molecules carried over in virus preparations from crude cell lysates.
10. The purpose of keeping the tube on its diagonal is to maintain cell viability by allowing for good gas exchange to occur and to make sure that the cells do not settle into a pellet at the bottom of the tube.
11. Remember when discarding materials that have come into contact with infectious virus, one needs to autoclave or treat waste with detergent to inactivate virus prior to disposal.
12. This step is optional, as the RsL11 T cells and CTLL used in later steps are minimally affected by vaccinia virus (22).
13. You may increase the antigen concentration to increase the stimulation response. We have used from 50 to 500 nM (8), but others have used up to μM concentrations (24). The GPMBP peptide fragment is less expensive than the whole protein.

14. Washing out the antigen is optional; however, using a defined time of antigen pulse may allow for better detection of changes in efficiency of antigen uptake or processing.
15. Background in the antigen presentation response will be defined as levels found in the absence of antigen or the absence of either cell type, as all three components are required for antigen presentation responses. The best control is with both cell types and no antigen, but multiple variations missing one component are useful to identify the source of any background.
16. Alternatively, one can add cells in 150 μl to each well to increase the volume of available supernatants (250 instead of 200 μl volume), as well as to give the cells more media for longer incubation periods.
17. The volume of fluid collected in each well for each time point should be held constant because the concentration of the biological effectors will be measured following this step. If volumes vary between samples, it will affect concentrations.
18. Check your pipetting technique by looking in the microscope for any evidence of cells in the supernatants you collected. We typically are able to collect a total of 150 μl (i.e., 50 μl three times, once for each time point) from wells containing 200 μl volume without aspirating cells, or a total 200 μl volume of supernatant from a well containing 250 μl (if an additional time point will be needed).
19. Take care that at later time points cell crowding and media depletion do not confound interpretation of results. The most strongly stimulated cells will proliferate the fastest and die the fastest creating a plateau effect.
20. Since Griess reagent measures nitrite, sodium nitrite can be used to make a standard curve if you want to quantify how much NO is in the actual samples.
21. The IL-2 concentrations used to maintain these cell lines will depend on the supplier. We often use an IL-2 in the supernatant from recombinant baculovirus infected cells. The cells are supplied by the ATCC, which recommends "10% T-STIM with Con A, available from Becton Dickinson."
22. For us, this typically takes 3–4 days of growth in the absence of IL-2. But it depends on how much IL-2 the cells had been growing in prior to putting them in media without IL-2.
23. For trypan blue staining, pipette 10 μl of cells into 90 μl of 0.4% Trypan Blue vital dye (1:10 dilution) and mix. Using a new pipette tip, put 10 μl of this solution onto a hemocytometer and count bright white cells (dead cells will appear gray or blue). Follow hemocytometer instructions for calculation of cells/ml. Usually (# cells/# squares) × dilution factor × 10^4 = cells/ml.

24. Important negative controls include the following: (a) media only control; three wells containing only 200 μl RPMI with no IL-2 present and (b) no IL-2 or supernatant control, three wells containing 150 μl CTLL cells, 50 μl RMPI (with no IL-2).
25. Important positive controls (CTLL and IL-2) include three wells containing 150 μl CTLL and 50 μl RPMI CTLL growth media with IL-2 present. An IL-2 standard curve can be constructed using known IL-2 concentrations.
26. The best data time point to graph is when the background is low in negative control wells and the positive groups of interest are near their peak absorbance reading but have not yet reached a plateau where the readings stop increasing. The optimal time point for analysis can be conveniently assessed by comparing the sample of interest that gives the highest reading to the negative control. The time point that gives the largest ratio of high reading value–negative control value is probably the best time point to analyze the data.

Acknowledgments

The author wishes to thank Dr. Mark Mannie, East Carolina University, for developing the rat antigen presentation system and for his generous help and advice. This work was supported by The North Carolina Biotechnology Center and NIH grant U54 AI057157 from Southeast Regional Center of Excellence for Emerging Infections and Biodefense.

References

1. Upton C, Slack S, Hunter AL, Ehlers A, Roper RL (2003) Poxvirus orthologous cluster: toward defining the minimum essential poxvirus genome. J Virol 77:7590–7600
2. Ludwig H, Mages J, Staib C, Lehmann MH, Lang R, Sutter G (2005) Role of viral factor E3L in modified vaccinia virus ankara infection of human HeLa cells: regulation of the virus life cycle and identification of differentially expressed host genes. J Virol 79:2584–2596
3. Guo ZS, Naik A, O'Malley ME, Popovic P, Demarco R, Hu Y et al (2005) The enhanced tumor selectivity of an oncolytic vaccinia lacking the host range and antiapoptosis genes SPI-1 and SPI-2. Cancer Res 65:9991–9998
4. Bradley RR, Terajima M (2005) Vaccinia virus K1L protein mediates host-range function in RK-13 cells via ankyrin repeat and may interact with a cellular GTPase-activating protein. Virus Res 114:104–112
5. Langland JO, Jacobs BL (2002) The role of the PKR-inhibitory genes, E3L and K3L, in determining vaccinia virus host range. Virology 299:133–141
6. Johnston JB, Barrett JW, Nazarian SH, Goodwin M, Ricuttio D, Wang G et al (2005) A poxvirus-encoded pyrin domain protein interacts with ASC-1 to inhibit host inflammatory and apoptotic responses to infection. Immunity 23:587–598
7. Jackson SS, Ilyinskii P, Philippon V, Gritz L, Yafal AG, Zinnack K et al (2005) Role of genes that modulate host immune responses in the immunogenicity and pathogenicity of vaccinia virus. J Virol 79:6554–6559
8. Rehm KE, Connor RF, Jones GJB, Yimbu K, Mannie MD, Roper RL (2009) Vaccinia virus decreases MHC class II antigen presentation, T cell priming, and peptide association with MHC class II. Immunology 128:381–392

9. Rehm KE, Jones GJ, Tripp AA, Metcalf MW, Roper RL (2010) The poxvirus A35 protein is an immunoregulator. J Virol 84:418–425
10. Stewart TL, Wasilenko ST, Barry M (2005) Vaccinia virus F1L protein is a tail-anchored protein that functions at the mitochondria to inhibit apoptosis. J Virol 79:1084–1098
11. Gomez CE, Vandermeeren AM, Garcia MA, Domingo-Gil E, Esteban M (2005) Involvement of PKR and RNase L in translational control and induction of apoptosis after Hepatitis C polyprotein expression from a vaccinia virus recombinant. Virol J 2:81
12. Wang G, Barrett JW, Nazarian SH, Everett H, Gao X, Bleackley C et al (2004) Myxoma virus M11L prevents apoptosis through constitutive interaction with Bak. J Virol 78:7097–7111
13. Roper RL, Payne LG, Moss B (1996) Extracellular vaccinia virus envelope glycoprotein encoded by the A33R gene. J Virol 70:3753–3762
14. Roper RL, Wolffe EJ, Weisberg A, Moss B (1998) The envelope protein encoded by the A33R gene is required for formation of actin-containing microvilli and efficient cell-to-cell spread of vaccinia virus. J Virol 72: 4192–4204
15. Nichols DB, Shisler JL (2006) The MC160 protein expressed by the dermatotropic poxvirus molluscum contagiosum virus prevents tumor necrosis factor alpha-induced NF-kappaB activation via inhibition of I kappa kinase complex formation. J Virol 80:578–586
16. Shisler JL, Jin XL (2004) The vaccinia virus K1L gene product inhibits host NF-kappaB activation by preventing IkappaBalpha degradation. J Virol 78:3553–3560
17. Guerra S, Lopez-Fernandez LA, Conde R, Pascual-Montano A, Harshman K, Esteban M (2004) Microarray analysis reveals characteristic changes of host cell gene expression in response to attenuated modified vaccinia virus Ankara infection of human HeLa cells. J Virol 78:5820–5834
18. Rubins KH, Hensley LE, Jahrling PB, Whitney AR, Geisbert TW, Huggins JW et al (2004) The host response to smallpox: analysis of the gene expression program in peripheral blood cells in a nonhuman primate model. Proc Natl Acad Sci USA 101:15190–15195
19. Wyatt LS, Earl PL, Eller LA, Moss B (2004) Highly attenuated smallpox vaccine protects mice with and without immune deficiencies against pathogenic vaccinia virus challenge. Proc Natl Acad Sci USA 101:4590–4595
20. Roper RL (2006) Characterization of the vaccinia virus A35R protein and its role in virulence. J Virol 80:306–313
21. Martina BE, van Doornum G, Dorrestein GM, Niesters HG, Stittelaar KJ, Wolters MA et al (2006) Cowpox virus transmission from rats to monkeys, the Netherlands. Emerg Infect Dis 12:1005–1007
22. Rehm KE, Connor RF, Jones GJ, Yimbu K, Roper RL (2010) Vaccinia virus A35R inhibits MHC class II antigen presentation. Virology 397:176–186
23. Mannie MD, Dawkins JG, Walker MR, Clayson BA, Patel DM (2004) MHC class II biosynthesis by activated rat CD4+ T cells: development of repression in vitro and modulation by APC-derived signals. Cell Immunol 230:33–43
24. Mannie MD, Norris MS (2001) MHC class-II-restricted antigen presentation by myelin basic protein-specific CD4+ T cells causes prolonged desensitization and outgrowth of CD4- responders. Cell Immunol 212:51–62
25. Campos-Neto A, Ovendale P, Bement T, Koppi TA, Fanslow WC, Rossi MA et al (1998) CD40 ligand is not essential for the development of cell-mediated immunity and resistance to *Mycobacterium tuberculosis*. J Immunol 160:2037–2041
26. Li P, Wang N, Zhou D, Yee CS, Chang CH, Brutkiewicz RR et al (2005) Disruption of MHC class II-restricted antigen presentation by vaccinia virus. J Immunol 175:6481–6488
27. Chahroudi A, Chavan R, Koyzr N, Waller EK, Silvestri G, Feinberg MB (2005) Vaccinia virus tropism for primary hematolymphoid cells is determined by restricted expression of a unique virus receptor. J Virol 79:10397–10407

Chapter 16

Characterization of Poxvirus-Encoded Proteins that Regulate Innate Immune Signaling Pathways

Florentina Rus, Kayla Morlock, Neal Silverman, Ngoc Pham, Girish J. Kotwal, and William L. Marshall

Abstract

Innate immune recognition of pathogens is critical to the prompt control of infections, permitting the host to survive to develop long-term immunity via an adaptive immune response. Poxviruses encode a family of proteins that inhibit signaling by Toll-like receptors to their downstream signaling components, severely limiting nuclear translocation of transcription factors such as IRF3 and NF-κB and thereby decreasing production of host interferons and cytokines. We describe bioinformatics techniques for identifying candidate poxviral inhibitors of the innate immune response based on similarity to the family of proteins that includes A52, A46, and N1. Robust luciferase assays can determine whether a given poxviral gene affects innate immune signaling, and in combination with other approaches can identify the cellular targets of poxviral innate immune evasion genes. Because apoptosis is an innate immune response of the cell to viral infection, assays for identifying poxviral genes that inhibit apoptosis can also be employed. Novel poxviral innate immune inhibitors are being identified via several approaches and these techniques promise to identify further complexities in the way that poxviruses interact with the host innate immune system.

Key words: Toll-like receptors, Toll/interleukin-1 receptor, Interferon regulatory factor-3, TANK-binding kinase 1, I-kB kinase complex, Interferon-stimulated response element, bcl-2, Vaccinia virus

1. Introduction

1.1. Toll-Like Receptors and Signaling Pathways

The Toll-like receptors (TLRs) are a key component of the innate immune response that protects against viral pathogens. The TLRs are a superfamily of pattern recognition molecules that respond to infection, as exemplified by the prototypic Toll receptor that is central to the *Drosophila* antifungal response. TLR signaling is triggered by binding of the components of pathogens to leucine-rich repeat regions of the TLRs. Following ligand-dependent engagement of the TLRs, signal transduction is thought to occur via

Stuart N. Isaacs (ed.), *Vaccinia Virus and Poxvirology: Methods and Protocols*, Methods in Molecular Biology, vol. 890, DOI 10.1007/978-1-61779-876-4_16,

homotypic interactions of TLRs and TIR-adaptor signaling proteins, which physically associate with the cytoplasmic domain of the receptors, the Toll/IL-1 receptor (TIR) domain (1, 2). This evolutionarily conserved signaling pathway is also implicated in signal transduction by the IL-1 receptor (IL-1R). Signaling involves the ligation of TLRs followed by clustering of TIR adapters, such as Mal/TIRAP, TRIF/TICAM-1, MyD88 (2–6), and TRAM/TICAM-2 (7, 8). Clustering of the TIR adapter, TRIF, leads to the apoptotic death of pathogen-infected cells (9). TIR adapter clustering also results in recruitment of the IL-1 receptor-associated kinases (IRAKs), tumor necrosis factor-associated factor 6 (TRAF6), and the I-kB kinase (IKK) complex (2). Utilization of this pathway at a minimum leads to NF-κB activation and the activation of various MAPK pathways, including p38, JNK, and ERK pathways (6).

Viral infections activate the innate immune response. Ligation of TLR3 and TLR4 by their respective ligands, e.g., double-stranded RNA (dsRNA) and lipopolysaccharide (LPS), clusters the TIR adapters TRIF and TRAM, which signal via the IKK complex to NF-κB and interferon regulatory factor-3 (IRF3) (8, 10). The downstream kinases involved in signaling to IRF3 include TANK-binding kinase 1 (TBK1) (11) and IKK-ε (12), which phosphorylate the IRF3 transcription factor that ordinarily exists in the cytoplasm in an unphosphorylated state. Upon phosphorylation, IRF3 is activated and translocates to the nucleus, where IRF3 binds to promoter elements such as those in the interferon-stimulated response element (ISRE) and IFN-β promoter. This eventually results in production of several antiviral cytokines, e.g., interferon-α and -β and the chemokines RANTES and IP-10 (8, 12).

1.2. Perturbation of the Host Innate Immune Signaling Pathways by the Poxviruses

Poxviruses perturb host innate immune signaling in many ways. For example, the soluble IFN-α/-β receptor of vaccinia virus (VACV) inhibits the type I interferon response at the receptor level (13). Moreover, proteins encoded by the E3L and K3L vaccinia virus genes bind dsRNA and inhibit IRF3 phosphorylation and IFN production (14). Although the mechanism of IFN signaling inhibition by soluble IFNR seems to overlap the E3- and N1-induced inhibition of IRF3 signaling to the ISRE, the soluble interferon receptors inhibit a distinct ISRE-binding transcription factor. Poxvirus inhibition of interferon signaling is likely critical since the TIR adapter, TRIF, signals to IRF3 (5, 8) and mediates a 20-fold reduction in poxvirus titer in vitro (15), which suggests a critical role for TLR-mediated innate immune response signaling via IRF3 in the control of poxvirus infections.

Poxviruses, and other viruses, encode viral homologs of human bcl-2 (vbcl-2s) that have critical roles in inhibiting innate immune responses like apoptosis. Apoptosis is an innate cellular immune response that aborts viral replication by engaging one of several pathways for programmed cell death. Studies of recombinant Epstein–Barr Virus (EBV) deleted for both their BHRF1 and

BALF1 vbcl-2s have revealed that while each EBV bcl-2 is not individually required for transformation, a deletion of both EBV bcl-2 homologs prevents survival of EBV via transformation (16). Vaccinia virus is the only virus other than EBV that is known to encode two vbcl-2s. Vaccinia virus vbcl-2s are designated N1 (17, 18) and F1 (19). F1 appears to be critical for preventing death of the infected cell by the innate apoptotic cellular response to viral infection, since cells infected with viruses deleted of F1 undergo apoptosis independently of apoptotic stimuli (19). In contrast, N1 inhibits innate immune signaling (20) and N1 is an antiapoptotic vbcl-2 structurally similar to cellular bcl-xl (17, 18). However, N1 functions distinctly from F1 in that cells infected with vaccinia viruses deficient in the N1 vbcl-2 do not undergo spontaneous apoptosis. Thus, vbcl-2s from vaccinia and several other viruses appear to be essential to certain innate immune evasion functions related to viral survival.

1.3. Bioinformatic Identification of Poxviral Innate Immune Inhibitors

Many pathogens have evolved mechanisms to inhibit TLR signaling pathways. Vaccinia viruses encode several proteins that inhibit signaling by the TLRs (20, 21), including A52, A46, and N1, which are all defined as mediating virulence in murine models of poxvirus infection (19–22). In cell-based experiments, N1 inhibits both NF-κB and IRF3 activation following TLR stimulation (20).

Swinepox virus (SPV) encodes five distinct "A52R-like" genes (23) that also display homology to N1. Our database searches demonstrated that four of these five A52 homologs encoded by SPV may have structural homologs located in similar positions on the vaccinia and variola virus genomes, including the prototype—A52, A46, and B22/C16 (see Table 1). The SPV-encoded "A52-like" proteins also displayed similarity to VACV proteins K7 and

Table 1
Identified vaccinia virus A52R/N1L gene family members with proposed functions

VACV Gene	Swinepox virus homolog	Transcription factor(s) inhibited	Mechanism	Function
A52R	SPV135	NF-κB (activates MAPK)	Targets IRAK2 and TRAF6	Inhibits IL-1, IL18, TLR signaling to NF-kB
N1L	Unclear	NF-kB and IRF3	Targets TBK1	Inhibits IL-1, TNF, LT, and TLR signaling
C16L/B22R	SPV001/SPV150	Unknown	Unknown	Structure "similar to TLR inhibitors"
A46R	SPV133	NF-kB&IRF3	Dominant negative TIR	Inhibits TL-1R, TLR signaling

Other poxviral innate immunomodulatory proteins include K1, which inhibits NF-κB translocation, and E3 and K3, each of which inhibits interferon responses via several proposed mechanisms

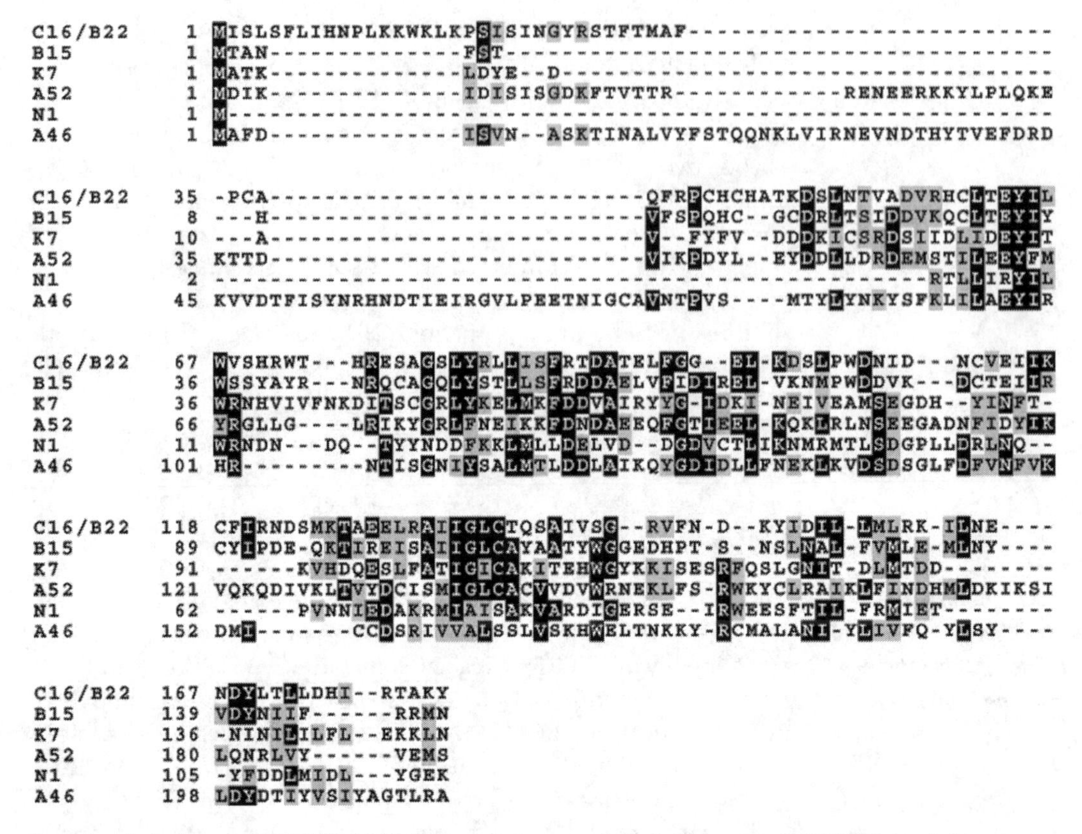

Fig. 1. Boxshade alignment "A52R-like" ORFs in vaccinia virus reveals similarities among 7 A52R/N1L innate immune evasion family members, some of unknown function. Alignment was performed using TCoffee and printed using the Boxshade 3.21 software at the URL http://www.ch.embnet.org/software/BOX_form.html.

B15, which are included in the alignment in Fig. 1. Our own searches detected that the N1 vaccinia virus protein is very similar (PHI-BLAST $e = 3 \times 10^{-12}$) to A52, which has been shown to inhibit Toll, IL-1, and IL-18 signaling to NF-κB (20). N1 also possesses remote homology to the SPV "A52-like" proteins (data not shown) and their homologs in vaccinia virus (Fig. 1). The structural similarity between the N1 and A52 inhibitors of Toll and IL-1 signaling led to an investigation of the effect of N1 on Toll and IL-1 signaling pathways. This investigation found that N1 modulates the innate immune response in a manner distinct from either A52 or A46 (20). Fold recognition ("threading") analysis (24) predicted that N1 had a similar fold to that of the bcl-2 family member, bid (unpublished observations). N1 was subsequently shown to also have the solution structure and certain functions of a bcl-2 protein (17, 18).

N1 appears to inhibit signaling by interacting with IKKs that are critically involved in NF-κB and IRF3 activation. Recently, three distinct proteins from Borna disease virus, hepatitis C virus, and rabies virus have subsequently been identified as using a similar

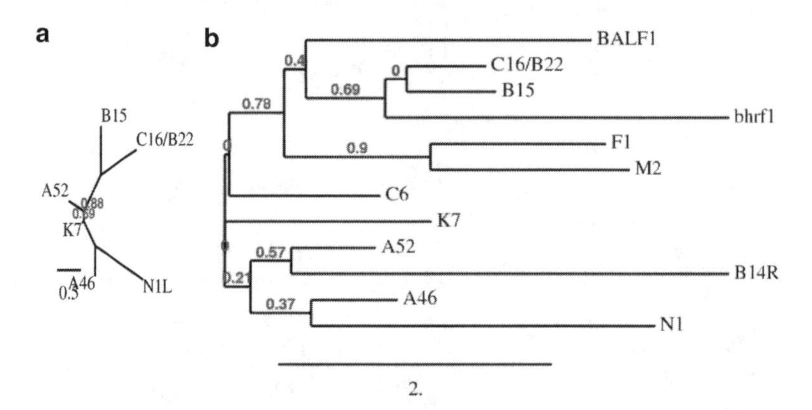

Fig. 2. Phylogenetic analysis was performed at http://www.phylogeny.fr (32) and displayed with the TreeDyn198.3 program (**a**) for VACV N1 and the VACV proteins similar to the 5 SPV "A52-like" proteins; (**b**) for all known VACV innate immune inhibitor family members, VACV vbcl-2, and EBV bcl-2s.

mechanism to N1 targeting the IKK complex (25–27). This suggests convergent evolution of viral inhibitors of the innate immune signal transduction pathways in these three viruses and vaccinia virus. This is in contrast to the evidence for divergent evolution of innate immune inhibitors from a common poxviral ancestral gene that is suggested by the data in the radial dendrogram in Fig. 2a. This evidence of divergent evolution of innate immune inhibitors from a common ancestral gene is compatible with the hypothesis that many poxviral genes evolved via duplication and subsequent recombination. In Fig. 2b, all the reported and predicted A52/N1 family members are aligned with the two vbcl-2s encoded by EBV and the two vbcl-2s encoded by vaccinia virus. Surprisingly, given the evidence for gene duplication in several poxviral genes, especially the inverted terminal repeats of the poxvirus genome, the N1 VACV vbcl-2 appears evolutionarily divergent from the F1 v-bcl-2, which is more closely related to the vbcl-2s of EBV.

Remarkably, poxviruses that are as evolutionarily distinct as vaccinia virus and SPV encode a family of similar proteins that inhibit host innate immune responses in vaccinia. These poxviral proteins inhibit TLR-mediated signaling, cytokine-dependent signaling, and/or apoptotic signaling. Several of the genes in this family of poxviral innate immune inhibitors are not characterized, such as C16/B22. However, other innate immune inhibitory proteins that display more remote similarity to the A52/N1 family shown in Fig. 2 are known to inhibit innate immune signaling. One example of such a vaccinia virus protein is the recently characterized B14 protein that inhibits IKKβ signaling and has the solution structure of a bcl-2-like protein. However, B14 displayed less amino acid sequence homology to A52/N1 in BLAST searches (28). The distinct roles of these A52/N1 family proteins and their functional relationship to other poxviral proteins with seemingly overlapping innate immune inhibitory functions (e.g., N1 and the F1 vbcl-2s of VACV) are still being investigated. This chapter provides methods for analyzing the effect of poxviral proteins on innate immune signaling.

2. Materials

2.1. PCR Cloning of Poxvirus Genes into Expression Plasmids

1. pCR3.1 (Invitrogen, Carlsbad, CA).
2. pfu Turbo polymerase (Agilent).
3. 10× PCR buffer (Agilent).
4. dNTP mix (Promega).
5. 200-μl Thin-walled capped PCR tubes.

2.2. Transfections of Cells

1. Cell lines: Wild-type HEK 293 cells (ATCC CRL-1573), HeLa cells (ATCC CCL2.2), and HEK 293 cells stably expressing CD14, MD2, and TLR2, TLR3, or TLR4 that have been described previously (29, 30) (see Note 1).
2. Tissue culture media: Dulbecco's modified Eagle's medium supplemented with 10% heat-inactivated bovine calf serum supplemented with penicillin/streptomycin and l-glutamine, and ciprofloxacin (10 μg/ml).
3. FLAG-tagged Mal, TRAM TRIF, IKKε, TBK1 constructs, and the pGL3-5×kB-luc plasmid were previously described (12). The plasmid encoding MyD88 was from M. Muzio (Mario Negri Institute, Milan, Italy).
4. ISRE luciferase plasmid (pISREluc; Stratagene/Agilent, Santa Clara, CA).
5. The plasmid thymidine kinase *Renilla* luciferase (pTk-renilla; Promega Inc., Madison, WI).
6. Poly I:C (GE Healthcare), stored at −20°C as a 2 mg/ml H_2O (100×) stock.
7. Human IL-1β and TNF-α (Cell Sciences, Canton, MA), used at 10 ng/ml.
8. TE buffer: 10 mM Tris–HCl, pH 7.2 and 1 mM EDTA.
9. Vaccinia virus strain Western Reserve (WR) (ATCC VR-1354).
10. Anti-V5 antibody (e.g., Sigma-Aldrich (Cat. # V8137)).
11. GeneJuice (Novagen, Madison, WI).
12. pEFBosTRAF6 expression plasmid, a gift of Dr. D. Golenbock and Brian Monks (Brian.Monks@umassmed.edu).
13. Commercial source of Renilla substrate (Promega, Madison, WI).
14. Passive lysis buffer (e.g., Part # E397A from Promega, Madison, WI).

2.3. Formulation of Firefly Luciferase Substrate from Lab Materials

1. 10 mM ATP solution (0.12 g ATP in 20 ml H_2O).
2. Acetyl CoEnzyme A, lithium salt. Buy 2 × 25 mg, add 1 ml H_2O to each vial to make 2 ml of 25 mg/ml.
3. To prepare 228 ml Firefly Luciferase Substrate (1×) final volume, the chemicals listed must be added in the following order (see Note 2): *0.817 g* Tricine (final concentration 20 mM), *1.21 ml* of 500 mM $MgSO_4$ (final concentration 2.67 mM), *45.6* µ*l* of 500 mM EDTA, pH 8.0 (final concentration 0.1 mM), *7.41 ml* of 1 M DTT (final concentration 32.5 mM), *12 ml* of fresh 10 mM ATP solution (final concentration 530 µM ATP), *1.89 ml* of 25 mg/ml fresh Acetyl CoEnzyme A solution (final concentration 270 µM), *30 mg* d-luciferin (e.g., purchased from Biosynth L-8200; see Note 3), add *199.86 ml* H_2O for a final volume up to 222.6 ml, and then add *570* µ*l* of 2 M NaOH (see Note 4) and *1.21 ml* of 50 mM magnesium carbonate hydroxide.

2.4. Co-immunoprecipitation and Immunoblot

1. Immunoprecipitation lysis buffer: 0.5% Triton X-100, 300 mM NaCl, 50 mM Tris–Hcl pH 7.6, 1 mM EGTA, 10 mM HEPES, 5 mM $MgCl_2$, 142 mM NaCl, and Complete™ protease inhibitor cocktail (Roche Diagnostic).
2. Protein G Sepharose beads (GE Healthcare).
3. Laemmli buffer.
4. 15% SDS-PAGE Tris–HCl Gel.
5. Gel running buffer: To make 10 l: 144.4 g glycine, 30 g Tris base, and 0.1% SDS dissolved in 10 l of water.
6. 0.45-µm PVDF membrane.
7. Blotting/transfer buffer: To make 10 l: 434 g glycine, 87 g Tris base dissolved in 10 l of water, and add SDS to a final concentration of 0.1%.
8. Anti-FLAG monoclonal antibodies (M5; Sigma-Aldrich).
9. 5% Milk: Carnation instant milk dissolved in PBS/T20.
10. Secondary donkey anti-mouse horseradish peroxidase (Amersham, Arlington Heights, IL).
11. Phosphate-buffered saline (PBS)/T20: 1× PBS/0.5% Tween 20.
12. ECL developer system (Perkin-Elmer).
13. Fuji-1000 Imager.

2.5. Flow Cytometric Assay for Apoptosis

1. GFP expression plasmid (pEGFPN1, Clontech).
2. 1 µM staurosporine (Sigma).
3. Binding buffer (Biovision, Mountain View, CA, e.g., Cat# K103).
4. Annexin V-Cy5 (Biovision, Mountain View, CA).

5. BD-LSR FACStar flow cytometer (Becton-Dickinson, San Diego, CA).
6. Flowjo software (version 4.1.1) (Tree Star, San Carlos, CA).

2.6. Heterodimerization Assay Between Cellular and Poxviral Proteins

1. Anti-bax, anti-bak, or anti-bad antibodies (Santa Cruz Biologicals, Santa Cruz, CA).
2. A 10% (v/v) slurry of Protein G Sepharose beads (GE Healthcare).

3. Methods

3.1. Construction of Plasmids Containing VACV ORFs that Encode Innate Immune Evasion Proteins

1. Constructs encoding vaccinia virus innate immune evasion genes can be created by PCR amplification of the candidate innate immune evasion protein-encoding open reading frame (ORF) of vaccinia virus followed by cloning into pCR3.1.
2. Template for the PCR reaction is vaccinia virus (strains COP or WR) at 1,000,000 pfu/μl in TE buffer and heat-treated for 5 min at 100°C.
3. To facilitate ultimate detection of the expressed VACV ORF from transfections, one primer should encode a C-terminal V5 tag. Thus, to facilitate cloning into pCR3.1, compatible 5′- and 3′-restriction sites are added to the primers. This is typically accomplished by amplifying with the 5′-primer containing a BamH1 site and the 3′-primer containing a V5 tag and Pst1 site (see Note 5).
4. Custom primers are purchased from Eurofins MWG Operon (Huntsville, Alabama).
5. Recipe for PCR mixture (piptetted into thin-walled 200-μl capped tubes) and PCR reaction program is shown in Table 2.
6. The resulting PCR product is digested with BamH1 and Pst1 and ligated into the BamH1 and Pst1 sites on similarly digested pCR3.1 (see Note 5).
7. The resulting VACV ORF cloned into pCR3.1 should be sequenced in both directions using BGH reverse- and T7-primers.
8. Expression of the V5-tagged construct should be confirmed on an immunoblot (see Subheading 3.3) using anti-V5 antibody at 1:1,000.

3.2. Luciferase Reporter Assays

1. HEK293 were typically seeded at 6×10^4 cells per ml in 96-well plates ~24 h prior to transfections (see Notes 6–8).
2. All HEK 293 cell transfections are performed using GeneJuice according to the manufacturer's recommendations. Briefly, to prepare a transfection mixture for addition to a single well in a

Table 2
PCR reaction mixture and amplification conditions[a]

Reagent	Volume (μl)
5′ primer (50 pmol)	1
3′ primer (50 pmol)	1
VACV DNA template	1
Puff Turbo polymerase	1
10× PCR buffer	5
dNTP mix	1
dH_2O	40
Total	50

[a]Amplification is performed on a programmable thermal cycler for 35 cycles of 30 s at 94°C, then 30 s at 55°C, and then 1 min/kb of VACV ORF insert at 68°C

96-well plate, 20 μl of DMEM is incubated with 0.5 μl of GeneJuice for 5 min, then 200 ng of DNA is added, and incubated for 15 min before addition to a well in a 96-well plate (see Note 9).

3. HEK cells are co-transfected with ISRE luciferase plasmid (25 ng/well) or NF-κB luciferase plasmid (25 ng/well), an equal amount of Tk-Renilla luciferase plasmid (25 ng/well), and 100 ng/well of the plasmid containing the poxviral innate immune inhibitor (made in Subheading 3.1) or empty plasmid (see Note 10).
4. 24–48 h post transfection (once the cells reach ~80% confluence), stimulation of signaling is achieved in one of several ways:
   (a) By ectopic overexpression in certain cases with 25–50 ng/well a component of the NF-κB signaling pathway (e.g., pEFBosTRAF6 expression plasmid)
   (b) Using specific TLR3 ligands (e.g., Poly I:C for TLR3-expressing cells), which leads to NF-κB or IRF3 signaling
   (c) Stimulation of signaling via NF-κB can be achieved via use of 10 ng/ ml IL-1β or TNF-α
5. To assay firefly and *Renilla* luciferase activity, cells were lysed using 50 μl of passive lysis buffer and the lysis proceeded for 15 min with rocking at room temperature (~20°C). Then, luciferase activity was immediately determined by adding 20 μl of luciferase reagent to 20 μl of lysate (3). Similarly, Renilla luciferase activity was assayed on a separate aliquot (3). Typical results are shown in Fig. 3 (see Note 11).

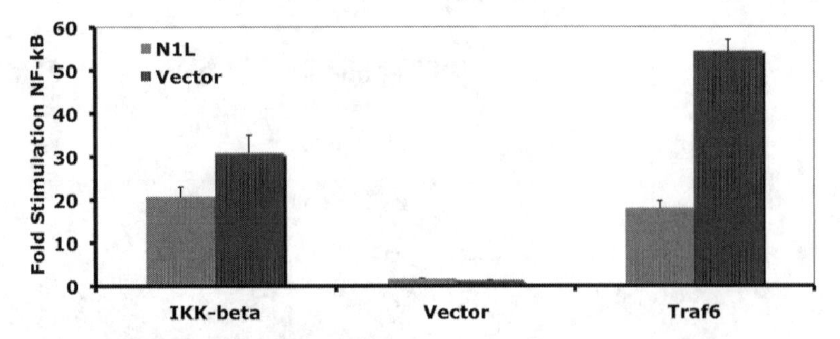

Fig. 3. N1 inhibits signaling to the NF-kB promoter induced by IKK-β or TRAF 6. HEK293 cells were co-transfected with 50 ng each of NF-kB luciferase and *Renilla* luciferase reporter plasmids and 0 or 100 ng/well of N1-encoding plasmid, supplemented with empty vector plasmid to a total of 200 ng DNA/well. After stimulation of signaling for 18 h, the cells were harvested and firefly luciferase activity was measured and divided by Tk-renilla luciferase activity to normalize for transfection efficiency. Fold luciferase stimulation was determined by dividing by the lowest mean value of the triplicate wells.

3.3. Co-immunoprecipitation and Immunoblot

1. HEK cells are co-transfected with V5-tagged poxviral innate immune suppressive gene and a FLAG-tagged component of the NF-κB or IRF3 signal transduction system (for examples of such FLAG-tagged components, see Fig. 4) (see Note 12).
2. For each transfection, a confluent T25 flask of HEK 293 cells was trypsinized and washed (see Note 8). Cells were then lysed on ice for an hour in 300 μl of immunoprecipitation lysis buffer.
3. Lysates were centrifuged at 4°C for 20 min at 14,000 × *g*, and proteins were immunoprecipitated with anti-V5 antibody bound to Protein G Sepharose beads.
4. Immunoprecipitated proteins were boiled in 20 μl of 2× Laemmli buffer for 5 min and separated by electrophoresis on a 15% SDS-PAGE gel.
5. Proteins were then transferred to a 0.45-μm PVDF membrane in a Biorad Mini Trans Blot apparatus for 20–30 min at 50–100 mA, followed by blocking the membrane in 5% milk.
6. The blot was probed with 1:1,000 anti-FLAG monoclonal antibodies overnight and washed four times for 5 min each in PBS/T20.
7. The blot was probed with a secondary donkey anti-mouse horseradish peroxidase for 1–3 h essentially as described (22).
8. Following the secondary antibody incubation steps, the PVDF membrane was washed 3–4 times with PBS/T20.
9. Resulting bands on the immunoblot were developed with the ECL developer system and visualized on a Fuji Imager or with X-ray film (see Note 13).
10. Immunoprecipitation with anti-FLAG-coated Protein G Sepharose beads followed by immunoblot with anti-V5 antibody is used to confirm the specificity of the results with IP using anti-V5 antibody.

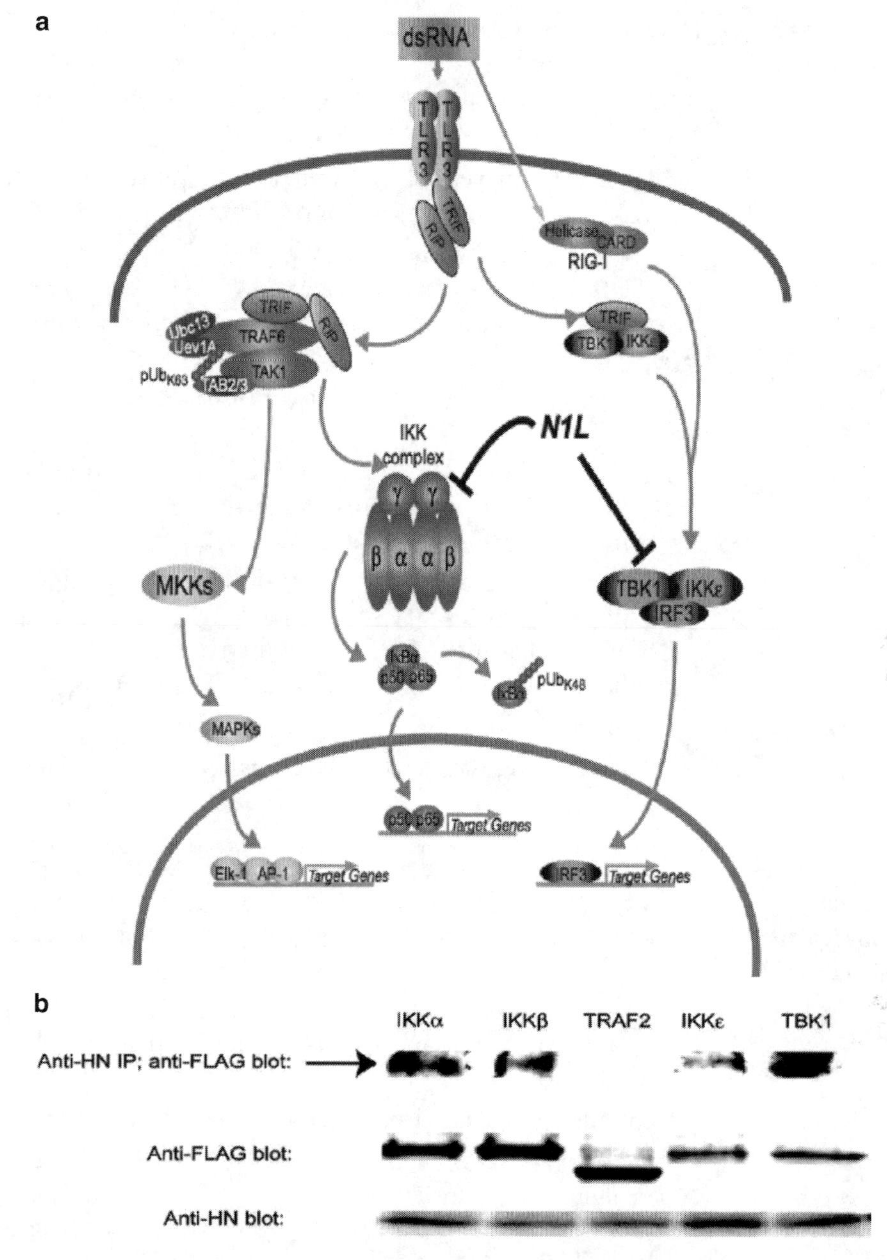

Fig. 4. N1 inhibition of signaling pathway by association with components of the IKK signaling complex. (**a**) Schematic of N1 inhibition of TLR-3-mediated signal transduction. This model depicts, in a simplified cartoon, the salient features of TLR3-induced NF-κB and IRF3 induction. N1L inhibits the activation of both of these signaling pathways by interacting with an unknown component of the IKKα/β/γ complex and interacting directly with the TBK1 kinase of the IKKε/TBK1 complex. (**b**) N1 associates with components of the IKK signaling complex (*top*), which is shown graphically at left. Histidine-arginine (HN)-tagged N1L was co-expressed in HEK cells with FLAG-tagged IKKα, IKKβ, TRAF2, IKKε, or TBK1 (lysates—bottom two blots). N1L was immunoprecipitated, with anti-HN antibody, and interacting proteins were detected by immunoblotting with anti-FLAG (top blot). The HN tag is 6XHN(His-Asn) C-terminal fusion peptide added to N1L cDNA via subcloning into the pDNR-dual vector, which encodes a C-terminal 6× histidine-arginine tag.

3.4. Flow Cytometric Assay for Apoptosis

Here, we treat HeLa cells co-transfected with VACV innate immune inhibitor expression plasmids and a GFP expression plasmid with a drug that induces apoptosis. Cells are then stained for Annexin V and analyzed by flow cytometry.

1. HeLa cells, in wells of a 24-well plate, were co-transfected with VACV innate immune inhibitor expression plasmids and GFP expression plasmid (pEGFPN1) and incubated.
2. Eighteen hours after transfection, some wells were treated with 1 μM staurosporine for 2 h to induce apoptosis (see Note 14).
3. At the same time, control cells are treated with vehicle alone (DMSO 0.2%).
4. Two hours after treatments, staurosporine-treated and control-treated wells were trypsinized and the cells collected by centrifugation (800 × g for 2 min).
5. The cells were resuspended in 500 μl of the Binding buffer, 5 μl of Annexin V-Cy5 was added, cells were incubated for 5 min, and analyzed by flow cytometry.
6. The cells were then analyzed on a BD-LSR FACStar flow cytometer.
7. The resulting data was analyzed using FlowJo software (version 4.1.1).

3.5. Heterodimerization Assay Between Cellular Proapoptotic Protein and Poxviral Candidate Antiapoptotic Proteins

1. HeLa cells plated in a 6-well plate were co-transfected with a V5-tagged VACV-encoded innate immune suppressive gene and cells processed in 300 μl of a immunoprecipitation lysis buffer as described in Subheading 3.3.
2. Proteins in the cell lysates were immunoprecipitated with 2 μg/sample anti-bax, anti-bak, or anti-bad antibodies bound to 20 μl of a slurry of Protein G Sepharose beads.
3. Immunoprecipitated proteins were boiled in Laemmli buffer for 5 min and electrophoresed on a 15% SDS-PAGE gel.
4. Proteins were transferred to a 0.45-μm PVDF membrane, followed by blocking and then washing the blot 3–4 times with PBS/T20.
5. The blot was probed with 1:1,000 anti-V5 antibody, washed 3–4 times with PBS/T20, and then probed with a donkey anti-mouse horseradish peroxidase for 30 min essentially as described (22). The blot was then washed 3–4 times with PBS/T20.
6. Resulting bands on the immunoblot were developed with the ECL developer system and visualized on a Fuji Imager or with film.

4. Notes

1. Stably transfected HEK293 cell lines are maintained in culture media containing geneticin sulfate (0.8 mg/ml).
2. The Firefly Luciferase Substrate solution should be aliquoted and stored at −20°C, in dark, being stable for multiple thaws.
3. After adding the luciferase substrate, the solution should be kept in dark from this point on.
4. After addition of NaOH, the solution should turn yellow.
5. Different restriction sites will need to be used if the VACV ORF contains either the BamH1 or Pst1 restriction enzyme sites.
6. Efficient transfection of the HEK cells is critical to the success of luciferase experiments. Key factors in achieving high transfection efficiency are the use of healthy, low passage number cells and the use of cells that are 50–80% confluent.
7. Cells should be allowed to recover overnight after splitting and before undergoing transfection and ideally are transfected the day after being plated.
8. HEK cells are very sensitive to EDTA, which should never be used in splitting HEK cells. Instead, just use trypsin.
9. Reaction components can be scaled up proportionally to make a master mix to add to all the wells. For example, to transfect an entire 96-well plate, 2,000 μl of DMEM incubated with 50 μl Genejuice and then 20 μg DNA would be added.
10. Transfection efficiency is normalized by inclusion of the plasmid encoding *Renilla* luciferase with all transfections. The amount of DNA transfected should be equalized among experiments by the addition of the appropriate amounts of the appropriate empty vector plasmid.
11. Low-level expression of VACV proteins can occur for a variety of reasons: (a) toxicity of the product, (b) transcription in the nucleus as opposed to the cytoplasm, or (c) the VACV protein being expressed may suppress transcription from NF-κB-dependent promoters such as the CMV promoter. Several possible solutions exist: (a) If the VACV protein is thought to be toxic to the cells, an inducible expression system can be used (e.g., Gene Switch™ (Invitrogen)). Alternatively, short time-course experiments could be done with transient transfectants that may more closely mimic the effects of VACV infection and protein expression. (b) Since VACV transcripts are normally made in the cytoplasm-form VACV promoters, one solution has been to clone VACV genes downstream of a T7 promoter and express these transcripts in the presence of T7 polymerase. Since these transcripts are uncapped, the cloning of an IRES

upstream of the ATG translational start site enhances translation several fold (31). (c) Since the CMV promoter is NF-κB dependent, many studies of NF-κB inhibitory proteins utilize the plasmid pEFBos to express such proteins since the pEFBos promoter contains an Sp1-driven promoter that does not have the potential NF-κB inhibitory disadvantage.

12. It may not be clear if the physical association between cellular and poxviral proteins observed in transfected cells actually represent associations that occur during an infection. Therefore, one can examine the association of innate immune inhibitors with the IKKs during an infection. With the aid of specific heterosera and antibodies to the IKK kinase such as IKKγ (Santa Cruz Biotech), such interactions can be investigated in infected cells.
13. When co-immunoprecipitation assays are performed, several problems may arise. First, the anticipated size of IKKγ and TANK is approximately 50 kDa, and the catalytic IKKs migrate at approximately 80 kDa. Thus, while many of these proteins should be detectable and easily distinguishable on an immunoblot, immunodetection of the 50 kDa IKKγ and TANK proteins may be obscured by the antibodies used for immunoprecipitation. In this case, one may use the TrueBlot™ secondary antibodies (eBioscience, San Diego, CA) that do not recognize the denatured primary antibodies on an immunoblot of the SDS-PAGE gel.
14. Use of alternate apoptotic stimuli (e.g., transfection of plasmids encoding proapoptotic proteins as an alternative to staurosporine) and more than one assay to confirm apoptosis is wisest.

Acknowledgments

We would like to acknowledge the invaluable advice of Eicke Latz, Kate Fitzgerald, Andrew Bowie, Brian Monks, and Doug Golenbock. This project has been funded in whole or part with funds from the National Institute of Allergy and Infectious Diseases, National Institutes of Health: R01 AI070940 to W.L.M. and R21AI069167 to N.S.

References

1. Akira S, Takeda K, Kaisho T (2001) Toll-like receptors: critical proteins linking innate and acquired immunity. Nat Immunol 2:675–680
2. Akira S, Hoshino K, Kaisho T (2000) The role of Toll-like receptors and MyD88 in innate immune responses. J Endotoxin Res 6: 383–387
3. Fitzgerald K, Palsson-McDermott E, Bowie A, Jefferies C, Mansell A, Brady G, Brint E, Dunne A, Gray P, Harte M, McMurray D, Smith D,

Sims J, Bird T, O'Neill L (2001) Mal (MyD88-adapter-like) is required for Toll-like receptor-4 signal transduction. Nature 413:78

4. Horng T, Barton G, Medzhitov R (2001) (2001) TIRAP: an adapter molecule in the Toll signaling pathway. Nat Immunol 2(9):835–41
5. Oshiumi H, Matsumoto M, Funami K, Akazawa T, Seya T (2003) TICAM-1, an adaptor molecule that participates in Toll-like receptor 3-mediated interferon-beta induction. Nat Immunol 4:161–167
6. Yamamoto M, Sato S, Mori K, Hoshino K, Takeuchi O, Takeda K, Akira S (2002) Cutting edge: a novel Toll/IL-1 receptor domain-containing adapter that preferentially activates the IFN-beta promoter in the Toll-like receptor signaling. J Immunol 169:6668–6672
7. Bin L, Xu L, Shu H (2003) TIRP, a novel Toll/interleukin-1 receptor (TIR) domain-containing adapter protein involved in TIR signaling. J Biol Chem 278:24526–24532
8. Fitzgerald KA, Rowe DC, Barnes BJ, Caffrey DR, Visintin A, Latz E, Monks B, Pitha PM, Golenbock DT (2003) LPS-TLR4 signaling to IRF-3/7 and NF-kappaB involves the toll adapters TRAM and TRIF. J Exp Med 198:1043–1055
9. Ruckdeschel K, Pfaffinger G, Haase R, Sing A, Weighardt H, Hacker G, Holzmann B, Heesemann J (2004) Signaling of apoptosis through TLRs critically involves toll/IL-1 receptor domain-containing adapter inducing IFN-beta, but not MyD88, in bacteria-infected murine macrophages. J Immunol 173:3320–3328
10. Doyle S, Vaidya S, O'Connell R, Dadgostar H, Dempsey P, Wu T, Rao G, Sun R, Haberland M, Modlin R, Cheng G (2002) IRF3 mediates a TLR3/TLR4-specific antiviral gene program. Immunity 17:251–263
11. McWhirter S, Fitzgerald K, Rosains J, Rowe D, Golenbock D, Maniatis T (2004) IFN-regulatory factor 3-dependent gene expression is defective in Tbk1-deficient mouse embryonic fibroblasts. Proc Natl Acad Sci USA 101:233–238
12. Fitzgerald K, McWhirter S, Faia K, Rowe D, Latz E, Golenbock D, Coyle A, Liao S, Maniatis T (2003) IKKepsilon and TBK1 are essential components of the IRF3 signaling pathway. Nat Immunol 4:491–496
13. Alcami A, Symons J, Smith G (2000) The vaccinia virus soluble alpha/beta interferon (IFN) receptor binds to the cell surface and protects cells from the antiviral effects of IFN. J Virol 74:11230–11239
14. Langland J, Jacobs B (2002) The role of the PKR-inhibitory genes, E3L and K3L, in determining vaccinia virus host range. Virology 299:133–141
15. Hoebe K, Du X, Georgel P, Janssen E, Tabeta K, Kim S, Goode J, Lin P, Mann N, Mudd S, Crozat K, Sovath S, Han J, Beutler B (2003) Identification of Lps2 as a key transducer of MyD88-independent TIR signalling. Nature 424:743–748
16. Altmann M, Hammerschmidt W (2005) Epstein-Barr virus provides a new paradigm: a requirement for the immediate inhibition of apoptosis. PLoS Biol 3:e404
17. Aoyagi M, Zhai D, Jin C, Aleshin AE, Stec B, Reed JC, Liddington RC (2007) Vaccinia virus N1L protein resembles a B cell lymphoma-2 (Bcl-2) family protein. Protein Sci 16:118–124
18. Cooray S, Bahar MW, Abrescia NG, McVey CE, Bartlett NW, Chen RA, Stuart DI, Grimes JM, Smith GL (2007) Functional and structural studies of the vaccinia virus virulence factor N1 reveal a Bcl-2-like anti-apoptotic protein. J Gen Virol 88:1656–1666
19. Wasilenko ST, Banadyga L, Bond D, Barry M (2005) The vaccinia virus F1L protein interacts with the proapoptotic protein Bak and inhibits Bak activation. J Virol 79:14031–14043
20. DiPerna G, Stack J, Bowie AG, Boyd A, Kotwal G, Zhang Z, Arvikar S, Latz E, Fitzgerald KA, Marshall WL (2004) Poxvirus protein N1L targets the I-kappaB kinase complex, inhibits signaling to NF-kappaB by the tumor necrosis factor superfamily of receptors, and inhibits NF-kappaB and IRF3 signaling by toll-like receptors. J Biol Chem 279:36570–36578
21. Bowie A, Kiss-Toth E, Symons J, Smith G, Dower S, O'Neill L (2000) A46R and A52R from vaccinia virus are antagonists of host IL-1 and toll-like receptor signaling. Proc Natl Acad Sci USA 97:10162–10167
22. Harte M, Haga I, Maloney G, Gray P, Reading P, Bartlett N, Smith G, Bowie A, O'Neill L (2003) The poxvirus protein A52R targets Toll-like receptor signaling complexes to suppress host defense. J Exp Med 97:343–351
23. Afonso CL, Tulman ER, Lu Z, Zsak L, Osorio FA, Balinsky C, Kutish GF, Rock DL (2002) The genome of swinepox virus. J Virol 76:783–790
24. Shi J, Blundell TL, Mizuguchi K (2001) FUGUE: sequence-structure homology recognition using environment-specific substitution tables and structure-dependent gap penalties. J Mol Biol 310:243–257
25. Brzozka K, Finke S, Conzelmann KK (2005) Identification of the rabies virus alpha/beta interferon antagonist: phosphoprotein P interferes with phosphorylation of interferon regulatory factor 3. J Virol 79:7673–7681

26. Otsuka M, Kato N, Moriyama M, Taniguchi H, Wang Y, Dharel N, Kawabe T, Omata M (2005) Interaction between the HCV NS3 protein and the host TBK1 protein leads to inhibition of cellular antiviral responses. Hepatology 41: 1004–1012
27. Unterstab G, Ludwig S, Anton A, Planz O, Dauber B, Krappmann D, Heins G, Ehrhardt C, Wolff T (2005) Viral targeting of the interferon-{beta}-inducing Traf family member-associated NF-{kappa}B activator (TANK)-binding kinase-1. Proc Natl Acad Sci USA 102: 13640–13645
28. Graham SC, Bahar MW, Cooray S, Chen RA, Whalen DM, Abrescia NG, Alderton D, Owens RJ, Stuart DI, Smith GL, Grimes JM (2008) Vaccinia virus proteins A52 and B14 Share a Bcl-2-like fold but have evolved to inhibit NF-kappaB rather than apoptosis. PLoS Pathog 4:e1000128
29. Kurt-Jones E, Mandell L, Whitney C, Padgett A, Gosselin K, Newburger P, Finberg R (2002) Role of toll-like receptor 2 (TLR2) in neutrophil activation: GM-CSF enhances TLR2 expression and TLR2-mediated interleukin 8 responses in neutrophils. Blood 100:1860–1868
30. Latz E, Visintin A, Lien E, Fitzgerald K, Monks B, Kurt-Jones E, Golenbock D, Espevik T (2002) Lipopolysaccharide rapidly traffics to and from the Golgi apparatus with the toll-like receptor 4-MD-2-CD14 complex in a process that is distinct from the initiation of signal transduction. J Biol Chem 277:7834–7843
31. Fuerst TR, Niles EG, Studier FW, Moss B (1986) Eukaryotic transient-expression system based on recombinant vaccinia virus that synthesizes bacteriophage T7 RNA polymerase. Proc Natl Acad Sci USA 83:8122–8126
32. Dereeper A, Guignon V, Blanc G, Audic S, Buffet S, Chevenet F, Dufayard J-F, Guindon S, Lefort V, Lescot M, Claverie J-M, Gascuel O (2008) Phylogeny.fr: robust phylogenetic analysis for the non-specialist. Nucleic Acids Res (Web Server Issue) 36:W465–W469

Chapter 17

Application of Quartz Crystal Microbalance with Dissipation Monitoring Technology for Studying Interactions of Poxviral Proteins with Their Ligands

Amod P. Kulkarni, Lauriston A. Kellaway, and Girish J. Kotwal

Abstract

Poxviruses are one of the most complex of animal viruses and encode for over 150 proteins. The interactions of many of the poxviral-encoded proteins with host proteins, as well as with other proteins, such as transcription complexes, have been well characterized at the qualitative level. Some have also been characterized quantitatively by two hybrid systems and surface plasmon resonance approaches. Presented here is an alternative approach that can enable the understanding of complex interactions with multiple ligands. The example given is that of vaccinia virus complement control protein (VCP). The complement system forms the first line of defense against microorganisms and a failure to appropriately regulate it is implicated in many inflammatory disorders, such as traumatic brain injury, Alzheimer's disease (AD), and rheumatoid arthritis. The complement component C3 is central to the complement activation. Complement regulatory proteins, capable of binding to the central complement component C3, may therefore effectively be employed for the treatment and prevention of these disorders. There are many biochemical and/or immunoassays available to study the interaction of proteins with complement components. However, protocols for many of them are time consuming, and not all assays are useful for multiple screening. In addition, most of these assays may not give information regarding the nature of binding, the number of molecules interacting with the complement component C3, as well as kinetics of binding. Some of the assays may require labeling which may induce changes in protein confirmation. We report a protocol for an assay based on quartz crystal microbalance with dissipation monitoring (QCM-D) technology, which can effectively be employed to study poxviral proteins for their ability to interact with their ligand. A protocol was developed in our laboratories to study the interaction of VCP with the complement component C3 using Q-sense (D-300), equipment based on QCM-D technology. The protocol can also be used as a prototype for studying both proteins and small-sized compounds (for use as anti-poxvirals) for their ability to interact with and/or inhibit the activity of their ligands.

Key words: Poxviral protein–ligand interaction, Quartz crystal microbalance with dissipation monitoring, VCP

Stuart N. Isaacs (ed.), *Vaccinia Virus and Poxvirology: Methods and Protocols*, Methods in Molecular Biology, vol. 890,
DOI 10.1007/978-1-61779-876-4_17, © Springer Science+Business Media, LLC 2012

1. Introduction

Poxviral outbreaks are common in different parts of the world (1–4). These viruses secrete many proteins which help these viruses in establishing infections in their host. Most of these proteins play an important role in the pathogenesis of these viruses by interacting with the immune system and/or by evading the immune response of the host. Study of interaction of these viral proteins with the proteins of the immune machinery of the host might prove beneficial for developing an effective treatment strategy to treat or prevent the progression of the pathogenesis of these viral infections. Some of the poxviruses also secrete complement regulatory molecules, which evade the complement-mediated immune response. These complement regulatory molecules may also be of great help in the treatment and prevention of a number of diseases, where the complement system is up-regulated. Vaccinia virus is one of the viruses belonging to poxviral family that secretes a complement control protein. The vaccinia virus complement control protein (VCP) is known to regulate the complement system (5–7), and by virtue of its complement regulatory activity, it has been shown to be effective in rodent models of traumatic brain injury (8), Alzheimer's disease (AD) (9, 10), spinal cord injury (11), and atherosclerosis (12), where up-regulation of the complement system is evident. VCP has previously been shown to interact with the C3b and C4b components of the complement system using surface plasmon resonance technology (SPR (13)). It has recently also been shown to bind to the complement component C3 using a relatively new quartz crystal microbalance with dissipation monitoring technology (QCM-D; (14)). The protocols in the aforementioned article are explained in greater detail here in this chapter.

The complement component C3 is one of the proteins of the complement system that has been found to be up-regulated in several neuroinflammatory disorders, such as AD. The level of C3 is known to be elevated in AD with mild to severe clinical symptoms with a low level of expression of complement regulatory molecules (15). Therefore, some believe that the activated complement components, especially C3 which is central to complement activation, need to be regulated in such disease states. The discovery and development of agents that prevent the detrimental effects of these complement components might have therapeutic potential. Techniques to identify compounds with an ability to bind to C3 and inhibit the activation of complement would be useful. However, immunoassays and other biochemical assays required to study the protein–protein interactions are costly and time consuming. Also, most of these assays do not give the information regarding the nature of binding of the two interacting moieties, number of molecules binding, as well as kinetics of binding. In addition, the assays need to be customized if one wants to study the interaction of small-sized compounds with the proteins.

There are many techniques available for protein adsorption or interaction (16). In order to carry out the protein binding study, we used QCM-D-based Q-sense (D-300), a Swedish-based technology thoroughly studied and developed for the protein adsorption studies at Chalmer's University of Technology, Goteberg, Sweden (17). QCM-D is a rapidly advancing technology used to study the protein–surface interactions and ligand–receptor interactions. This technology has recently been used to study the change in conformation of the protein molecules during co-precipitation (18), deposition kinetics of nanoparticles (19), screening membrane-active antimicrobial peptides (20), and changes in viscoelastic properties during the phase transitions between the two states of cowpea chlorotic mottle virus (21).

The QCM-D technique offers the advantage of real-time monitoring, rapidity, simplicity, sensitivity and economy, compared to the other routinely used techniques. The potential of QCM as a biosensor and its sensitivity was found to be comparable to that of SPR (22, 23), a commonly used technique for protein binding studies. Using bare gold crystals of QCM, it was shown to be as sensitive as ELISA when compared using gp41-based HIV-ELISA (24). It offers an advantage of real-time monitoring and rapidity (10 min to 2 h versus several hours to 1–2 days in ELISA) without losing sensitivity. It does not involve labeling of protein molecules and, therefore, decreases concern about possible changes in conformation when protein needs to be labeled. It also takes into account dissipation (D) values or the ratio of change in dissipation to change in frequency (dD/dF) values at the same time point. These values are explained later in the text, and can be used to study the nature (rigid vs. reversible) as well as kinetics of protein binding. The many advantages of QCM-D technology over the other routinely employed techniques as discussed above led us to employ Q-sense D-300 (Q-sense) to study the interaction of VCP with complement components.

QCM-D has previously been employed to study the interaction of the complement system with biomaterials (25). The technique was also used to study the interaction of the complement components with complement inhibitory molecules. QCM-D was compared to the ELISA technique to show that biomaterial surfaces activate the complement system and this activation can be inhibited by complement regulatory molecules, such as factor H (26). Using QCM-D, it was also shown that the polystyrene surface activates the alternative pathway (AP) of complement activation. C3 deposited on the polystyrene surface can form C3 convertase and activates the AP of complement activation (25). Thus, Q-sense based on QCM-D may be employed to investigate the complement regulatory molecules with an ability to bind the activated form of C3. In this chapter, we describe the application of the technique of using the polystyrene sensor (PS) crystals to study the interaction of VCP with the complement component C3 adsorbed on PS crystals.

The overall objective of applying the QCM-D technique is to specifically compare the interactions of VCP and a truncated version of VCP (tVCP) with C3 on the PS surfaces.

Objectives

(a) Compare the nature of binding of VCP and tVCP with that of C3 using PS.

(b) Compare the adsorption kinetics of interaction of VCP and tVCP with PS.

(c) Compare the interaction of C3 with the aforementioned surfaces as well as with VCP and tVCP adsorbed onto these surfaces.

2. Materials

1. Q-sense, model D-300.
2. Polystyrene sensor crystals (PS; Q-sense).
3. Piranha solution: 1:1:5 of HCl:H_2O_2 (30%):H_2O.
4. 1% Hellmanex-II solution (a cleaner concentrate for cuvette washing).
5. Nitrogen gas.
6. UV chamber (e.g., Bioforce, nanoscience) (see Note 1).
7. Phosphate-buffered saline (PBS).
8. Purified proteins of interest, passed through a 0.22-μM filter and degassed prior to use (see Note 2).

3. Methods

3.1. Basic Principles of QCM, Terminologies Used, and Description of Q-Sense

QCM is used to measure the interaction between protein or adsorbing molecules and that of the surface as well as with each other, employing a piezoelectric sensor crystal. For the QCM studies, this piezoelectric sensor crystal is coated with a very thin layer of gold on both the sides. The bottom surface serves as electrodes to which AC voltage is applied. This results in oscillation of the sensor crystal at its resonant frequency, f. This inherent f of the quartz crystal changes upon adsorption of a thin layer of an adsorbing moiety (AM). In QCM, this change in frequency (dF) is correlated with the change in mass (ΔM or dM). The amount of a substance deposited on the crystal surface is determined from this change in frequency using Sauerbrey's equation as discussed by Höök in his thesis and the references mentioned therein (17). As discussed by Höök, the Sauerbrey's equation is based on the assumptions that the AM should be rigidly deposited on the crystal, distributed evenly, and the added mass is smaller than the weight of the crystal. "Fit analysis" as outlined in the manual of q-sense should be carried out to check whether the experiments follow the assumptions based on the Sauerbrey's equation.

Dissipation factor or D factor in QCM-D gives information regarding the energy dissipated during one oscillation after the adsorption of the AM on the crystal. The ratio $\mathrm{d}D/\mathrm{d}F$ in turn gives information regarding the rigidity of binding at the interfaces (27–29). Further information on this factor can be obtained from the research articles published in the literature (14, 27–31).

3.2. Q-Sense: Practical Considerations (Q-Sense Reference Manual (14, 17)

1. The basic layout of the Q-sense (model D-300) is as shown in Fig. 1. It consists of four basic components: sensor crystal, measurement chamber, electronic unit, and acquisition software.
2. Inside the chamber (Fig. 1b), the PS crystals (Fig. 1d) are operated in thickness shear mode at 5 MHz frequency and are mounted as per the instructions in the manual.
3. The smooth surface of the crystal is used for the measurements, and the opposite side of the crystal is connected to the A/C supply of the electrical unit.
4. The measurement chamber also consists of a sample holder (Fig. 1c), where the solution of the AM is kept. The sample holder can be directed to either a "temperature loop" or to the "sensor crystal" using a controller knob.

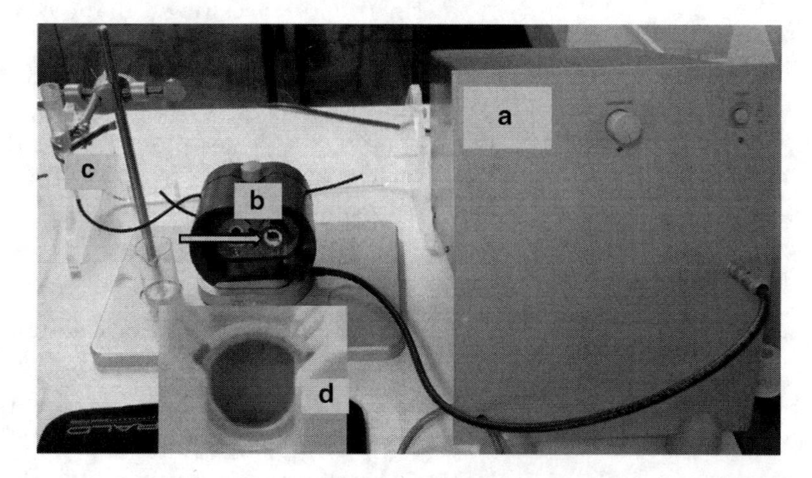

Fig. 1. The Q-sense (D-300) apparatus and its components. Shown are the electronic unit (**a**), measurement chamber (**b**), and the sample holder (**c**). The PS sensor crystal mounted on the 'O' ring of the measurement chamber is shown by an *arrow near* (**b**) and the smooth surface of the sensor crystal where AMs are adsorbed is shown in the *inset* (**d**).

5. The measurement chamber is connected to a computer to measure the changes in f and D values at different overtones. These changes in f and D values are monitored online using q-soft/QTools software supplied with the equipment and are used for further analysis.

3.3. Explanation of Terminology

1. AM1 = AM bound to the surface of the sensor crystal. Therefore, AM1 is indicated by putting the first letter of the surface (ps = polystyrene) in lower case, next to AM1. For example, when the AM is VCP or C3 and when the AM is bound to the surface of the sensor crystal, the abbreviations are VCPps and C3ps.
2. AM2 = a second adsorbing moiety that is bound to AM1. For example, when C3 is bound to VCPps, the abbreviation would be C3-VCPps or when VCP is bound to C3ps, the abbreviation would be VCP-C3ps.
3. dF_{30} and dD_{30} refer to dF and dD values after 30 min, respectively. dF_{30} and dD_{30} values for AMs are calculated as follows.
   (a) $dF_{30} = dF$ at 30 min $- dF_{initial}$, where $dF_{initial}$ refers to the baseline or initial f or D values prior to the adsorption of AMs.
   (b) dF_{fin} or dD_{fin} values indicate the final dF and dD values recorded after washing the surface with PBS after the adsorption is over. dF_{fin} = Final dF (or dD) values after washing with PBS $- dF_{initial}$ (or $dD_{initial}$).
   (c) Using these dD and dF values, calculate dD/dF ratios to get the information regarding the rigidity (affinity) or viscoelasticity of binding.
   (d) Calculate the number of molecules of AM adsorbed (n) onto the crystal surface or onto the other AM using Avogadro's formula ($n = (6.022 \times 10^{23} \times \text{mass adsorbed} (\text{ng/cm}^2))/\text{molecular weight of the AM}$).

3.4. Studying Interactions of Poxviral Proteins with Their Ligands Using QCM-D Technology

1. Just before starting the experiment and after completion of the experiment, clean the surface of the crystal and the chamber according to the protocol provided in the Q-sense instruction manual by washing the PS crystal with piranha solution and washing the chamber with 1% Hellmanex-II solution (see Note 3).
2. After each washing step, dry the sensor crystals and chamber using nitrogen gas.
3. Expose the sensor crystal to UV light for 5–10 min to destroy impurities on the sensor surface (see Note 1).
4. Ensure that resonant frequencies of the crystals match the values provided in the instruction manual (see Notes 4 and 5).
5. Obtain baseline readings using PBS before starting each experiment as well as prior to each adsorption (see Note 6).
6. Prior to the adsorption of AM1, obtain a baseline for the resonance frequency of the crystal by adding PBS to the surface.

Table 1
The outline of the Q-sense experiments

Sr	Surface	AM 1	AM 2
1	PS	VCP	C3
2	PS	C3	VCP
3	PS	tVCP	C3

Sr serial number, *PS* polystyrene, *AM* adsorbing moiety (AM1, AM2)

7. Add PBS until there is no further drop in the frequency. An example of the surfaces, AMs, and the sequence of adsorbing moieties is shown in Table 1.
8. Adsorb AM (e.g., VCP, tVCP, or other poxviral protein or C3) on the PS crystal surface. This is AM1 (see Notes 7–9).
9. After adsorption of the first AM on the sensor crystal for a particular period of time (approximately 30-50 min), wash with PBS to remove any unadsorbed molecules.
10. Add the second AM (AM2) to the sensor surface to adsorb to AM1 and allow adsorption for the same period of time (approximately 30–50 min) (see Notes 7 and 8).
11. Wash with PBS to remove any unadsorbed molecules.
12. Measure the changes in f and D values at different overtones using QTools software.
13. Analyze the data captured using q-soft/QTools software (see Notes 10–12).
14. Confirm whether the adsorption of the AM follows the Sauerbrey's equation using fit analysis and plot the data as shown in Fig. 2.
15. Calculate the dD/dF ratios to get information regarding rigidity of binding of AMs.
16. Plotting the dD values (Y axis) at different time intervals against dF values at the same interval reveals information regarding the kinetics of binding.
17. Calculate the number of molecules of AMs adsorbed onto the PS surface and/or to each other using Avogadro's formula (see Note 13).

3.5. Analysis and Graphical Representation of the Data

The data can be analyzed manually or using the software provided with the equipment, and is either graphically presented as shown in Figs. 3 and 4 using Excel/other graphing programs or tabulated as shown in Tables 2 and 3.

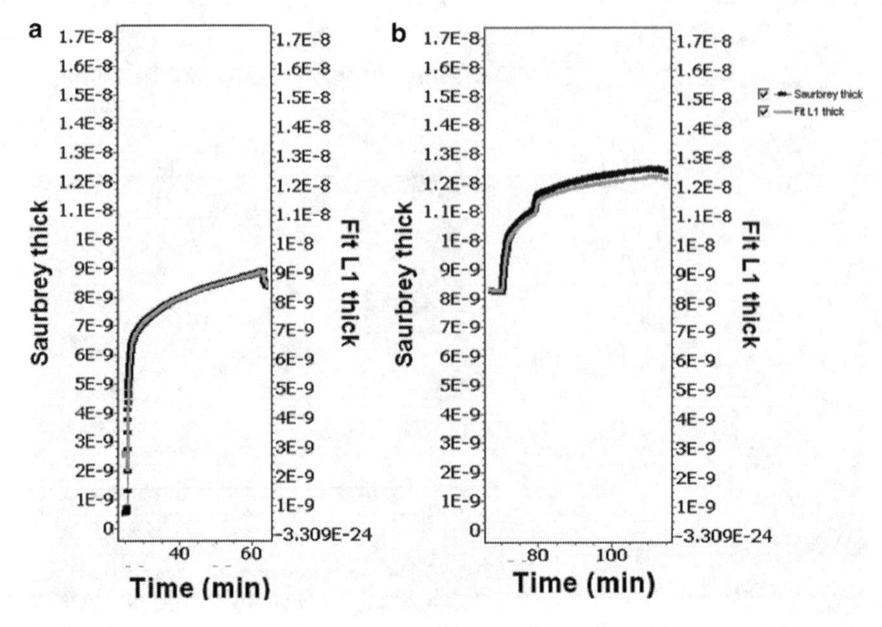

Fig. 2. Fit analysis for the adsorption of C3 onto VCP adsorbed onto PS. The values of Saurbrey thickness (*Y*1-axis) and Fit L1 thickness (*Y*2-axis) obtained after Fit analysis done using the QTools software coincide with each other indicating that the adsorption of VCP on PS (**a**) and C3 on VCP (**b**) obeys Sauerbrey's equation.

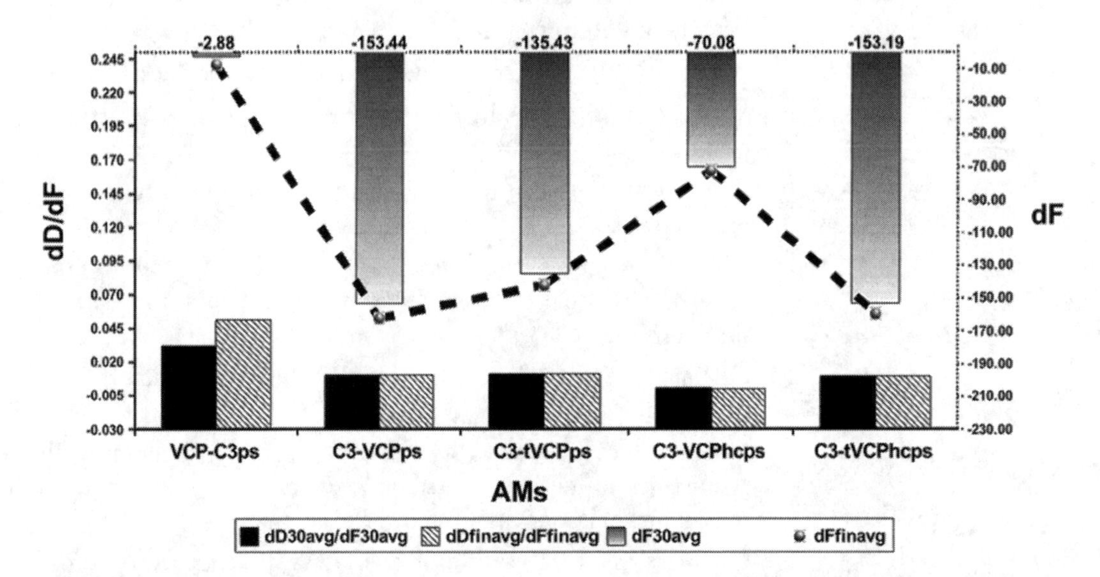

Fig. 3. d*D*/d*F* ratios (the primary *X* and *Y* axis) and d*F* values (the secondary *Y* axis [*right*] and the secondary *X* axis shown [*top*]) for binding of VCP, tVCP, and C3 on PS. d*D*/d*F* ratios at a 30-min interval are shown by *solid black bars*, whereas final d*D*/d*F* ratios are shown by *hatched bars*. The dF_{30} values are shown by the *gray gradient bars*. The dF_{fin} values are shown by a *black broken line with black dots*. The d*F* and d*D* values shown indicate the average values obtained from the two experiments unless specified in Table 1. "hc" and "ps" in "hcps" stand for high concentration and polystyrene surface, respectively. See Note 14. Figure modified from ref. 4 published in the Open Journal of Biochemistry.

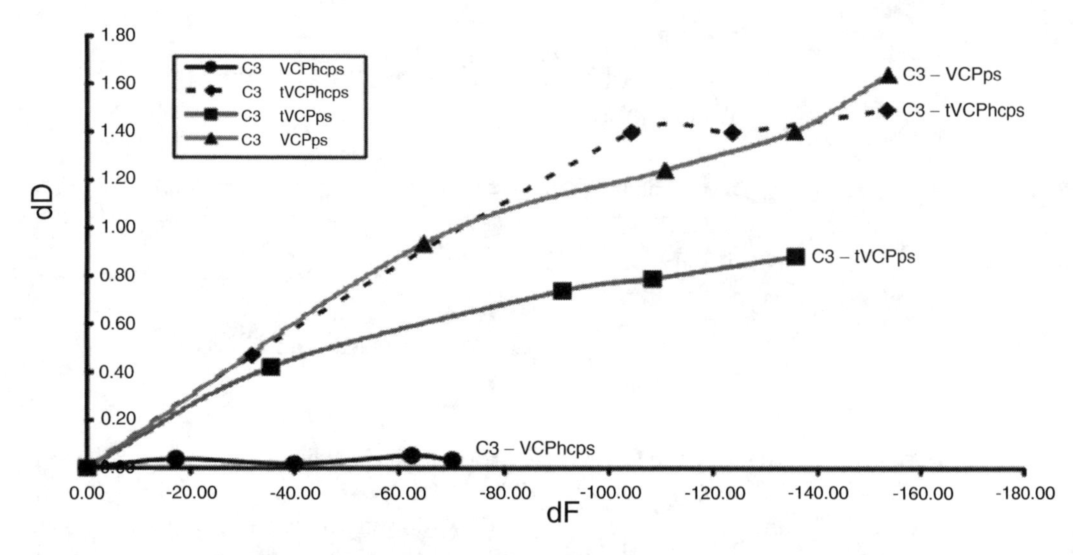

Fig. 4. Absorption of C3 on VCP and tVCP absorbed on PS crystal surface. d*F* values for the interaction of VCP and tVCP adsorbed on the PS surfaces with C3 (*X*-axis) were plotted against d*D* values (*Y*-axis) at five different time points. The data label for each series indicates AMs (Series: C3-VCPhcps, C3-tVCPhcps, C3-tVCPps, and C3-VCPg indicate adsorption of C3 on VCPhcps, tVCPhcps, tVCPps, and VCPg, respectively. "hc" prior to ps in the case of VCP denotes VCP at high concentration) (see Notes 14 and 15). Figure modified from ref. 4 published in the Open Journal of Biochemistry.

Table 2
The number of molecules of AMs adsorbed for interaction of VCP and tVCP with C3 on PS

Sr	AMs	Final dD/ dF ratio	Final dF	dF corrected	Mass (ng/cm²)	No. of molecules adsorbed (×1,013)
1.	VCP-C3ps	0.0517	-7.3670	-5.1569	30.4257	0.6362
2.	VCPps	0.0141	−24.7845	−17.3492	102.3600	2.1403
3.	tVCPps	0.0137	-35.4495	-24.8147	146.4064	4.6897
4.	VCPhcps	0.0340	−140.8510	-98.5957	581.7146	12.1635
5.	tVCPhcps	0.0124	-35.9400	-25.1580	148.4322	4.7546
6.	C3ps	0.0105	−177.9875	−124.5913	735.0884	2.3298
7.	C3-VCPps	0.0099	−162.2495	−113.5747	670.0904	2.1238
8.	C3-tVCPps	0.0110	−142.2435	-99.5705	587.4657	1.8620
9.	C3-VCPhcps	0.0002	-72.4925	-50.7448	299.3940	0.9489
10.	C3-tVCPhcps	0.0094	−159.2420	−111.4694	657.6695	2.0845

Sr serial number, *AM* adsorbing moiety, *PS* polystyrene, hc in "hcps" stands for high concentration (see Note 14), and d*F* corrected refers to the d*F* values corrected for water bound to the protein molecules (see Notes 13, 16, and 17)

Table 3
The ratio of binding of the number of molecules of AM1 to AM2

Sr	Binding moieties (AM1:AM2)	Ratio AM1:AM2
1.	VCPps:C3-VCPps	12.818:1
2.	tVCPps:C3-tVCPps	2.519:1
3.	C3ps:VCP-C3ps	3.662:1

3.5.1. Nature of Binding

To study the viscoelastic properties of binding (rigidity or strongness of binding), the dD/dF ratio at a particular time point (30 min and/or final values) and dF values at the same time point can be plotted (Fig. 3). In addition to the viscoelastic properties of interaction of compounds with the complement component C3, the dF values give information regarding the amount of AMs adsorbed at that time point.

Low dF values indicate no or little adsorption (e.g., VCP-C3ps in Fig. 3). dD/dF ratio indicates rigidity of binding. As shown in Fig. 3, the order for dD/dF ratios for the different AMs is (C3 vs. VCPhcps) Đ (C3 vs. VCPps) = (C3 vs. tVCPps) = (C3 vs. tVCPhcps) < (VCP vs. C3ps). The ratio was the lowest for C3 vs. VCPhcps with dD/dF ratio approaching zero, suggesting very strong binding (see Note 14).

3.5.2. Kinetics of Binding

Figure 4 shows the kinetics of binding of AMs. In this graph, the dD values (Y-axis) at five different time points (0, 1, 5, 15, and 30 min) are plotted against dF values (X-axis) at the same time points (see Note 15). The dD values change linearly with dF values during the first phase of adsorption (i.e., the adsorption of C3 on PS, or on VCP and tVCP). In most other cases, this first phase is followed by the second phase, where dD values either increase or decrease with the change in dF values. In some cases, the adsorption of AMs may constitute a single phase (e.g., the adsorption of C3 on VCPhcps as shown in Fig. 4). A single-phase adsorption may be due to formation of a monolayer of the AM and the second phase may indicate the formation of a bilayer.

3.5.3. Number of Molecules of AMs Adsorbed onto the Surface and/or on the Other AM

Calculation of the mass of the AMs adsorbed onto the crystal surface or on the other AM using dF values is shown in Table 2. Calculate the number of molecules of the AMs adsorbed onto the surface of the crystal or onto the other AM using the Avogadro's formula and mass adsorbed (see Notes 13 and 16–18). Calculation of the ratio of the AMs adsorbed onto the surface or the other AMs as shown in Table 3.

4. Notes

1. UV chamber generates UV light at the specific wavelengths of 185 and 254 nm, and is used to remove organic contaminants on the crystal surface.
2. As examples for this chapter, we show data with VCP and complement component C3. Purified VCP and an N-terminal truncation of VCP (tVCP) expressed in *Pichia pastoris* yeast expression system, were purified by using a Hi-trap heparin column. tVCP lacks the first short consensus repeat (SCR) domain of VCP and, thus, is a histidine-tagged protein comprising the 2nd, 3rd, and 4th SCR domains of VCP. In the work described in this chapter, the final concentration of tVCP was 1.8767 times higher than that of VCP supplied and its molecular weight (18.8 kDa) is 1.5319 times less than that of VCP (28.8 kDa). Thus, in order to prepare tVCP at the same concentration of VCP, 40.82 to 76.06 μl of tVCP was dissolved in 2 ml of PBS solution prior to experimentation. To prepare C3, 15 μl of 2 mg/ml C3 was diluted in 2 ml PBS prior to the experiment. C3 was kindly provided by Prof. Krishna Murthy's laboratory, University of Alabama, Birmingham, AL, USA.
3. The crystals can also be cleaned using 1:1:5 parts of ammoniac:H_2O_2(30%):H_2O. Water used in the experiment and/or to wash the sensor crystal surface was of deionized double-distilled Millipore grade, and it was degassed using nitrogen gas prior to use.
4. When the resonant frequencies of the sensor crystal do not match the values provided in the instruction manual, the sensor crystal needs to be discarded. However, sometimes the resonant frequencies may not match the values provided in the manual due to inadequate washing and drying steps. So before discarding, first try to rewash the sensor crystal and the chamber. Then, dry them with the nitrogen gas, subject it to the UV light as mentioned in the protocol, and check the resonant frequencies again.
5. For all the experiments, only the crystals showing comparable and similar pattern of f and D values were used.
6. In most of the Q-sense-based experiments, temperature needs to be controlled appropriately by use of the controller knob on the electronic unit. Experiments can be carried out at various temperatures between 15 and 40°C. Extremes of temperature may induce aberrant changes in frequency. For the protocol described here, the experiments were carried out at 25°C. Before adding the solution onto the sensor crystal, make sure

that the temperature of the solutions of all the AMs is brought to the room temperature and is nearly the same.

7. The position of the sample holder (Fig. 1c) can be adjusted to optimize the flow rate. The same flow rate should be maintained in all the experiments.
8. It is recommended that the final volume of each AM to be used should be about 2 ml for an accurate measurement.
9. To prepare VCP solution, approximately 50 μl of 1.2 mg/ml of endotoxin-free purified VCP was diluted to 2 ml using ice-cold PBS solution at the time of experiment. The tVCP used in the study was purified by desalting and by washing with PBS and concentrated by passing the solution several times through a 5-kDa cutoff filter.
10. If the adsorption does not follow Sauerbrey's equation, a method called Z-match and/or viscoelastic modeling is recommended. All the instructions regarding modeling are provided with the software.
11. Orientation of AMs may be important for binding. For example, we found that VCP does not bind to C3 when C3 is first adsorbed onto to the crystal surface. Under such circumstances, before making any conclusion, the sequence of addition of AMs should be reversed to check the binding orientation.
12. The surface of the sensor crystal may also be coated with a monolayer of specific antigen, antibody, or any other protein/peptide under investigation. Many types of sensor crystals are also available for specific use. For example, PEG-COOH and/or amine coupling reactions can be used with certain surfaces.
13. If the water binding capacity of AMs is not known, the number of molecules bound to each other and with the surface will not be a true reflection, but an approximate value. In order to get exact values of the number of molecules of AMs binding to each other or to the sensor crystal surface, the water bound to proteins should be determined experimentally. Optical techniques, such as ellipsometry (ELM) or optical wave guidance lightweight microscopy (OWLS), give information regarding the "dry mass" of the protein adsorbed. QCM-D gives information regarding "wet mass (Protein plus water bound)." So the information regarding water bound to the protein adsorbed could be calculated by combining the optical techniques with QCM-D as described by Höök et al. (32). Alternatively, one can estimate that approximately 30% of the mass of protein molecules accounts for the water bound to them, and therefore for the current example, it was deducted from the final mass of AMs interacting with the surfaces or other AMs.

14. VCPhc means VCP at higher concentration. The concentrations of VCPhc and tVCPhc were 2.5 times higher than those of VCP and tVCP, respectively.

15. The zero values used in the dD vs dF graphs as shown in Fig. 4 do not indicate a time point or a data point, but only the starting point of the graph. These graphs are plotted to avoid time dependency and reveal the shapes of the graph in order to study the phases of adsorption of AMs either on the surface or on the other AMs.

16. For calculating the number of molecules of AMs adsorbed onto the surface of the crystal or onto the other AM using Avogadro's number, the molecular weights of the AMs must be expressed in nanograms because the mass of the AMs adsorbed is expressed in the same unit. For the current example, the molecular weights of C3, tVCP, and VCP in nanograms (ng) were $1,900\times10^{11}$, 188×10^{11}, and 288×10^{11}, respectively. These values were taken either from the Calbiochem catalog (C3) or from the literature (VCP and tVCP).

17. When the number of molecules of water bound to the AM is not known, approximate values of water of hydration of the protein molecules based on literature values are used to calculate the mass deposited using Sauerbrey's equation, and the number of molecules adsorbed on the surface using Avogadro's equation.

18. Although not shown here, to get more information regarding the viscoelastic properties of the AMs, modeling with QTools software provided with the equipment is an option.

References

1. Silva DC, Moreira-Silva EA, Gomes JA, Fonseca FG, Correa-Oliveira R (2010) Clinical signs, diagnosis, and case reports of Vaccinia virus infections. Braz J Infect Dis 14: 129–134
2. Abrahão JS, Silva-Fernandes AT, Lima LS, Campos RK, Guedes MI, Cota MM, Assis FL, Borges IA, Souza-Júnior MF, Lobato ZI, Bonjardim CA, Ferreira PC, Trindade GS, Kroon EG (2010) Vaccinia virus infection in monkeys, Brazilian Amazon. Emerg Infect Dis 16:976–979
3. Prasad VG, Nayeem S, Ramachandra S, Saiprasad GS, Wilson CG (2009) An outbreak of buffalo pox in human in a village in Ranga Reddy District, Andhra Pradesh. Indian J Public Health 53:267
4. Formenty P, Muntasir MO, Damon I, Chowdhary V, Opoka ML, Monimart C, Mutasim EM, Manuguerra JC, Davidson WB, Karem KL, Cabeza J, Wang S, Malik MR, Durand T, Khalid A, Rioton T, Kuong-Ruay A, Babiker AA, Karsani ME, Abdalla MS (2005) Human monkeypox outbreak caused by novel virus belonging to Congo Basin clade, Sudan. Emerg Infect Dis 16:1539–1545
5. Kotwal G, Moss B (1988) Vaccinia virus encodes a secretory polypeptide structurally related to complement control proteins. Nature 335:176–178
6. Kotwal G, Isaacs S, McKenzie R, Frank M, Moss B (1990) Inhibition of the complement cascade by the major secretary protein of vaccinia virus. Science 250:827–830

7. McKenzie R, Kotwal G, Moss B, Hammer C, Frank M (1992) Regulation of complement activity by vaccinia virus complementcontrol protein. J Infect Dis 166:1245–1250
8. Hicks RR, Keeling KL, Yang MY, Smith SA, Simons AM, Kotwal GJ (2002) Vaccinia virus complement control protein enhances functional recovery after traumatic brain injury. J Neurotrauma 19:705–714
9. Kulkarni AP, Pillay NS, Kellaway LA, Kotwal GJ (2008) Intracranial administration of vaccinia virus complement control protein in Mo/Hu APPswe PS1δE9 transgenic mice at an early age shows enhanced performance at a later age using a cheese board maze test. Biogerontology 9:405–420
10. Pillay N, Kellaway L, Kotwal G (2008) Early detection of memory deficits in an Alzheimers disease mice model and memory improvement with Vaccinia virus complement control protein. Behav Brain Res 192:173–177
11. Reynolds DN, Smith SA, Zhang YP, Mengsheng Q, Lahiri DK, Morassutti DJ et al (2004) Vaccinia virus complement control protein reduces inflammation and improves spinal cord integrity following spinal cord injury. Ann N Y Acad Sci 1035:165–178
12. Thorbjornsdottir P, Kolka R, Gunnarsson E, Bambir SH, Thorgeirsson G, Kotwal GJ, Arason GJ (2005) Vaccinia virus complement control protein diminishes formation of atherosclerotic lesions: complement is centrally involved in atherosclerotic disease. Ann N Y Acad Sci 1056:1–15
13. Smith S, Sreenivasan R, Krishnasamy G, Judge K, Murthy K, Arjunwadkar S, Pugh D, Kotwal G (2003) Mapping of regions within the vaccinia virus complement control protein involved in dose-dependent binding to key complement components and heparin using surface plasmon resonance. Biochim Biophys Acta 1650:30–39
14. Kulkarni AP, Randall PJ, Murthy K, Kellaway LA, Kotwal GJ (2010) Investigation of interaction of vaccinia virus complement control protein and curcumin with complement components C3 and C3b using quartz crystal microbalance with dissipation monitoring technology. Open J Biochem 4:9–21
15. Zanjani H, Finch C, Kemper C, Atkinson J, McKeel D, Morris J, Price J (2005) Complement activation in very early Alzheimer disease. Alzheimer Dis Assoc Disord 19:55–66
16. Ramsden JJ (1993) Experimental methods for investigating protein adsorption kinetics at surface. Q Rev Biophys 27:41–105
17. Höök F (2004) Development of a novel QCM technique for protein adsorption studies. Thesis, Chalmers University of Technology, Goetberg University, Goetberg, Sweden
18. Giamblanco N, Yaseen M, Zhavnerko G, Lu JR, Marletta G (2011) Fibronectin conformation switch induced by coadsorption with human serum albumin. Langmuir 27:312–319
19. Jiang X, Tong M, Li H, Yang K (2010) Deposition kinetics of zinc oxide nanoparticles on natural organic matter coated silica surfaces. J Colloid Interface Sci 350:427–434
20. McCubbin GA, Praporski S, Piantavigna S, Knappe D, Hoffmann R, Bowie JH, Separovic F, Martin LL (2011) QCM-D fingerprinting of membrane-active peptides. Eur Biophys J 40: 437–446
21. Rayaprolu V, Manning BM, Douglas T, Bothner B (2010) Virus particles as active nanomaterials that can rapidly change their viscoelastic properties in response to dilute solutions. Soft Matter 6:5286–5288
22. Ayela C, Roquet F, Valera L, Granier C, Nicu L, Pugnière M (2007) Antibody–antigenic peptide interactions monitored by SPR and QCM-D. A model for SPR detection of IA-2 autoantibodies in human serum. Biosens Bioelectron 22:3113–3119
23. Vikinge T, Hansson K, Liedberg P, Lindahl T, Lundström I, Tengvall P, Höök F (2000) Comparison of surface plasmon resonance and quartz crystal microbalance in the study of whole blood and plasma coagulation. Biosens Bioelectron 15:605–613
24. Aberl F, Wolf H, Kößlinger C, Drost S, Woias P, Koch S (1994) HIV serology using piezoelectric immunosensors. Sensors Actuators B 18–19:271–275
25. Andersson J, Ekdahl KN, Larsson R, Nilsson UR, Nilsson B (2002) C3 Adsorbed to a polymer surface can form an initiating alternative pathway convertase. J Immunol 168:5786–5791
26. Andersson J, Larsson R, Richter R, Ekdahl K, Nilsson B (2001) Binding of a model regulator of complement activation (RCA) to a biomaterial surface: surface bound factor H inhibits complement activation. Biomaterials 22:2435–2443
27. Rodahl M, Höök F, Kasemo B (1996) QCM operation in liquids: an explanation of measured variations in frequency and Q factor with liquid conductivity. Anal Chem 68: 2219–2227
28. Höök F, Rodahl M, Brzezinski P, Kasemo B (1998) Energy dissipation kinetics for protein and antibody-antigen adsorption under shear oscillation on a quartz crystal microbalance. Langmuir 14:729–734
29. Höök F, Rodahl M, Kasemo B, Brzezinski P (1998) Structural changes in hemoglobin during adsorption to solid surfaces: effects of pH, ionic strength, and ligand binding. Proc Natl Acad Sci USA 95:12271–122276

30. Rodahl M et al (1997) Simultaneous frequency and dissipation factor QCM measurements of biomolecular adsorption and cell adhesion. Faraday Discuss 107:229–246
31. Brash JL, Horbett TA (1995) In proteins at interfaces II. In: Brash JL, Horbett TA (eds) ACS symposium series 602, Washington DC, 1995
32. Höök F, Vörös J, Rodahl M, Kurrat R, Böni P, Ramsden J, Textor M, Spencer N, Tengvall P, Gold J, Kasemo B (2002) A comparative study of protein adsorption on titanium oxide surfaces using *in situ* ellipsometry, optical waveguide lightmode spectroscopy, and quartz crystal microbalance/dissipation. Colloids Surf B 24:155–170

Chapter 18

Central Nervous System Distribution of the Poxviral Proteins After Intranasal Administration of Proteins and Titering of Vaccinia Virus in the Brain After Intracranial Administration

Amod P. Kulkarni, Dhirendra Govender, Lauriston A. Kellaway, and Girish J. Kotwal

Abstract

Poxviral proteins are known to interact with the immune system of the host. Some of them interact with the transcription factors of the host, whereas others interact with the components of the immune system. Vaccinia virus secretes a 28.8-kDa complement control protein (VCP), which is known to regulate the complement system. This protein helps the virus to evade the immune response of the host. Such viral proteins might also prove beneficial in the treatment and prevention of the progression of the disorders, where up-regulation of the complement system is evident. VCP has been shown experimentally to be effective in protecting tissues from inflammatory damage in the rodent models of Alzheimer's diseases (AD), spinal cord injury, traumatic brain injury, and rheumatoid arthritis. Not only VCP, but also other poxviral proteins could be used therapeutically to treat or prevent the progression of the brain disorders, where the immune system is inadequately controlled. However, being a protein that cannot traverse the brain barrier because of its size, delivery of such proteins to the central nervous system (CNS) could be a limiting factor in their usefulness as CNS therapeutics. In this chapter, we show methods for the intranasal route of administration of a protein and show ways to detect its distribution in the cerebrospinal fluid (CSF) and to the different parts of the brain. These protocols can be extended to examine the distribution of viral antigens in the brain. A protocol is also included to quantitate vaccinia virus in different segments of the brain after intracranial administration of the virus.

Key words: Vaccinia virus protein, Poxvirus, Intranasal administration, Cerebrospinal fluid, Immunohistochemistry, Brain, Central nervous system, Intracranial administration, Viral brain distribution

1. Introduction

In many inflammatory disorders, the immune system is known to be inappropriately activated. Up-regulation of the complement system, which forms an important part of innate immune response,

Stuart N. Isaacs (ed.), *Vaccinia Virus and Poxvirology: Methods and Protocols*, Methods in Molecular Biology, vol. 890,
DOI 10.1007/978-1-61779-876-4_18, © Springer Science+Business Media, LLC 2012

is evident in many disorders of the brain such as spinal cord injury (SCI), traumatic brain injury (TBI), and Alzheimer's disease (AD) (1–3). The complement system is also inappropriately activated in peripheral disorders, such as rheumatoid arthritis and atherosclerotic lesion formation (4, 5). In addition to the complement system, the cytokines and chemokines, which also form an important part of the immune system, are known to be up-regulated in many inflammatory disorders (6). The immune system is also activated during organ transplantation. In such inflammatory disorders, regulation of the components of the immune system is desirable. Many novel therapies currently being developed for the treatment and prevention of progression of these disorders are, therefore, aimed at modulation of the immune system.

Poxviruses secrete many proteins that enable them in establishing infection in the host through their ability to evade the immune response of the host. Some of these proteins like cytokine response modifier E (CrmE; 18 kDa) encoded by cowpox virus are known to mimic the human cytokine receptors and thereby evade the immune response (7). Vaccinia virus encodes a 28.8-kDa vaccinia virus complement control protein (VCP), which is known to regulate the host complement system (8–10). Some of these immunomodulatory proteins of poxviral origin may offer potential in developing novel strategies to modulate the immune system.

The peripheral disorders, such as atherosclerosis and rheumatoid arthritis, are relatively easy to treat as the drugs do not have to cross the blood–brain barrier (BBB) to reach the site of action. The poxviral proteins could be used to treat these disorders without any modification. Indeed, VCP was found to be effective in the treatment of rheumatoid arthritis and atherosclerosis after being administered systemically without any modification. VCP and other poxviral proteins could be utilized for their ability to modulate the immune system in many disorders of the CNS in addition to these peripheral disorders. VCP was found to be effective in the prevention and treatment of neuroinflammatory disorders in a number of rodent models after being intrathecally or directly administered into the brain (8–14). However, the direct administration of substances into the brain is an invasive procedure and thus we determined if VCP could be administered noninvasively and detected in CSF and brain tissue. Many proteins and peptide molecules have previously been delivered to the brain using the intranasal route of administration (15–17). NGF, a 20-kDa protein, when delivered intranasally, was found to improve cognitive performance in mice (18). Horseradish peroxidase (HRP; 40 kDa) as well as HRP coupled to wheat germ agglutinin (WGA-HRP; 62 kDa) have previously been delivered to the CNS using this route of administration (19–21). VCP (28.8 kDa) is smaller than HRP (40 kDa) and some other proteins that have been delivered to the CNS via this route. Therefore, since the relative size of many poxviral proteins is small, there is good likelihood that intranasally administered proteins would reach the brain. However, one of the limiting factors for using the intranasal route of administration is the degradation of proteins by the intranasal enzymes. Kulkarni et al. recently reported delivering VCP to the CNS via intranasal administration (22). The intranasal administration of truncated VCP (tVCP), optimization of the protocol to collect the cerebrospinal fluid (CSF), transcardial perfusion, collection of the brain tissue, optimization of an ELISA, as well as the immunohistochemistry protocol for the detection of VCP are discussed in this chapter. The protocols outlined might prove useful in studying the CNS distribution of poxviral proteins.

In addition to intranasal administration of protein and protocols to detect the protein in the CNS, this chapter also describes intracranial administration of vaccinia virus and quantification of the virus in different segments of the brain.

2. Materials

1. VCP or tVCP previously expressed in *Pichia pastoris* and purified.
2. Rats: Wistar rats (300 g).
3. Thermoregulatory heating pad.
4. Thin flexible gel-loading pipette tips.
5. Dental needles (27G × 1/2"[0.4 × 13 mm]).
6. Thick Technicon tubing with a small inner diameter that fits tightly on a dental needles.
7. Stereotaxic apparatus for rats.
8. The Rat brain atlas (Paxinos and Watson 1982).
9. Microscope.
10. Ketamine/xylazine anesthetic: Mixture of ketamine hydrochloride (Anaket-V; 90 mg/kg) and xylazine (Rompun 2%; 10 mg/kg) prepared by mixing 9 ml of 100 mg/ml of ketamine with 5 ml of 20 mg/ml of xylazine.
11. Endotoxin-free normal saline and endotoxin-free double-distilled deionized water.
12. Hamilton syringe (50 μl).
13. Syringe pump model 341A (e.g., SAGE thermoregulatory instruments).
14. ELISA plates.
15. Wash buffer: 0.1% Tween-20 in PBS.
16. Blocking buffer: 0.5% Tween-20 in PBS.

17. Endogenous peroxidase blocking solution: 0.6% of a 30% H_2O_2 in 2% methanol in PBS, pH = 7.2 (see Note 1).
18. TMB substrate.
19. 1 M phosphoric acid solution.
20. Sodium pentobarbitone.
21. 50-ml syringes.
22. Blunted 18-gauge needle (see Note 2).
23. 4% paraformaldehyde solution: Dissolve 4 g of paraformaldehyde in 100 ml of 0.10 M PBS, pH 7.2, along with one pellet of sodium hydroxide pellets and heat to 60–70°C with constant stirring (see Note 3).
24. Transcardial perfusion set, clamps, artery forceps and cotton (see Note 4 and Subheading 3.5.1).
25. Wax pastilles, melting point 56–58°C (Saarchem supplied by Merck).
26. Microtome.
27. Coated histopack glass slides.
28. Xylol or xylene.
29. 2–3% H_2O_2 in PBS.
30. 50% methanol in PBS.
31. 0.05% trypsin: 50 mg of bovine pancreas trypsin in 10 ml of double-distilled deionized water and stored at –20°C.
32. 1% calcium chloride: 1 g of calcium chloride in 1 L of double-distilled deionized water.
33. Trypsin in calcium chloride solution. This solution should be prepared freshly just before the experiment by diluting mixture of 1 ml of 0.05% trypsin and 1 ml of 1% calcium chloride to 10 ml with water after adjusting pH to 7.4 with NaOH.
34. Goat serum.
35. Wash buffer II: 0.05% Tween-20 in PBS.
36. Avidin or streptavidin.
37. DAB substrate kit.
38. 5% copper sulfate solution: 5 g of copper (II) sulfate or cupric sulfate in 100 ml deionized double-distilled water. (The solution needs to be prepared freshly just 30 min prior to use.)
39. Mayer's Hematoxylin: 1 g hematoxylin, 50 g potassium or aluminum alum dissolved in 1 L of deionized double-distilled water.
40. Scott solution: 2 g sodium hydrogen carbonate and 20 g magnesium sulfate in 1 L of deionized double-distilled water.
41. Entellan mounting medium.
42. BALB/c mice.
43. Methoxyflurane.

44. Insulin syringe, 2/10-cc 29.5-gauge needle.
45. Wild-type vaccinia virus and vGK5 (Gift from Bernard Moss).
46. BSC-1 cells: African green monkey cells (ATCC).

3. Methods

Since there is uncertainty about the uptake or distribution of the poxviral protein after intranasal administration, the protein can directly be administered into various brain structures for comparison. It is important that such positive controls are available for ELISA and immunohistochemistry work. Thus, when a CNS distribution pattern is not known, rats should be divided into four groups for intracranial injections. In the current protocol, VCP was administered directly into the left lateral ventricle (four rats), the olfactory lobe (one rat), and into the hippocampus (one rat). For the intranasal administration of protein, rats were given the protein (e.g., VCP and tVCP) or saline intranasally.

3.1. Intranasal Administration of Proteins into Rats

1. Anesthetize the rat with 1 μl/g of body weight (either i.m./ i.p.) with ketamine/xylazine anesthetic. For example, for a rat weighing 300 g, the volume of anesthetic mixture will be 300 μl.
2. Once the rat is surgically anesthetized, position it on its back on a heater pad with the head positioned 70–90° angle to the surface (23) (see Note 5).
3. Using thin flexible gel-loading pipette tips, administer 50–70 μl of VCP, tVCP, or saline into both nostrils sequentially over a period of 20 min to the respective treatment groups (see Note 6).
4. Collect the CSF and brain samples as described in the sections below.

3.2. Intracranial Administration of VCP into the Rat Brain (see Note 7)

1. Anesthetize the rats using ketamine/xylazine anesthetic (see Subheading 3.1, step 1).
2. Once the rat is surgically anesthetized and its head shaven, fix its head in the stereotaxic apparatus with the ear bars. Once the rat's head is positioned in the stereotaxic apparatus, mark the position of the interaural line (IAL) on the rat cranium as a reference point (see Note 8).
3. Remove the needle holder from the stereotaxic apparatus and fix a dental needle into a specific holder attached to the manipulator so that it can be accurately guided to the cisterna magna. Connect tubing connected to a Hamilton syringe fixed to a specified syringe pump.

Table 1
Stereotaxic coordinates for VCP administration

Sr	Part of the brain	AP	ML	DV
1.	Lateral ventricle	7.7	1.6	3.1
2.	Hippocampus	4.48	2.2	2.8
3.	Olfactory lobe	13.7	1.6	3.1

Brain areas, and the stereotaxic coordinates (Anteroposterior [AP], Mediolateral [ML], and Dorsoventral [DV]) used for the direct administration of VCP via 27G dental needle into various parts of the brain using IAL as a reference line (Sr represents serial number)

4. Draw normal saline into the syringe via the needle followed by a small amount of air to serve as a marker and to prevent mixing.
5. Draw solution of protein (or saline) to be administered intracranially into the syringe.
6. Using the brain coordinates shown in Table 1 and with reference IAL, slowly administer up to a 10 μl volume of protein solution (in this example, VCP at (25 μg/10 μl) into the left lateral ventricle of each rat. The minute bubble in the syringe/tubing serves as a visual marker to track delivery of the drug solution.
7. If the CNS distribution of the protein is not known, directly administer the protein into different parts of the brain such as hippocampus and olfactory lobe using the brain coordinates shown in Table 1. Intracranial administration of the protein into various regions serves as a control to compare the distribution pattern of the compound once it has been administered intranasally.

3.3. Collection of CSF Samples

1. After the intracranial or intranasal administration of proteins, locate the posterior ridge of the dorsal cranium. This can be felt underneath the skin using the blunt end of the scalpel or seen once the skin of the scalp is reflected laterally from the midline incision.
2. Similar to Subheading 3.2, step 3, set up the stereotaxic apparatus with a dental needle, but use thick Technicon tubing and tuberculin syringe (see Note 9).
3. Position the dental needle 1–1.2 mm posterior to this point, 0.8–1 mm mediolaterally, and insert needle ventrally to a depth of about 9 to 13 mm (depending upon the stereotaxic setup and position of the head; see Note 10) or until it punctures the atlanto-occipital membrane above the cisterna magna and CSF

appears in the tubing with gentle suction. Collect the CSF very slowly (see Notes 9, 11, and 12).

4. To collect samples at various time points, maintain the rat on anesthesia, and allow the needle to remain in the same position while on the stereotaxic apparatus (see Notes 13 and 14).
5. Transfer the CSF to a microfuge tube, immediately spin the tube at 3,200 rpm for a minute to spin down any contaminating blood, and remove the clear supernatant.
6. Store the samples at –20°C (or –80°C for long-term storage) for further analysis.
7. After collection of the CSF samples from the rats belonging to different treatment groups at various time points are collected from the rat, one can sacrifice the rat and harvest the brain as described in Subheading 3.5.

3.4. Detection of VCP and tVCP in CSF Using ELISA

One needs to determine the sensitivity of the technique, optimize the dilutions of primary and secondary antibodies, check the antibodies for cross-reactivity in the CSF samples, and increase the specificity with reduced background staining.

3.4.1. Optimization of the ELISA Protocol

1. Dilute the stock solutions of the proteins under investigation (e.g., VCP or tVCP) in PBS or other solvent to obtain a range of concentration (e.g., 1, 0.1, 0.2, 0.5, 0.01, 0.02, 0.05, and 0.005 ng/μl).
2. Coat the ELISA plate with 100 μl of each solution and incubate overnight at 4°C with agitation.
3. To check the cross-reactivity of the primary antibodies with the control CSF samples, coat individual ELISA plate wells with at least three control CSF samples collected from different rats (100 μl each) and incubate overnight at 4°C with agitation.
4. After overnight incubation (~16 h), discard the coating solution, and remove any drops by gently tapping the plates on the filter paper.
5. Wash the wells with wash buffer three times (see Note 15).
6. Treat plates with blocking buffer and incubate for 60–90 min to reduce the background or nonspecific binding (see Notes 16 and 17).
7. Wash the wells three times and remove any drops of the wash solution by gentle tapping on the filter paper.
8. If there is concern about endogenous peroxidase activity in the CSF, one should inactivate endogenous peroxidase activity by washing the plates three times with 200 μl endogenous peroxidase blocking solution (see Note 1 and Subheading 3.4.2).

9. Incubate the wells with the primary antibodies (e.g., rabbit-anti-VCP (R-Ig) or chicken-anti-VCP (Ch-Ig)) at different concentrations (e.g., 1:1,000, 1:2,000, and 1:4,000 in 0.1% Tween-20) for about 4 h at room temperature (RT) (see Notes 18 and 19).
10. Remember to include wells that do not have primary antibody added to check for cross-reactivity of secondary antibodies with the proteins under investigation (e.g., VCP and tVCP).
11. Wash the wells with wash buffer three times and then add the appropriate conjugated horseradish peroxidase secondary antibodies (1:10,000) and incubate for about 2 h at RT (see Note 19).
12. Add 100 μl TMB substrate to each well.
13. Allow blue color to develop for about 15–20 min, and then add 100 μl of 1 M phosphoric acid to the solution to stop the reaction. Mix gently and thoroughly.
14. Transfer the yellow-colored solution (200 μl) to another 96-well plate, measure the optical density at 450 nm (OD_{450}), and plot the results using a graph as shown in Fig. 1.

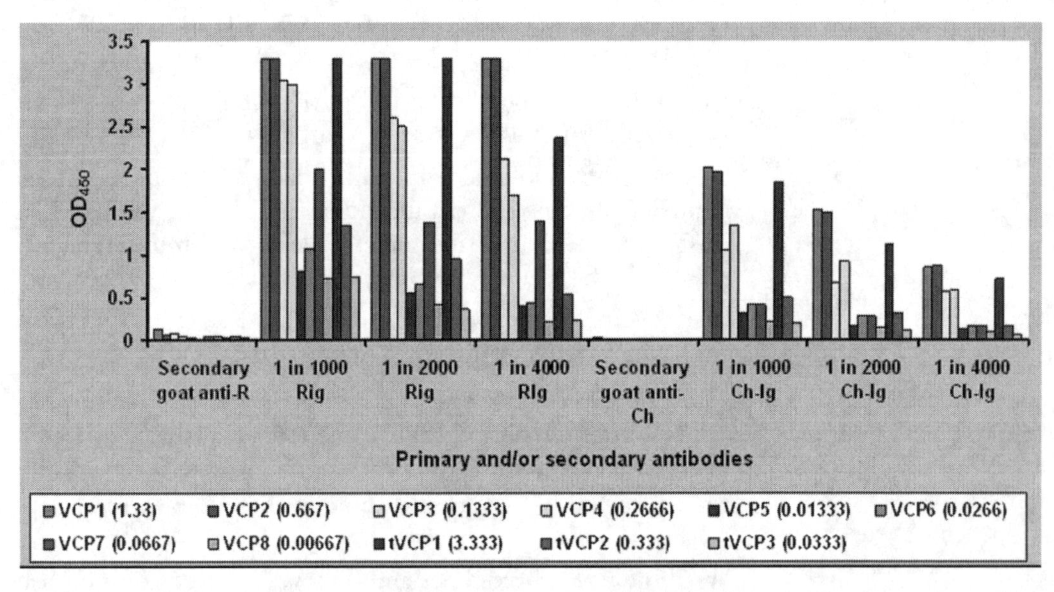

Fig. 1. Optimization of the ELISA protocol for VCP and tVCP detection in CSF. OD_{450} values as an index of peroxidase activity (*Y*-axis) are plotted against 1:10,000 dilutions of the secondary antibodies (goat-anti-R and goat-anti-Ch) and the concentrations of R-Ig and Ch-Ig (1:1,000, 1:2,000, and 1:4,000). VCP1 to VCP8 and tVCP1 to tVCP3 shown in the legend represent different dilutions of VCP and tVCP (*R* rabbit, *Ch* chicken). Their concentrations (ng/μl) are shown in the parentheses of legends. As shown in this figure, the R-Ig is more sensitive than Ch-Ig, but the sensitivity is attributed to several fold higher concentrations of R-Ig used. The concentration below 0.00666 ng/μl can be detected using the primary antibodies at all dilutions. There is nonspecific adsorption of the secondary antibody on VCP and tVCP to a very small extent and the adsorption can be directly correlated to the concentration of VCP and tVCP.

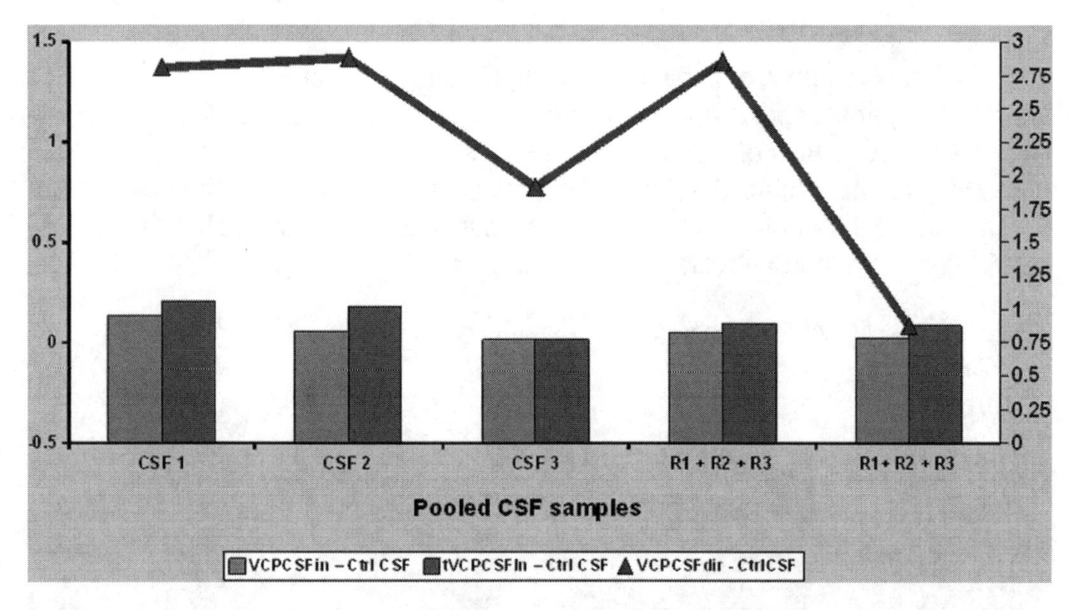

Fig. 2. Detection of VCP and tVCP in CSF samples after intranasal and intracranial inoculations. OD_{450} values (left *Y* axis) of intranasally administered protein (VCPCSFin and tVCPCSFin) of pooled CSF samples at different time intervals (CSF1, CSF2, and CSF3) as well as from two sets of three rats each (R1 + R2 + R3 and R4 + R5 + R6) from each group (*X* axis). The right-sided *Y*-axis shows the OD_{450} values of VCP in CSF after direct intracranial inoculation of protein (VCPCSFdir; line graph). The left- and right-sided *Y*-axes values have different scales since the values of VCPCSFdir were much higher than those of VCPCSFin and tVCPCSFin. The values shown in the graph are corrected OD_{450} obtained by subtracting OD_{450} values of Ctrl CSF samples at the same interval or same individual group sets. OD_{450} for the detection of the VCP and tVCP in CSF samples after ELISA suggested that the OD_{450} of the treatment groups differs from that of control groups. After intranasal administration, the OD_{450} was high at the first time interval, and then decreased at the subsequent time intervals. At all intervals, the OD_{450} of CSF from animals that were given a protein treatment was higher than that of Ctrl CSF. These trends were similar for tVCPCSFin, but the difference between the OD_{450} at the first and second intervals was higher than that of VCPCSFin at the same time intervals. For VCPCSFdir, the OD_{450} was several fold higher than that of the intranasal treatment groups at all time intervals. For the first two time intervals after intracranial inoculation, the OD_{450} values were off scale indicating that the concentrations were much higher during the first two intervals. While not shown here, when the OD_{450} values of the CSFctrl were subtracted from the OD_{450} values of the treatment groups, it was found that the difference was higher for tVCP than that for VCP, which suggested that tVCP was delivered to a greater extent than VCP. This figure is reproduced from ref. 22 with permission from Bentham Science Publishers, copyright ©2011.

3.4.2. Detection of VCP and tVCP in CSF by ELISA

1. The CSF samples drawn from the three rats belonging to the same treatment group are pooled and 100 μl of the pooled sample is added to wells of an ELISA plate and the optimized ELISA protocol followed. See Fig. 2 for example of results.
2. The CSF samples can be checked for the endogenous peroxidase activity of samples by adding 25 μl of pooled CSF samples to the ELISA plate wells (see Note 20).
3. As a control, add 20 μl of control CSF in triplicate to wells and then add 5 μl of PBS or the proteins under investigation (e.g., VCP or tVCP at 0.5 ng/μl).
4. Incubate the plates at 30°C for 30 min.
5. Add 100 μl of TMB substrate mix to each well as in Subheading 3.4.1, step 13.

Table 2
Endogenous peroxidase activity of CSF samples and effect of VCP and tVCP on peroxidases. When tVCP and VCP were added to CSFCtrl and OD450 were measured, OD450 of VCP-treated CSFCtrl samples was less than that of the untreated samples (Table 2). When tVCP tt the same concentration, the OD450 of CSFCtrl samples treated with tVCP was also less than that of CSFCtrl samples, but greater than that of VCP-treated CSFCtrl samples

Sr	CSF samples	CSF ctrl	CSF ctrl + VCP (5 μl/25 μl)	CSF ctrl + tVCP (5 μl/25 μl)	VCPCSFin	tVCPCSFin	VCPdir
1.	CSF1	0.649	0.579	0.650	0.440	0.230	0.428
2.	CSF2	0.498	0.504	0.515	0.473	0.155	0.217
3.	CSF3	NA	NA	NA	0.016	0.103	0.286
4.	R1 + R2 + R3	0.859	0.795	0.813	0.348	0.046	0.290
5.	R4 + R5 + R6	0.851	0.782	0.788	0.822	0.899	0.249

The OD_{450} after the peroxidase assay to measure the peroxidase activity of CSF samples from the different treatment groups. Some of the samples not subjected to analysis are indicated as NA

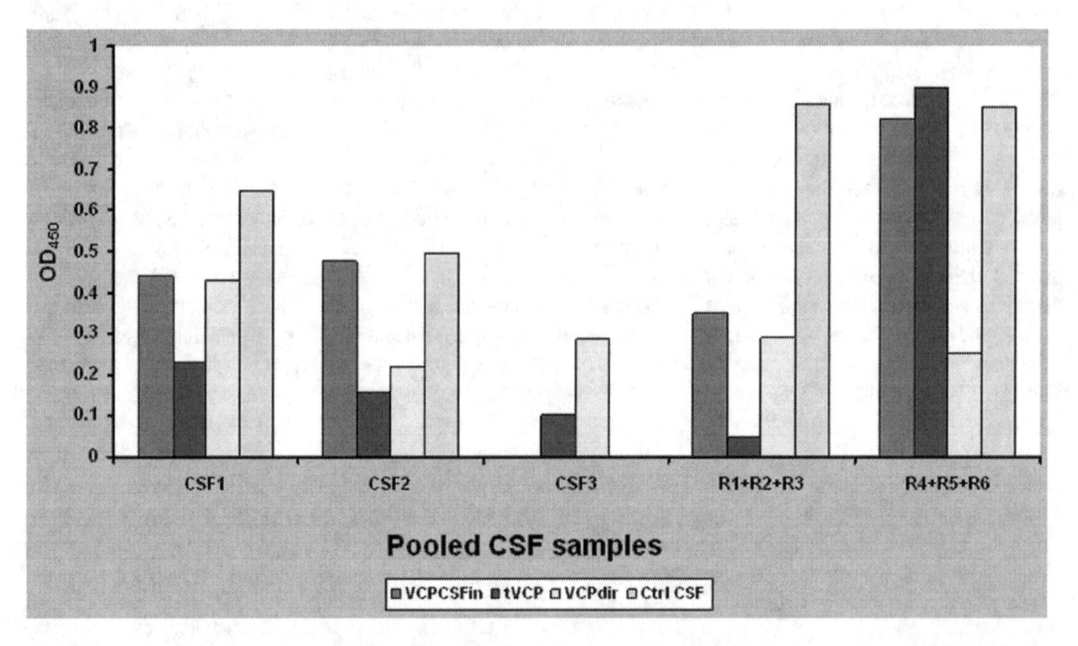

Fig. 3. Endogenous peroxidase activity of CSF samples from treatment groups. OD_{450} indicating endogenous peroxidase activity (*Y* axis) of pooled CSF samples collected at different time intervals (CSF1, CSF2 and CSF3) and pooled CSF samples from two different sets of three rats each (R1 + R2 + R3 and R4 + R5 = R6) from the treatment groups (*X* axis). The order of OD_{450} of pooled samples from different treatment groups was Ctrl CSF > VCPCSFin > VCPdir > tVCPCSFin.

6. Stop the reaction after 10–15 min by adding 75 μl of 1 M phosphoric acid to each well, and record the OD at 450 nm. See Table 2 and Fig. 3 for example of results.
7. If endogenous peroxidase activity is found, then proceed with detection of VCP in CSF by including an endogenous peroxidase blocking step.

Table 3
Quenching of endogenous peroxidases in CSF samples

Sr	CSF samples	CSF ctrl	VCPCSFin	VCPdir	tVCPCSFin
1.	CSF1	0.005	0.004	0.004	0.002
2.	CSF2	0.007	0.003	0.001	0.001
3.	CSF3	0.006	0.003	0.001	0.002
4.	R1, R2, R3	0.005	0.005	0.001	0.003
5.	R4, R5, R6	0.004	0.003	0.002	0.003

OD_{450} of CSF samples subjected to peroxidase assay after treatment with endogenous peroxidase blocking solution. After such treatment, the OD_{450} was reduced almost to zero indicating complete blocking of inherent or endogenous peroxidase activity

8. Add 40 μl of CSF samples into wells of an ELISA plate and incubate at 4°C overnight.
9. Discard the samples after the overnight incubation, and wash the well three times with wash buffer.
10. Incubate the wells with 200 μl of endogenous peroxidase blocking solution for 10 min at RT to block the endogenous peroxidase activity (see Note 1).
11. Wash the wells with wash buffer three times.
12. Add 100 μl of TMB substrate to each well.
13. Measure the ODs at 450 nm after a 15-min incubation (see Table 3 for example of results).

3.5. Preparation of Brain Tissue for Immunohistochemistry

3.5.1. Transcardial Perfusion

1. After the collection of CSF samples, the rats can be subjected to transcardial perfusion to collect the brains for further analysis. For this, anesthetize the rats with the lethal dose of sodium pentobarbitone (100 mg/kg). Before starting the perfusion, ensure that the rat is not responsive to the footpad pinch test.
2. A perfusion system was constructed from an infusion set. Three pieces of tubing are needed: one about 15–20-cm long and the other two about 10-cm long. Connect these tubes to a three-way stopcock. Connect the open end of the long tubing to the hub of a blunted 18-gauge needle (see Note 2). Connect one open end of the tube to a 50-ml syringe containing PBS, and connect the open end of the other short tube to another 50-ml syringe containing 4% paraformaldehyde. Both these tubes are connected to a three-way stopcock. Set aside two to four 50-ml syringes filled with PBS and three to four 50-ml syringes filled with 4% cold paraformaldehyde (see Note 21).

3. Set the stopcock to inject PBS from appropriate syringe into the long tubing. Fill the tube with the PBS up to the stopcock, and then flush tubing with PBS to clear any paraformaldehyde and/or air bubbles.
4. Using a scalpel, incise to the skin aiming midline above the sternum and through this midline incision to expose the heart (see Note 22).
5. Hold apart the ribs and the skin with forceps to adequately expose the heart. Retractors could be used for this purpose.
6. While the heart is still beating, insert the blunted needle into the left ventricle of the heart and clamp it in position (see Notes 2 and 23).
7. First, perfuse the rat transcardially using 100–150 ml of 0.1 M PBS, pH 7.4, to remove the blood. It is best if the flow of PBS is slow and steady. The pressure on the syringe is applied until the heart swells with PBS. Then, immediately cut the right atrium to allow blood and saline to escape.
8. While maintaining steady pressure on the syringe, once the blood is cleared using the PBS infusions, immediately administer 200–250 ml of 4% paraformaldehyde slowly over 10–15 min. Spontaneous movements of the rat paws and tail (fasciculations) appear once 250 ml of formaldehyde is delivered. Round eyes and stiffening of body are signs of good perfusion (see Note 3).
9. Decapitate the rat. Expose the brain in situ, reposition in stereotaxic apparatus, and cut into appropriate blocks as shown in Fig. 4 for further processing (see Note 24).

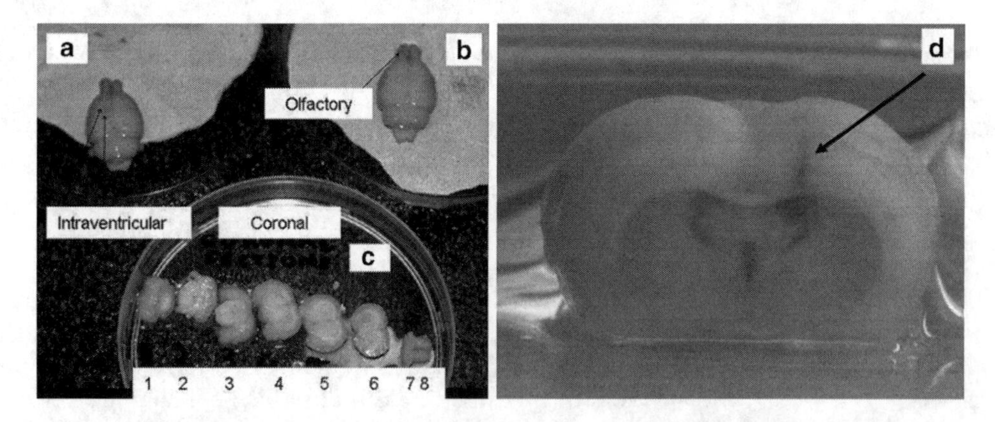

Fig. 4. Direct administration of VCP in the rat brain. Based on the needle track (*arrows*), shown are rat brains after stereotactic left lateral intraventricular (**a**) or olfactory lobe (**b**). The fixed rat brain was divided into different parts that are numbered from 1 to 8 (**c**) and later embedded in wax. Parts 7 and 8, which represent the olfactory lobes, were used as a positive control for studying the pattern of distribution in the rats to which VCP was administered intranasally. Brain block section showing direct administration of VCP into the left lateral ventricle is shown in (**d**). This figure is reproduced from ref. 22 with permission from Bentham Science Publishers, copyright ©2011.

Table 4
Brain tissue processing

Sr	Treatment	Exposure time (min)	Temperature (°C)
1.	96% Ethanol	10	31–40
2.	96% Ethanol	10	31–40
3.	Absolute alcohol	15	31–40
4.	Absolute alcohol	15	31–40
5.	Absolute alcohol	15	31–40
6.	Absolute alcohol	15	31–40
7.	Xylol	15	31–40
8.	Xylol	15	31–40
9.	Wax	20	60
10.	Wax	20	60
11.	Wax	25	60
12.	Wax	25	60

Dehydration in alcohol at different concentrations, xylol treatment, and processing in wax (*Sr* serial number)

10. Fix the whole rat brain in 4% paraformaldehyde buffer overnight and then cut into different blocks of tissue (see Note 3).
11. Process each block of tissue as shown in Table 4.
12. Embed the brain blocks in wax until the time they are used for the immunohistochemical staining. To embed the tissue in wax, the dehydrated tissue blocks are placed into molds for subsequent cutting on a microtome. The appropriate orientation of these blocks is important while placing them into the molds. One must ensure that all the blocks have the same orientation.
13. Molten wax is then poured into the mold, which hardens at room temperature.
14. These blocks are then used to cut 4–10 μm thin sections with a microtome.

3.5.2. Immunohistochemical Staining

1. Cut 4-μm-thick coronal brain sections from brain blocks in wax from different parts of the brain (Fig. 4) of different treatment groups (see Note 25).
2. Place cut sections on coated histopack glass slides and fix these sections on the slides by keeping them in oven at 42°C for an hour.

3. Dewax the sections by incubating in xylol for 10 min. Repeat.
4. Rehydrate section with alcohol with incubating sections in alcohol with increasing concentrations of water. Incubate in absolute alcohol for 1 min. Repeat two more times.
5. Then, incubate in 90% alcohol for 1 min. Repeat two more times.
6. Then, incubate in 70% alcohol for 1 min. Repeat two more times.
7. After this, subject the sections to 2–3% H_2O_2 in PBS or 50% methanol in PBS.
8. Treat these sections with trypsin in calcium chloride solution for the antigen retrieval (see Note 26).
9. Block sections using goat serum diluted 1 in 20 for 10 min (see Note 27).
10. After draining off the goat serum, incubate the sections with the primary antibody (in this example, chicken-anti-VCP diluted 1 in 166.33 (3 μl/500 μl) at RT for several hours (e.g., ~5 h) or overnight at 4°C in a humidity-controlled chamber.
11. After blocking step, wash sections three times with wash buffer II and then rinse once with PBS (0.1 M, pH 7.2).
12. Incubate the sections with the secondary antibody (in this example, goat-anti-chicken diluted 1 in 350–400) and incubate in a humidity chamber at RT for 1.5–2 h.
13. Wash sections three times in wash buffer II, and rinse with PBS (0.1 M, pH 7.2).
14. Incubate sections with avidin (1:400 to 1:600) or streptavidin (1:400).
15. Add DAB substrate kit (freshly prepared by mixing DAB with substrate diluent), and allow the color to develop for 10–15 min.
16. Wash the sections and treat with 5% copper sulfate solution for 3 min and counterstain with hematoxylin for 30 s to 1 min.
17. Treat with Scott solution for 1 min and wash under tap water.
18. Dehydrate section by going through successive incubations with alcohol containing less and less water (70%, 90%, and absolute EtOH, and then finally xylol).
19. Mount coverslips using entellan mounting medium.
20. Take digital photographs at various magnifications (see Figs. 5 and 6).

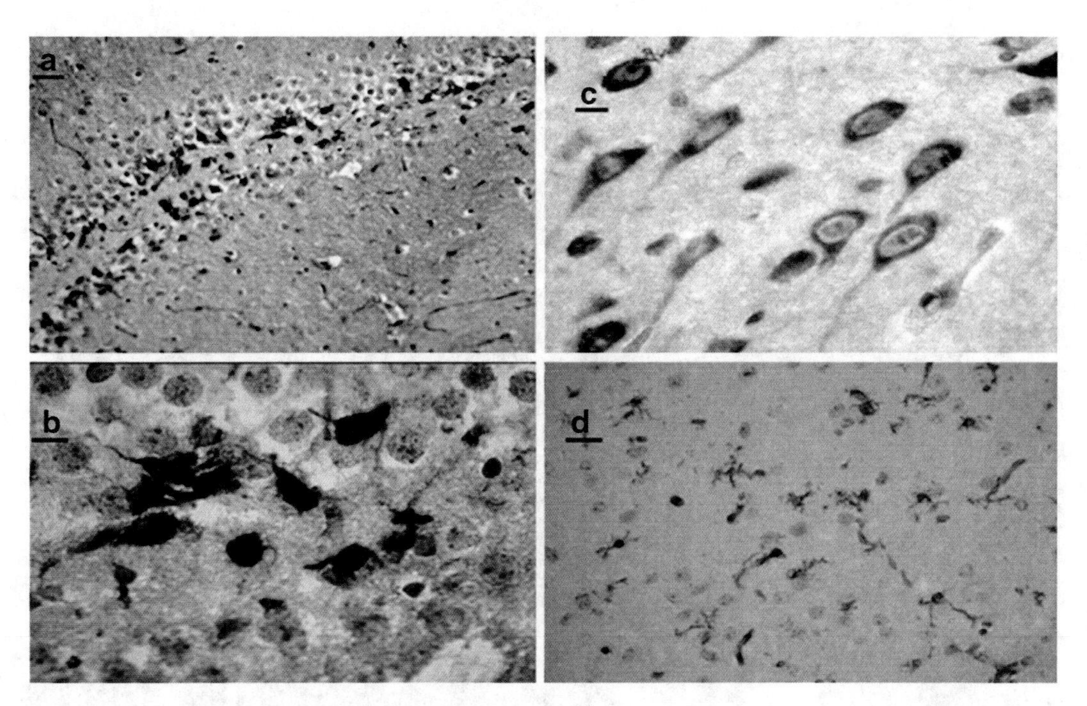

Fig. 5. Immunostaining of brain sections from direct inoculation of VCP into the rat brain. Shown is the immunostaining of brain sections from the hippocampus at various magnifications: (**a**) 10×; (**b**, **c**) 40×; (**d**) is brain area 5 (Fig. 4) at 20×. Dark brown staining of the hippocampus or other brain region against much lighter background is indicative of positive staining. Panel (**b**) shows dark brown staining of the whole neuron and panel (**c**) shows light staining that probably suggests some intraneuronal staining. This figure is reproduced from ref. 22 with permission from Bentham Science Publishers, copyright ©2011.

3.6. Evaluation of Vaccinia Virus Replication in Brain Tissue and Spread in the Brain

In this final section, we show how to inoculate vaccinia virus intracranially into mice and titer the virus from brain tissue. As an example, we use vaccinia viruses with differing virulence (determined previously by LD50 studies (24)). One virus is the wild-type vaccinia virus (strain WR). The attenuated virus is a recombinant vaccinia virus with the N1L gene deleted (vGK5) (25). Deleting N1L gene expression results in a significant decrease of vaccinia virus replication in mouse brain (24). The protocol describes intracranial injection of virus and then removing the brain and dissecting it into four quarters and obtaining the titers in each quarter.

1. Anesthetize 3-week-old BALB/c mice (n=6–8).
2. Remove the hair around the site of the injection on the head with a clipper and swab with 70% alcohol.
3. Holding a Hamilton syringe by wrapping all the fingers around it in a vertical position, inject 20 μl (100 pfu) of the accurately titered virus intracranially into the left cerebral hemisphere.
4. Euthanize the mice at various time points post infection (24, 48, and 96 h), and perfuse pericardially with 10 ml of PBS as described in Subheading 3.5.1, step 7.

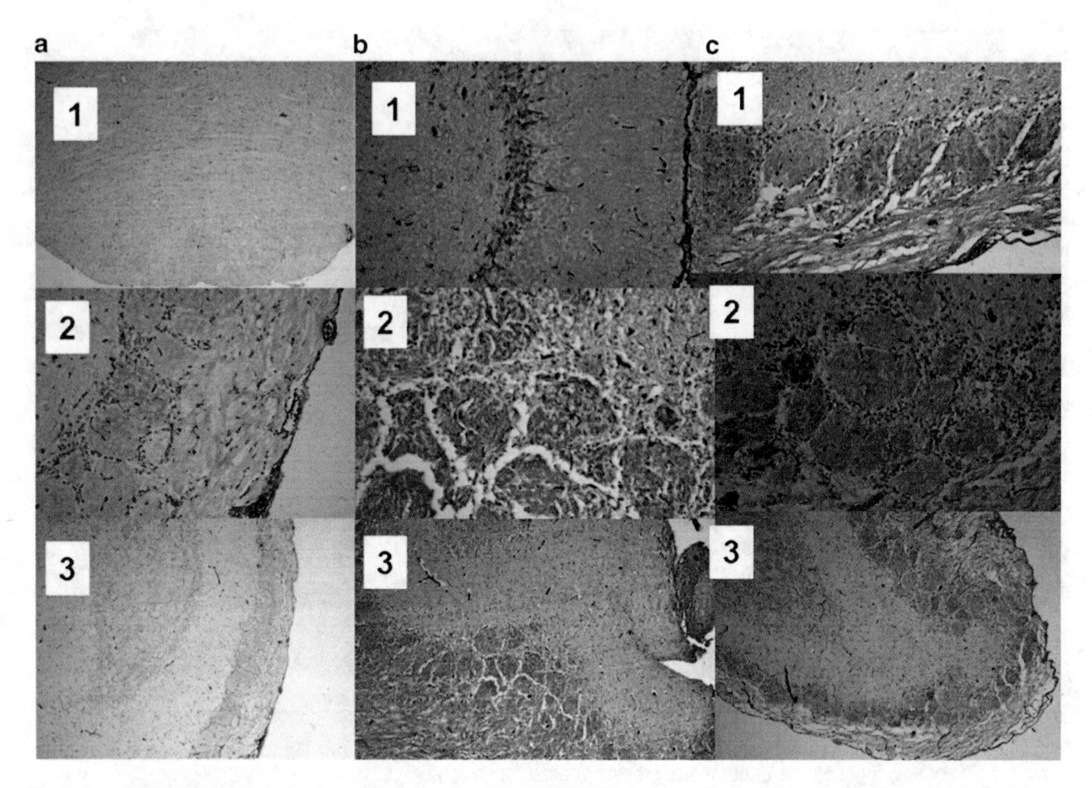

Fig. 6. Immunohistochemical staining of brain sections from VCP after intranasal administration. Immunostaining of 4-μm sections from the tissue block areas shown in Fig. 4. Panel (**a**), intranasal saline-treated control, where **a1** is area 6 (Fig. 4) [×4], and **a2** and **a3** are the olfactory lobe [×40] and [×10], respectively. Panel (**a**) serves as a negative staining control. Panel (**b1**) is a section of the rat brain, where VCP was directly administered into the hippocampus [×10] and serves as a positive staining control. **b2**–**c3** are area 7, 8, or olfactory lobes (Fig. 4) of the rat brain after VCP was administered intranasally. Different magnifications are shown (**b2**, **c1**, and **c2** = ×40; **b3** and **c3** = ×20). **b2** and **b3** and **c1**, **c2**, and **c3** represent glomerular staining of the area 7, 8, or olfactory lobe. A similar pattern of staining in the glomerular cell layer of the olfactory lobes of tVCP-treated rats was observed. The rats in which VCP was administered directly into the olfactory lobe of the brain showed a similar pattern of staining of the glomerular cell layer. This figure is reproduced from ref. 22 with permission from Bentham Science Publishers, copyright ©2011.

5. Remove the brains and slice into four quarters along the long axis vertically and horizontally about half way from the two ends (see Fig. 7a).
6. Homogenize each quarter in 2 ml of PBS and then sonicate.
7. Determine the titers as described before (26).
8. Plot the results as shown in Fig. 7.

Fig. 7. (continued) post infection of vGK5 were found to be significantly less in all brain areas (area 1, $^{*}P<0.0001$; area 2, $^{**}P<0.000008$; area 3, $^{***}P<0.0001$; and area 4, $^{****}P<0.00001$). (**d**) Virus titers of brain areas, (log 10)/g brain tissue, 4 days post infection. Viral titers 4 days post infection of vGK5 were found to be significantly less in all brain areas (area 1, $^{*}P<0.004$; area 2, $^{**}P<0.0009$; area 3, $^{***}P<0.001$; and area 4, $^{****}P<0.002$). Standard deviations are indicated on the graphs of **b**, **c**, and **d** (6–8 animals per group). This figure is reproduced from ref. 24 with permission from John Wiley and Sons, copyright ©2004.

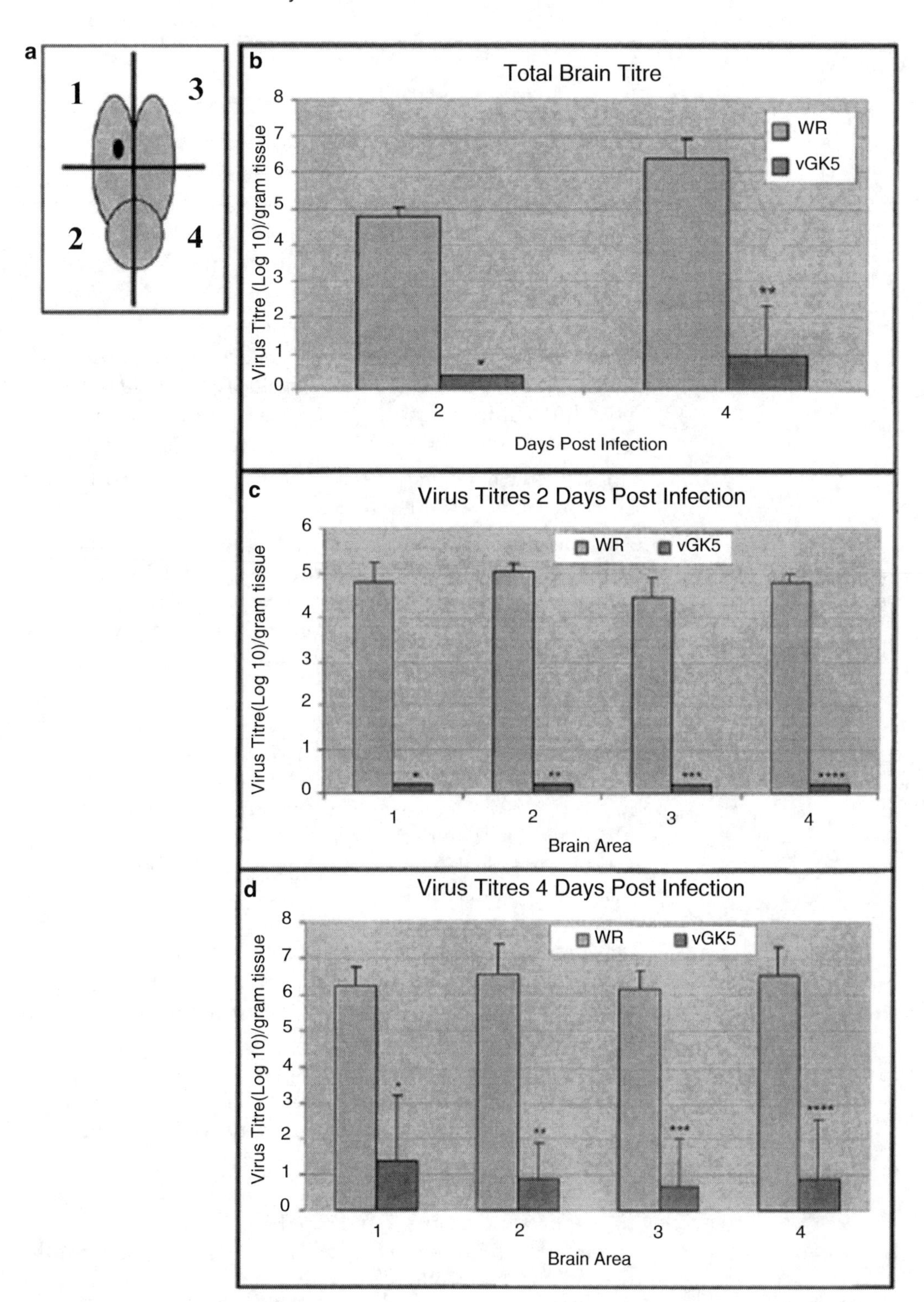

Fig. 7. Viral titers in mouse brain after intracranial inoculation. (**a**) Diagram of the location of injection (*black dot*) and the areas in which the brain was sectioned for virus titration. (**b**) Total virus titer, (log 10)/g brain tissue, 2 and 4 days post infection with either vGK5 or WR. Viral titers of vGK5 were found to be significantly less on days 2 and 4 post infection (*$P<0.00003$ and **$P<0.0008$, respectively). (**c**) Virus titers of brain areas, (log 10)/g brain tissue, 2 days post infection. Viral titers 2 days

4. Notes

1. Endogenous peroxidase activity can also be blocked without using methanol (27).
2. For transcardial perfusion, cut, smooth, and polish the end of the needle so that it does not puncture the other heart chambers. It is easy to insert such a blunted needle into the left ventricle of the heart with some practice and is the preferred method used by most of the laboratories.
3. Wear gloves while administering paraformaldehyde. It is recommended that blood and paraformaldehyde should be appropriately discarded in a waste container for proper disposal (and not flushed down the sink).
4. Antiseptic cream/spray is required if the rats are to be kept alive after the CSF withdrawal.
5. According to van den Berg et al. (23), the position of the rat is important while administering the protein and collecting the samples. The supine position with head angulated between 70 and 90° while administering the drugs intranasally is, thus, strongly recommended.
6. The intranasal administration of proteins could be a bit tricky. It is advisable to administer the protein (or control saline) very slowly and alternate between the left and right nostrils. To do this, 10 μl is loaded into a pipette at a time and administered into one nostril very slowly. The rat needs to be monitored for breathing. Once you have ensured that administering into the nostril is not blocking the airways and the rat is breathing normally, one can start administering the next 10 μl volume to the other nostril, adding a total volume of 50–70 μl. The tip of the gel-loading pipette should not be inserted deep into the nostril. The aim is for the exposure of the protein to the olfactory epithelium of the rats. If the pipette tip is inserted deep into the nostril, the solution is more likely to enter the airways.
7. The weight of rats in each group should be approximately 300 g for accurate comparison of brain structures using stereotaxic apparatus using brain atlas such as the one by Paxinos and Watson (28).
8. The IAL is defined as the imaginary line between the tips of the ear bars and its midpoint is used as the zero coordinate for the anterior/posterior plane and thus the zero position of the needle. For more information regarding the IAL and other reference points (Bregma and Lamda), please refer to the atlas by Paxinos and Watson (28). The midpoint of the IAL can be marked on the rat cranium for ease of marking the zero position of the needle in the midline.

9. Ensure that the tubing used to collect the CSF is sufficiently long so as to observe CSF entry. The thick tubing must fit exactly onto the 27-gauge dental needle as well as the needle of the insulin syringe. Thick Technicon tubing of the required diameter is needed to maintain appropriate pressure while withdrawing the CSF.
10. The length of the needle required to reach the cisterna is fairly constant from rat to rat (approximately 9–10 mm); however, it depends upon the degree of lowering of the incisor bar relative to the ear bar, hence tilting the head at an angle. The angle may change from rat to rat as the distance between the posterior ridge and cisterna magna may be different depending upon the size of the rat. Lower the incisor bar to an optimal level to maximally expose the atlanto-occipital membrane covering the cisterna magna for facilitating withdrawal of CSF. There are numerous variations of the method of CSF withdrawal depending upon the design of the experiment (23, 29). The chronic cannulation of the cisterna magna is possible for the serial collection of the samples; however, this method has a drawback of being susceptible to infection, blockade, and high risk of contamination with the red blood cells (23).
11. Collect the CSF very slowly. If the CSF is withdrawn too quickly, the rat may suddenly go into coma and die.
12. The maximum volume of CSF that can be withdrawn from the rats is 40–60 μl from each rat at a time. The volume that can be collected is variable over time. The collection of CSF should be stopped after the first appearance of blood in the CSF.
13. Maintaining anesthesia by constant monitoring of the rat is required so that there is no movement.
14. For the ELISA analysis below, the CSF samples at different time intervals can be pooled from different rats from the control and treatment groups. The pooling of the samples increases the sample volume and reduces the number of samples, thereby making the analysis a bit easier. In the current protocol, in order to make the volume up to 100 μl, the samples from different time intervals (0–1 h and 1–2 h, 2–3 h and 3–4 h, and 4–5 h and 5–6 h or 5–8 h) were further pooled to give three different samples representing three different time intervals (0–2 h, 2–4 h, and 4–6 or 4–8 h). The remaining CSF samples at each time interval from each rat were mixed and the CSF samples mixed in such a manner were further pooled from three different rats (Rl = rat 1 + rat 2 + rat 3; R2 = rat 4 + rat 5 + rat 6) to give two different sets of CSF samples for the estimation of VCP level in the individual rats. The levels of VCP and tVCP in the CSF samples can be plotted in a graph as shown in Fig. 2.

15. 1% Tween-20 reduces the sensitivity, whereas 0.1 and 0.01% are known to increase the sensitivity as shown in a previous study (30). Therefore, the concentration of Tween-20 in the wash buffer or diluent buffer should be from 0.01 to 0.1%.
16. Use 0.5% BSA in 0.1% of Tween-20 as a blocking buffer to further reduce the nonspecific binding and retain the sensitivity.
17. Increase the time of incubation in blocking buffer from 60–75 min to 120–140 min to further block nonspecific binding sites.
18. We have found that our rabbit-anti-VCP antibody (R-Ig) is an excellent primary antibody due to its high sensitivity, and the optimum concentration of this R-Ig is 1:3,000. This keeps the sensitivity intact and minimizes nonspecific binding.
19. To further decrease background, the incubation time with the primary antibody can be reduced by an hour and that of the secondary antibody by half an hour.
20. Since peroxidase is very stable and thus can withstand the freeze and thaw cycles, the CSF samples can be stored at –20°C and subjected to a peroxidase assay for measuring the inherent peroxidase activity of these samples at any time.
21. Other cardiac perfusion protocols call for buffer containing HEPES and protease inhibitor cocktail.
22. Ensure that the heart does not get damaged due to sharp surgical instruments used to cut the ribs and skin to expose the heart.
23. While inserting the blunt end of the needle into the left ventricle, ensure that it does not puncture the intraventricular septum.
24. Make sure that the olfactory lobes, which are tiny, are intact. The proteins delivered intranasally may accumulate in this part of the brain, and special care must be taken while removing the olfactory lobes of rats.
25. While cutting, check the sections specifically for the appearance of the needle tract to confirm accurate placement within lateral ventricle or particular part of the brain. This is to confirm appropriate delivery of protein to the intended target area (Fig. 4b) and the olfactory lobe (Fig. 4d).
26. Formalin or paraformaldehyde treatment results in protein cross-linking. It is, therefore, necessary to expose the protein epitopes. Trypsin breaks the formalin or paraformaldehyde cross-linking and unmasks the antigens/epitopes, which can later be detected by the antibodies.

27. If the background staining is high, sections can be treated with 0.03% Triton X-100 to further reduce the nonspecific background staining. 1% BSA can also be added to normal serum and Triton X-100 to reduce the background staining.

Acknowledgments

The protocols were approved by the Faculty of Health Sciences Animal Ethics Committee of South Africa. The authors gratefully acknowledge Morea Petersen, Barbara Young (Human Biology), and Nafisa Ali (Anatomical Pathology) for their help with the immunohistochemistry protocol and cutting sections. LK is the recipient of the project funded by the South African Medical Research Council. APK is funded by the Claude Leon Foundation, UK, for the Postdoctoral research at the University of Cape Town. APK was supported through various fellowships by UCT for his Ph.D. studies.

References

1. Anderson AJ, Robert S, Huang W, Young W, Cotman CW (2004) Activation of complement pathways after contusion-induced spinal cord injury. J Neurotrauma 21:1831–1846
2. Stahel PF, Morganti-Kossmann MC, Kossmann T (1998) The role of the complement system in traumatic brain injury. Brain Res Brain Res Rev 27:243–256
3. Pasinetti GM (1996) Inflammatory mechanisms in neurodegeneration and Alzheimer's disease: the role of the complement system. Neurobiol Aging 17:707–716
4. Schubart AF, Ewald RW, Schroeder WC, Rothschild HJ, Bhatavadekar DN, Pullen PK (1965) Serum complement levels in rheumatoid arthritis. A longitudinal study of 43 cases with correlation of clinical and serological data including rheumatoid factor and thermolabile inhibitor of the F-II L.P. test. Ann Rheum Dis 24:439–450
5. Torzewski J, Bowyer DE, Waltenberger J, Fitzsimmons C (1997) Processes in atherogenesis: complement activation. Atherosclerosis 132:131–138
6. McGeer EG, Klegeris A, McGeer PL (2005) Inflammation, the complement system and the diseases of aging. Neurobiol Aging 1:94–97
7. Saraiva M, Alcami A (2001) CrmE, a novel soluble tumor necrosis factor receptor encoded by poxviruses. J Virol 75:226–233
8. Kotwal G, Moss B (1988) Vaccinia virus encodes a secretory polypeptide structurally related to complement control proteins. Nature 335:176–178
9. Kotwal G, Isaacs S, McKenzie R, Frank M, Moss B (1990) Inhibition of the complement cascade by the major secretary protein of vaccinia virus. Science 250:827–830
10. McKenzie R, Kotwal G, Moss B, Hammer C, Frank M (1992) Regulation of complement activity by vaccinia virus complementcontrol protein. J Infect Dis 166:1245–1250
11. Hicks RR, Keeling KL, Yang MY, Smith SA, Simons AM, Kotwal GJ (2002) Vaccinia virus complement control protein enhances functional recovery after traumatic brain injury. J Neurotrauma 19:705–714
12. Reynolds DN, Smith SA, Zhang YP, Mengsheng Q, Lahiri DK, Morassutti DJ et al (2004) Vaccinia virus complement control protein reduces inflammation and improves spinal cord integrity following spinal cord injury. Ann N Y Acad Sci 1035:165–178
13. Pillay N, Kellaway L, Kotwal G (2008) Early detection of memory deficits in an Alzheimers

disease mice model and memory improvement with Vaccinia virus complement control protein. Behav Brain Res 192:173–177

14. Kulkarni AP, Pillay NS, Kellaway LA, Kotwal GJ (2008) Intracranial administration of vaccinia virus complement control protein in Mo/Hu APPswe PS1δE9 transgenic mice at an early age shows enhanced performance at a later age using a cheese board maze test. Biogerontology 9:405–420
15. Gozes I (2001) Neuroprotective peptide drug delivery and development: potential new therapeutics. Trends Neurosci 24:700–705
16. Thorne RG, Frey WH (2001) Delivery of neurotrophic factors to the central nervous system: pharmacokinetic considerations. Clin Pharmacokinet 40:907–946
17. Talegaonkar S, Mishra PR (2004) Intranasal delivery: an approach to bypass the blood brain barrier. Indian J Pharmacol 36:140–147
18. Capsoni S, Giannota S, Cattaneo A (2002) Nerve growth factor and galantamine ameliorate early signs of neurodegeneration in anti-nerve growth mice. Proc Natl Acad Sci USA 99:12432–12437
19. Kristensson K, Olsson Y (1971) Uptake of exogenous proteins in mouse olfactory cells. Acta Neuropathol 19:145–154
20. Broadwell RD, Balin BJ (1985) Endocytic and exocytic pathways of the neuronal secretory process and trans-synaptic transfer of wheat germ agglutinin-horseradish peroxidase *in vivo*. J Comp Neurol 242:632–650
21. Thorne RG, Emory CR, Ala TA, Frey WH (1995) Quantitative analysis of the olfactory pathway for drug delivery to the brain. Brain Res 692:278–282
22. Kulkarni AP, Govender D, Kotwal GJ, Kellaway LA (2011) Modulation of anxiety behavior by intranasally administered vaccinia virus complement control protein and curcumin in a mouse model of Alzheimer's disease. Curr Alzheimer Res 8:95–113
23. van den Berg MP, Romeijn SG, Verhoef JC, Merkus FW (2002) Serial cerebrospinal fluid sampling in a rat model to study drug uptake from the nasal cavity. J Neurosci Methods 116:99–107
24. Billings B, Smith SA, Zhang Z, Lahiri DK, Kotwal GJ (2004) Lack of N1L gene expression results in a significant decrease of vaccinia virus replication in mouse brain. Ann N Y Acad Sci 1030:297–302
25. Kotwal GJ, Hugin A, Moss B (1989) Mapping and insertional mutagenesis of a vaccinia virus gene encoding a 13,800-Da secreted protein. Virology 171:579–587
26. Kotwal GJ, Abrahams MR (2004) Growing poxviruses and determining virus titer. Methods Mol Biol 269:101–112
27. Jha P, Smith SA, Justus DE, Kotwal GJ (2005) Vaccinia virus complement control protein ameliorates collagen-induced arthritic mice. Ann N Y Acad Sci 1056: 55–68
28. Paxinos G, Watson C (1982) The rat brain in stereotaxic coordinates, 2nd edn. Academic/Harcourt Brace Jovanovich publishers
29. Frankmann SP (1986) A technique for repeated sampling of CSF from the anesthetized rat. Physiol Behav 37:489–493
30. Halim ND, Joseph AW, Lipska BK (2005) A novel ELISA using PVDF microplates. J Neurosci Methods 143:163–168

INDEX

A

Accidents. *See* Biosafety
Activation 53, 60, 135, 199, 200, 260, 261, 274–276, 283, 289–291
Alignment. *See* Bioinformatics
Alternative selection
  blasticidin resistance gene 94
  gpt (xanthine-guanine-phosphoribosyl-transferase) 94
  hemagglutinin (A56) 94
  host range 38, 72–74, 94
  thymidine kinase (J2R) 94
  zeocin 94
ALVAC. *See* Viruses
Anesthesia
  avertin 156, 160, 164, 172, 220, 222, 228
  CO_2/O_2 195
  isoflurane 149, 150, 164, 165, 173
  ketamine 156, 179, 186, 187, 195, 196, 307, 309
  xylazine 156, 179, 186, 187, 195, 196, 307, 309
Annotation. *See* Bioinformatics
Antigen presentation 259–270
Antigen presenting cell (APC) 197, 213, 259–261
Apoptosis 260, 261, 267, 273, 274, 279, 284, 286
Area under the curve (AUC). *See* Calculation(s)
Avogadro's number. *See* Calculation(s)

B

Bacterial artificial chromosome (BAC) 37–56, 94
BALB/c. *See* Mice
Bioassay 265–267
Bioinformatics 233–257, 275
  alignment 233–235, 238–241, 243–249, 252, 253
  annotation 235–238, 241, 246, 247, 253, 254
  base-by-base (BBB) 235, 254
  BLAST 233, 237, 238, 248–250, 255
  dotplot 233, 235, 244, 245
  DOTTER 235, 244, 245
  genomic analysis 233–235, 237, 241, 242, 244, 254
  graphical user interfaces (GUI) 234
  homology 24, 25, 37, 47–49, 136, 250, 275–277
  JAVA 235–238, 241, 246, 248, 253, 255
  modeling 147, 250, 251, 300, 301
  next generation (nucleotide) sequencing (NGS) 253
  sequence query language (SQL) 235, 237, 238, 255
  tool 233–257
  viral genome organizer (VGO) 236, 241–244
  viral orthologous clusters (VOC) 233, 234, 237–241
Bioluminescence. *See* Luciferase
Biosafety
  accident 1–19, 238
  cabinet 1, 2, 18, 52, 150, 156, 173, 183, 193, 216
  decontamination 174, 183
  disinfect 2, 5, 8, 11, 18, 164, 178, 183, 202, 216
  disinfectants 2, 5, 13, 178, 183
  goggles 3
  mask 150, 173, 324
Bleeding, submandibular 188
Blood brain barrier (BBB) 235, 237, 241, 252–254, 306
Borna disease virus. *See* Viruses

C

Calculation(s)
  area under the curve (AUC) 167, 169-171, 175
  Avogadro's number 141, 294, 295, 298, 301
  molecular weight 80–82, 84, 115, 141, 224, 225, 237, 294, 299
  tissue culture infectious dose 50 ($TCID_{50}$) 50, 88, 89
Carboxymethylcellulose (CMC) 149, 155, 157, 179, 182, 186, 191, 195

Stuart N. Isaacs (ed.), *Vaccinia Virus and Poxvirology: Methods and Protocols*, Methods in Molecular Biology, vol. 890, DOI 10.1007/978-1-61779-876-4, © Springer Science+Business Media, LLC 2012

Cardiac perfusion 324
C57/C57Bl/6. *See* Mice
Cell-associated enveloped virus (CEV). *See* Viruses
Cell lines
 BHK-21 (baby hamster kidney cells) 61, 64, 67, 69, 71, 72, 74, 77, 81, 85, 87, 89, 93, 94, 96, 97, 99, 100, 103–110
 BSC 2, 3, 13, 17, 25, 27, 119, 126, 127, 178, 181–186, 191, 194, 201, 286, 309
 CEF (chick embryo fibroblasts) 61, 64, 65, 67–69, 72, 74, 77, 84–87, 89, 93, 110
 DF-1 (chicken fibroblast cell line UMNSAH/DF-1) 61, 64, 67–69, 74, 77, 81, 87, 89
 HEK293 .... 135, 140, 282, 285
 HeLa 61, 87, 123, 124, 126, 128, 129, 131, 137, 143, 221, 224–226, 229, 278, 284
 L929 .... 178, 183
 RK 61, 64, 72–74, 76, 77, 89, 91, 119
 RK-13 .... 61, 64, 72–74, 76, 77, 89, 91, 119
 293T 124, 126, 127
Cell markers
 CD4 (helper T lymphocytes) 259, 260, 262, 264
 CD8 .... 182, 191–193, 197–217, 260
 CD16/CD32 (Fc block) 202
 CD62L (l-selectin) 199, 200, 202, 207, 208
 CD107 200
 major histocompatibility complex class I molecules (MHC-I) 199
Cellular responses 275
Centrifugation
 centrifuge 2, 13, 27, 43, 48, 64, 65, 67, 80–82, 96, 99, 107, 108, 116, 117, 183–185, 202–207, 209, 211–213, 216, 221, 261, 264, 266
 cesium chloride (CsCl) 89, 115, 117-120
 cushion 67, 68, 116, 117, 148, 184, 185, 194, 195, 202, 227
 density 113, 115–117, 119, 120, 128, 143, 312
 equilibrium centrifugation 114, 117, 118, 131
 gradient 63, 88, 89, 117, 118, 120, 128, 131, 136, 185, 194, 195, 296
 purification 5, 33, 34, 37, 38, 40, 42, 55, 59, 61, 62, 67, 72, 74, 76, 77, 79–81, 88–90, 93, 101, 102, 104, 106, 107, 126, 127, 131, 136, 179, 185, 194, 196
Cesium chloride (CsCl). *See* Centrifugation
CFSE. *See* Flow cytometry
Cherry. *See* Fluorescent markers
Chloramphenicol 40, 41, 43, 48–50
Chorioallantois vaccinia virus Ankara (CVA) 38, 39
CMC. *See* Carboxymethylcellulose (CMC)
Collagenase 203, 209
Color markers
 galactosidase 33, 94, 162
 glucuronidase 94, 100, 109
Conformational epitope 231
Cowpoxvirus. *See* Viruses
Cross-link 222, 226, 230, 324
Crystal violet 96, 108, 149, 155, 156, 179, 182, 186, 191, 193
CVA. *See* Chorioallantois vaccinia virus Ankara (CVA)
Cytokines 200, 205, 260, 261, 265, 267, 268, 273, 274, 306
 inteferon (IFN) 191, 200, 205, 215, 261, 267, 274
 interleukin 2 (IL-2) 261, 262, 273
 tumor necrosis factor (TNF) 205, 261, 274, 275, 278, 281
Cytotoxicity
 in vitro 203, 210–213
 in vivo 208, 210, 216

D

Decontamination. *See* Biosafety
Density. *See* Centrifugation
Dianisidine 64, 87
DimerX. *See* Tetramers
Disinfectants. *See* Biosafety
Dissemination 161–175
DNA polymerase (VVpol). *See* Genes
DNase 40, 62, 137, 141, 142, 203, 209
Dryvax. *See* Viruses

E

Ear pinnae 147, 148, 150–152, 154, 155
Ectromelia virus (ECTV). *See* Viruses
Effector cells 199, 211, 212, 260
Electrocompetent 28, 39, 42, 47, 48
Electron microscopy (EM). *See* Microscopy
ELISA 177, 181, 190, 229, 291, 307, 309, 311–313, 315, 323
Entry 123–131, 144, 323
Enveloped virus (EV). *See* Viruses
Epitope 25, 197, 199–201, 204, 207, 212, 220, 231, 324
Epstein Barr virus (EBV). *See* Viruses
Equilibrium centrifugation. *See* Centrifugation
Error. *See* Statistics
Ethidium bromide 26, 29, 31, 138, 141
Euthanasia 163, 180, 188
Extracellular virus (EV/EEV). *See* Viruses

F

Flow cytometry................................192, 197, 200, 204–208, 210, 212, 284
CFSE 182, 192, 208–213, 216, 217
FACS................................197, 202, 204, 205, 207, 209, 212, 280, 284
intracellular cytokine staining (ICS)........................199, 200, 203, 205
Fluorescent markers
cherry..123–131
fluorescent............................46, 71, 74, 76, 79, 94, 101, 103, 105, 109, 123, 124, 129, 131, 162, 197, 224, 261
GFP.......................................41, 71, 74, 76, 79, 82, 83, 90, 95, 100, 108, 135, 140, 144, 215, 279, 284
RFP 71, 74, 76–80, 90, 91
YFP 126, 128, 130, 131
Footpad. *See* Routes of innoculation
Fowlpox. *See* Viruses
Fugene. *See* Transfection

G

GalK selection 41, 46, 54
Genes
A10 (core)...220, 225, 239
A14 (MV envelope protein) 220, 225
A22...53
A33/A33R (EV envelope protein)............118, 220, 225
A46..273, 275, 276
A52..201, 273, 275–277
A56/A56R (hemagglutinin) 94, 118, 220, 225
A4L (EV107) 123, 125–127, 193
A24L (rpo132; the large subunit of the DNA dependent RNA polymerase)..136
A32L 44, 45
A34R...118
A35R...260
A36R...118
A50R...243
B5/B5R (EV envelope protein) 116, 118, 119, 220, 225
B15/B15R (IL-1beta binding protein)................41, 46, 47, 49, 276
B8R 197
B14R...49
B22R 275
B25R 44, 45
C3 (secreted complement control protein, VCP) 220, 225, 289–299, 320
C7L (host range gene).......................220, 222, 227, 228
C16L 275
C19L 44, 45
D8 (MV envelope protein)..............................220, 225
D2L...240
D13L (non-structural protein) 220, 225
D3R...240
E3L 71, 201, 226, 274, 275
EV107 (A4L) 193
F13/F13L (EV envelope protein).....................220, 225
F10L..44, 45
F12L...118
G1L..63, 71
G5R..250, 251
G8R...250
H3 (MV envelope protein)...............................220, 225
H5 71
H7R...240
I3 (DNA binding protein)........................24, 26, 28–30, 32, 33
I3L-I4L..42
I8R 63, 71
J2R (thymidine kinase)......................9, 42, 94, 163, 278
J4R..243
J6R..44, 45
K7..275
K1L (host range gene).............................63, 72, 75–77, 79, 90, 91, 94, 220, 222, 227, 228
K3L 274
K4R...243
L1 (MV envelope protein)......................219, 220, 296
L4R 116, 118, 119
mc129..136
N1L...275, 276, 283, 319
N2L...24, 25, 27, 31, 32
rpo132 (A24L; the large subunit of the DNA dependent RNA polymerase)...136
topoisomerase..23, 24
VACWR144...281
Vvpol (DNA polymerase).........................24–26, 29–32
WR148 (non-structural protein) 220, 225
Genomic analysis. *See* Bioinformatics
GFP. *See* Fluorescent markers
Glycerine 64, 85
Glycosylase..239, 240, 249, 251
Goggles. *See* Biosafety
gpt. *See* Alternative selection
Gradient. *See* Centrifugation
Granzyme B (GzmB)....................................199, 200, 208
GST-fusion 231

H

Helper virus. *See* Viruses
Hemocytometer.......................202, 204, 221, 223, 228, 269
Hematoxylin...308, 318
Hepatitis C virus. *See* Viruses
Homology. *See* Bioinformatics
Host range. *See* Alternative selection

Human diseases
Alzheimer's disease 290, 306
rheumatoid arthritis 306
spinal cord injury 290, 306
traumatic brain injury 290, 306

I

IFN. *See* Cytokines
Immunohistochemistry 307, 309, 315–319, 325
Immunology 59, 148
Immunomodulatory 275, 306
Immunoprecipitation 225–227, 230, 279, 282, 284, 286
Innate 260, 273–286, 305
Insertion sequence primers *(E. coli)* 50
Interleukin 2 (IL-2). *See* Cytokines
Intracellular cytokine staining (ICS). *See* Flow cytometry
Intracellular enveloped virus (IEV). *See* Viruses
IPTG 32, 222, 227
Irradiate/irradiator 182, 192, 197

J

JAVA. *See* Bioinformatics

L

LD50 .... 228, 319
Library/libraries 32, 96, 138, 162, 253
Limiting dilution 105–107, 110
Lipofectamine. *See* Transfection
Low melting point agarose 62, 87, 96
Luciferase
bioluminescence 162, 163, 167
firefly 139, 144, 162, 164, 172, 279, 281, 282, 285
luciferin 162, 279
renilla 139, 144, 278, 281, 282, 285
Luminex 261
Lysis 26, 27, 44, 137, 140, 200, 202, 203, 208, 210, 212–213, 217, 221, 225, 226, 260, 278, 279, 281, 282, 284

M

Macrophages 260–263
Major histocompatibility complex class I molecules (MHC-I). *See* Cell markers
Markerless 38
Mask. *See* Biosafety
Mass spectrometry 47, 52, 226–227, 230
Mature virion (MV/IMV). *See* Viruses
Metabolic label/radio-labeling 115
Mice 8, 17, 19, 147–158, 162–175, 178, 182, 193, 196, 200, 201, 203–206, 208–212, 214–216, 220, 228, 260, 306, 308, 319
Microscopy 45, 74, 94, 103–104, 123, 128–129, 140, 144, 300
electron microscopy (EM) 68, 88, 128, 131, 145
Model antigen 261, 263–265
Modeling. *See* Bioinformatics
Molluscum. *See* Viruses
Monkeypox (MPXV). *See* Viruses
Mousepox. *See* Viruses
Mouse strains
AKR 178
BALB/c 161–178, 200–214, 220–228, 308, 319
C57BL/6 .... 156, 157, 178, 195–198, 200–203, 210, 216
C3H 178
DBA 178, 203
MVA. *See* Viruses

N

Neutralization 182, 191, 220
Nitric oxide (NO) 261, 265, 267, 269
NYVAC. *See* Viruses

O

Overlay. *See* Carboxymethylcellulose (CMC); Low melting point agarose

P

Paraformaldehyde 203, 206, 207, 209, 212, 216, 221, 224, 308, 315–317, 322, 324
Passage 50, 60, 61, 72, 81, 82, 86, 89, 90, 93, 97, 102–104, 126, 144, 183, 194, 285
Pathogenesis
virulence 260, 275
weight loss 170, 171, 175
PCR 23, 39, 62, 124, 136, 181, 222, 278
Peptide 25, 162, 199–202, 204–206, 208–217, 262, 263, 267, 268, 283, 291, 300, 306
Peroxidase substrate solution 62
Phage 38, 46, 54, 64, 84
Plasmids
p240 (poxvirus firefly luciferase reporter virus) 138, 139
p300 (poxviral EGFP reporter plasmid) 138–140
pCR3.1 .... 278, 280
phRG-TK (p238; Transfection control plasmid expressing renilla luciferase under a herpes promoter) 138, 139
pMVA-βGus 95, 102, 109
pMVA-rsGFP 95, 98, 109
pRB21 .... 97, 109, 138, 139

pRB21-E-Koz-EGFP-X-flag-strepII-N 139
pRB21-E-Koz-firefly luciferase-H 139
pRB21-pE/L-EGFP-SFX (poxviral EGFP reporter plasmid) 138, 139
pRB21-pE/L-FF luciferase (poxvirus firefly luciferase reporter virus) 138, 139
pyMCV1-E-C (PCR MCV control plasmid) 138, 139, 141–143
Primers 24–26, 32, 33, 40, 41, 44, 45, 47, 49–52, 55, 63, 79–84, 90, 125, 137, 181, 193, 194, 235, 253, 280
Promoter..................................32, 33, 41, 54, 60, 71, 95, 97, 100, 108, 109, 126, 136, 139, 144, 162, 163, 237, 238, 248, 252, 253, 256, 274, 282, 285, 286
Psoralen inactivated virus. *See* Viruses
Purification. *See* Centrifugation

R

Rabies virus. *See* Viruses
Radioactive 18, 114, 120, 197, 200, 230
Receiver operating characteristic (ROC) curve analysis. *See* Statistics
Recombination 3, 38, 41, 42, 46, 54, 55, 60, 71, 76, 95, 97–101, 103, 104, 109, 251, 277
Recombineering 37–56
Red recombineering 39, 46
Renilla. *See* Luciferase
Rescue..38–41, 44–47, 49–56
RFP. *See* Fluorescent markers
Rheumatoid arthritis. *See* Human diseases
Rhodamine filter set 103
RNA...32, 53, 55, 136, 253–254, 274
Routes of innoculation
footpad 178, 196
intracranial... 307, 305–321
intradermal 148–156, 178, 187
intranasal 148, 166–169, 175, 178, 180–187, 206, 305
intravenous 163, 187–188, 209, 222
subcutaneous... 19, 178, 187

S

Sequence query language (SQL) 235, 237, 238, 255
Shope fibroma virus. *See* Viruses
Single cross over 99, 102–105, 131
Sonicate/sonicator...13, 14, 62, 67, 69, 72, 74, 76, 77, 79, 81, 82, 86–88, 90, 95, 97, 99, 100, 103–108, 126, 131, 139, 149, 154, 158, 173, 179, 184–187, 194, 195, 222, 227, 320
Spinal cord injury. *See* Human diseases
Spleen.................................... 166, 168–171, 188, 189, 191, 192, 203, 208–211, 213, 215, 216, 220, 222, 223, 229
ST-246 (Tecovirimat) 114, 115, 118
Statistics
error 32, 52, 217
receiver operating characteristic (ROC) curve analysis 163, 168
t-test 168, 169, 174
Strain. *See* Viruses
Subcutaneous. *See* Routes of innoculation
Sucrose. *See* Centrifugation
Syringe 12, 61, 65, 66, 148–152, 155–157, 163–165, 179, 180, 187, 188, 195, 203, 209, 213, 223, 307–310, 315, 316, 319, 323

T

T cell.. 4, 7, 182, 192, 193, 197, 199–201, 207, 208, 210, 213, 214, 262, 266, 267
Tecovirimat (ST-246). *See* ST-246
Termination (TTTTTNT). *See* Termination signal
Termination signal.......................... 53, 54, 89, 242, 244, 256
Tetramers.. 200, 204
DimerX 204
Thymidine kinase (TK). *See* Genes, J2R
Tissue culture infectious dose 50 (TCID50). *See* Calculation(s)
Titer/titering 3, 5, 17, 68, 70, 82, 87, 89, 91, 100, 108–110, 126, 143, 148, 155, 173, 179, 186, 190–191, 194, 195, 202, 229, 268, 274, 305–325
Tool. *See* Bioinformatics
Topoisomerase. *See* Genes
Transcription reaction.. 145
Transfection...6, 143, 144, 285
fugene 62, 95, 100
lipofectamine 126, 127, 139
Traumatic brain injury. *See* Human diseases
TROVAC. *See* Viruses
Trypan blue 221, 228, 266, 269
Tumor necrosis factor. *See* Cytokines

U

UV-inactivated virus. *See* Viruses

V

VAC-BAC. *See* Viruses
Vaccination 2, 4–8, 11, 14–19, 113, 168, 174, 177, 228, 230
Vaccinia immune globulin (VIG) 7, 9, 163–175
VCP. *See* Genes, C3
Viral DNA 67, 74, 76, 79, 84, 89, 90, 224, 225, 230
Virulence. *See* Pathogenesis
Viruses
ALVAC 14
Borna disease virus 276
cell-associated enveloped virus (CEV) 114, 117

Viruses *(Continued)*
chorioallantois vaccinia virus Ankara (CVA) 38, 39
Cowpox virus (CPXV) 9, 38, 114, 201, 243, 306
CVA (*see* chorioallantois vaccinia virus Ankara)
Dryvax 163, 167–168, 174
ectromelia virus (ECTV) 113, 177, 200, 201, 239, 243–244, 246
Epstein Barr virus (EBV) 274
extracellular virus (EV) or (EEV) 114, 116, 117–120, 219, 220, 230
fowlpox 38, 40, 44, 45, 52, 53
helper virus 38, 41, 44, 53
hepatitis C virus 276
IHD-J 115, 119, 120
intracellular enveloped virus (IEV) 114
mature virion (MV/IMV) 117, 118, 124, 128, 145, 219, 220, 229
modified vaccinia virus Ankara (MVA) 3, 14, 17, 38, 59, 93, 202
Molluscum 135–145
monkeypox (MPXV) 14, 18, 113, 177, 244, 245
Mousepox 177–197
MVA (*see* modified vaccinia virus Ankara)
NYVAC 3, 14, 59
psoralen inactivated virus 44
rabies virus 44
shope fibroma virus 44
temperature-sensitive mutant 44
TROVAC 14
UV-inactivated 222, 223
VAC-BAC 38, 39, 44, 53
variola virus (VARV) 1, 2, 14, 18, 113, 177, 233, 244, 245, 253, 254, 275
Western Reserve (WR) 2, 4, 5, 7–9, 17, 27, 38, 72, 96, 109, 119, 124, 126, 127, 136, 140, 142, 143, 150, 153, 158, 163, 201, 202, 215, 220, 222–227, 278, 280, 319, 321

W

Web links
http 33, 40, 46, 137, 235–237, 244, 248, 249, 252
www 9, 141, 235, 236, 251–253, 255, 276, 277
Weight loss. *See* Pathogenesis
Western blot 85–86, 119, 227, 231
Western Reserve (WR). *See* Viruses

Y

YFP. *See* Fluorescent markers

Printed by Publishers' Graphics LLC
SO20120614